Dr. Seuss and the Art of War

Secret Military Lessons

Edited by

Montgomery McFate
US Naval War College

ROWMAN & LITTLEFIELD
Lanham • Boulder • New York • London

Executive Acquisitions Editor: Michael Kerns
Assistant Editor: Elizabeth Von Buhr
Sales and Marketing Inquiries: textbooks@rowman.com

The views expressed here are those of the author or contributors alone and do not necessarily represent the views, policies, or positions of the US Department of Defense or its components, to include the Department of the Navy or the US Naval War College.

Credits and acknowledgments for material borrowed from other sources, and reproduced with permission, appear on the appropriate pages within the text.

Published by Rowman & Littlefield
An imprint of The Rowman & Littlefield Publishing Group, Inc.
4501 Forbes Boulevard, Suite 200, Lanham, Maryland 20706
www.rowman.com

86-90 Paul Street, London EC2A 4NE

Copyright © 2024 by Montgomery McFate
Contributors retain copyright to their respective contributions.

All rights reserved. No part of this book may be reproduced in any form or by any electronic or mechanical means, including information storage and retrieval systems, without written permission from the publisher, except by a reviewer who may quote passages in a review.

British Library Cataloguing in Publication Information Available

Library of Congress Cataloging-in-Publication Data

Names: McFate, Montgomery, editor.
Title: Dr. Seuss and the art of war : secret military lessons / edited by Montgomery McFate, US Naval War College.
Description: Lanham : Rowman & Littlefield, 2024. | Includes bibliographical references and index.
Identifiers: LCCN 2024005880 (print) | LCCN 2024005881 (ebook) | ISBN 9781538193617 (cloth) | ISBN 9781538193624 (paperback) | ISBN 9781538193631 (epub)
Subjects: LCSH: War (Philosophy) | Military art and science. | Seuss, Dr.—Criticism and interpretation. | Strategy. | Art and war.
Classification: LCC U21.2 .D727 2024 (print) | LCC U21.2 (ebook) | DDC 355.02—dc23/eng/20240208
LC record available at https://lccn.loc.gov/2024005880
LC ebook record available at https://lccn.loc.gov/2024005881

Taryn Chance Carlough and Hunter Courage Carlough
deserve credit (and ice cream!) for inspiring this book.
Their secret decoder ring revealed the hidden wisdom of Dr. Seuss.

Contents

Foreword ... xi
Acknowledgments ... xix

Part I. Introduction

1 Learning the Art of War from Dr. Seuss | Montgomery McFate ... 3
   Dr. Seuss in Military Art and Science ... 4
   Life in Uniform ... 7
   Hidden in Plain Sight ... 10
   Lessons from the Master ... 13
   Conclusion ... 16

Part II. Bounding the Subject

2 *I Had Trouble in Getting to Solla Sollew* and Strategy | Antulio J. Echevarria II ... 23
   Troubles ... 25
   Psychological Resilience ... 27
   Misadventure ... 29
   Isolationism/Anti-Isolationism ... 30
   Military Leadership ... 32

Friction	33
End States	33
Utility of Force	36
In Conclusion	38

3 *Oh, the Places You'll Go!* and Grand Strategy in Outer Space | Saadia M. Pekkanen — 42

Grand Strategy	43
Three Uncertainties	46
Bright Places with Boom Bands	53
In Conclusion: The Great Balancing Act	54

4 *Horton Hears a Who* and International Human Rights Law | John Hursh — 61

The Making of Horton	62
Listening	64
The Encounter	65
Do No Harm	67
Tolerance	67
Development	69
Torture	71
Collective Action	73
Conclusion	74

Part III. Specialized Domains of Warfare

5 *Thidwick the Big-Hearted Moose* and Environmental Security | Rebecca Pincus and Montgomery McFate — 85

Capacity and Collapse	86
Short-Term Interests, Long-Term Consequences	88
Common Pool Resource Dilemmas	90

Social Norms and Vulnerability	92
Nasty, Brutish, and Short	93
Mass Extinction	95
Deep Ecology	97
In Conclusion: "All Stuffed, as They Should Be"	98

6 *The Cat in the Hat* and Cyber Warfare | Jon R. Lindsay and Michael Poznansky — 102

Unwitting Cooperation	103
Unintended Consequences	107
Staged Operations	109
Leaving No Trace	110
Offense-Defense Balance	113
Deviance and Authority	115
In Conclusion: Warnings to Ponder	116

7 *Private Snafu* and Political Propaganda | Kevin P. Eubanks — 122

Hollywood at War	123
Propaganda and Its Discontents	125
Propaganda and Democracy	127
Anti-Authoritarianism	132
Individual Freedom	135
Civil Society	139

8 *Yertle the Turtle* and Authoritarianism and Resistance | Katherine Blue Carroll — 146

Staying in Power	148
Explaining Mass Protest	152
Lone Dissidents (Explaining Mack)	157
Conclusion: Plea for a Seussian Research Agenda	161

Contents

9 *Hunches in Bunches:* Intelligence and National Security Decision-Making | Genevieve Lester, John Nagl, and Montgomery McFate — 170

- The Wrong Hunch — 172
- The Happy Hunch — 173
- Real Tough Homework Hunch — 174
- The Better Hunch — 177
- The Sour Hunch — 178
- Very Odd Hunch — 179
- The Unhelpful Hunch — 180
- The Spookish Hunch — 181
- The Nowhere Hunch — 182
- The Up Hunch — 182
- The Down Hunch — 184
- Wild Hunches in Big Bunches — 185
- Super Hunch — 186
- In Conclusion: Munch Hunch — 187

Part IV. Theories of Warfare

10 *Horton and the Kwuggerbug* and Deception in International Relations | Chris C. Demchak — 195

- The Kwuggerbug and the Deceptive False Flag State — 199
- The Grinch Deceives the Hoobubs . . . Again — 203
- Marco Comes Late and Spins a Tale . . . — 205
- Resilience in "How Officer Pat Saved the Whole Town" — 207
- Conclusion: Deception Is a Strategy That Pays — 207

11 *Did I Ever Tell You How Lucky You Are?* and Luck in Warfare | Erich Henry Wagner and Montgomery McFate — 212

- The Bunglebung Bridge and the Dangers of War Technology — 215
- Zayt Highway Eight and the Dangers of Roads — 218

The Schlottz and the Wounds of War	219
Hawtch-Hawtcher Bee-Watcher and Military Intelligence	220
Poogle-Horn Players and the Danger of Broken Tools	223
Harry Haddow's Shadow: Changing One's Luck	223
The Brothers Ba-Zoo: Lucky Comrades in War	226
That Forest in France: Luck and the Environment in Warfare	227
The Seventeenth Radish: Luck and the Randomness of Survival	229
In Conclusion: A Game of Cards	230

12 *The Butter Battle Book* and Deterrence and Escalation | Sam J. Tangredi — 235

Controversy and Reality	237
Arms Race	238
Tech Pace	239
Escalation and Hesitation	240
Rational or National	241
Renegades and Provocateurs	243
Deterrence and Nonoccurrence	245
Punishment or Denial	247
War Termination	250
In Conclusion: Guns or Butter	252

Part V. Consequences of Warfare

13 *How the Grinch Stole Christmas* and Traumatic Stress | Montgomery McFate — 261

The Grinch as Allegory	262
Wartime Trauma	263
Diagnostic Criteria	264
Resilience and Healing of the Grinch	272
Trauma and Resilience of Who-Ville	275
In Conclusion: Redemption in "War Stories"	277

Bibliography 287
Index 329
About the Contributors 345

Foreword

While Dr. Seuss has often been mistaken for children's literature, a number of senior officers in the US military have long recognized Dr. Seuss as a great strategic thinker! And with the publication of *Dr. Seuss & the Art of War*, the significance of Dr. Seuss to national security will finally be made apparent to the American public. This wonderfully unique and highly entertaining volume conclusively demonstrates the contemporary relevance of Dr. Seuss for such diverse topics as cyber warfare, intelligence, grand strategy, and the role of "luck" in combat. Moreover, this book also illuminates a convergence between the core values in Dr. Seuss books and the core values of the US military, both of which originate in the US Constitution.

The foundation of Dr. Seuss's military knowledge comes from his own uniformed service during World War II. Theodor Geisel—the name of the man behind the nom de plume Dr. Seuss—firmly believed that WWII was not an elective conflict. As he said, "we were going to have no choice in the matter."[1] Despite the fact that Geisel, at the age of thirty-nine, could have been considered too old for service, he volunteered nonetheless. Assigned to the US Army Signal Corps, Geisel served under Major Frank Capra in Hollywood at Fox Studios, which for the duration of the war was known as "Fort Fox."

Geisel's Hollywood film projects supported the nation's efforts to educate its soldiers about their roles and responsibilities while in uniform. His cartoon *Private Snafu* (Situation Normal, All [Fouled] Up), was shown to the troops at the end of the weekly newsreels that were popular at that time. By watching *Private Snafu* cartoons, soldiers internalized important lessons, such as the benefits of wearing mosquito repellant, investing in savings bonds, and maintaining their weapons. His subsequent film, *Your Job in Germany*, explained to soldiers that the postwar occupation had equal importance to that of the fighting to bring the war to an end. It conveyed that only by reconstructing Germany's society, economy, and political system could a resurgence of fascism be prevented. As the nation relearned in Afghanistan and Iraq—which Geisel knew all along—"winning the peace" during the post-conflict period is what achieves the conditions necessary for a stable and secure long-term future.

During his US Army service, Geisel refined the basic skills that all soldiers acquired, developing an understanding of US wartime and postwar strategy, and honing his propaganda skills (or what might now be called "information operations"). In the European theaters in 1944, Geisel observed accounts of combat leadership, deception operations, troop morale, combined operations, and other aspects of contemporary warfare while visiting General Omar Bradley on the Siegfried Line. Then, while trapped behind German lines at Bastogne, Geisel shared the experiences of soldiers throughout history—existential dread, trouble reading a map, and damp socks.

The books of Dr. Seuss reflect these diverse military experiences—generally concealed and unacknowledged until now—in their prose, illustrations, and meaning. And I will elaborate on a few thoughts about ordinary military life, strategy and tactics, special operations, and core values that are found in Dr. Seuss books.

EVERYDAY LIFE IN THE PROFESSION OF ARMS

Surprisingly, many Dr. Seuss books contain illustrations of ordinary, day-to-day military service.

Since time immemorial, soldiers have stood watch on the perimeter of camps, fortifications, and bases, which is represented by vigilant creatures in several Dr. Seuss books (such as *The Butter Battle Book* [1984], *How Officer Pat Saved the Whole Town* [1950], and *The 500 Hats of Bartholomew Cubbins* [1938]). Perhaps the best example of standing watch can be found in *Horton Hatches the Egg* [1940], in which a mother bird asks Horton to sit on her egg while she vacations on Palm Beach. Because "an elephant's faithful / One hundred percent!" Horton refuses to leave the nest, enduring rain, boredom, frost, taunting, and frustration. "He sat all that day / And he kept the egg warm. . . . / And he sat all that night / Through a *terrible* storm. / It poured and it lightninged! / It thundered! It rumbled! / 'This isn't much fun,' / The poor elephant grumbled." Like Horton, soldiers stand watch regardless of the weather. Although that task is not much fun, it is part of a soldier's job and Horton's performance of it with respect to the egg provides an exemplary illustration of the discipline, devotion, and dedication to duty that are required.

Physical adversity (another common fact of military service) is also exemplified in many Dr. Seuss books. In *I Had Trouble in Getting to Solla Sollew* (1965), the narrator sprains his "tail bone," is pecked by vicious birds, stubs his toe, and experiences severe insomnia while sheltering from a fierce storm in an empty house.[2] "I ran in the house and I fell in a heap. / I needed my rest, but I just couldn't sleep. / Did *you* ever sleep, when your feet were like ice, / With a family of owls and a family of mice?" Similarly, in *Oh the Places You'll Go!*, the narrator gets "all hung up / in a prickle-y perch" when his hot air balloon malfunctions, becomes trapped in a "Slump," and finds himself "mixed up with many strange birds." But despite physical fear, difficult terrain, and physical hardship, the narrator perseveres: "On you will go / though the Hakken-Kraks howl. / Onward up many / a frightening creek, / though your arms may get sore / and your sneakers may leak." In real life, of course, soldiers will encounter only metaphorical Hakken-Kraks; nevertheless, those require the same grit and determination the narrator emphasizes.

Unappetizing food (which is another constant of military life, especially on deployment) features prominently in *Green Eggs and Ham* (1960). Despite Sam-I-Am's entreaties, the narrator rejects Sam's polite offer. "I do not like them, / Sam-I-am. / I do not like / green eggs and ham." Indeed, a large serving platter of fried eggs with green yolks and a large hunk of green meat hardly seem appealing regardless, of whether they are served in a different location (such as "in a house," "in a box," or "in a car"); with a different companion (such as "with a mouse," "with a fox," or "with a goat"); or under different conditions (such as "in the dark," or "in the rain"). Yet, as anyone who has served in uniform knows, when soldiers get hungry enough, they will devour even green eggs and ham.[3]

Weaponry, another common feature of military life, appears in quite a few Dr. Seuss books, as well. In *Horton Hatches the Egg*, hunters sneak up on Horton from behind with rifles while he guards a nest in a tree. "He heard the men's footsteps! / He turned with a start! / *Three rifles were aiming / Right straight at his heart*!" (Instead of shooting Horton, he is taken captive and sold into the circus.) Another Dr. Seuss story featuring rifles is *Thidwick the Big-Hearted Moose* (1948), in which hunters pursue Thidwick to take his horns as a trophy. (The lucky moose escapes by shedding his horns.) In *I Had Trouble in Getting to Solla Sollew*, the soldiers carry lances taller than themselves, and at the end of the book the protagonist wields a large bat for dealing with his troubles. Beyond small arms and light weapons, *The Butter Battle Book* features lethal heavy weapons (such as the Kick-a-Poo Kid; the Eight-Nozzled, Elephant-Toted Boom-Blitz; and the Utterly Sputter) and even an atomic weapon (the Big-Boy Boomeroo). Although the US military does not yet possess an Eight-Nozzled, Elephant-Toted Boom-Blitz in its arsenal, DARPA may very well be working on something like that.[4]

STRATEGY AND TACTICS

Many Dr. Seuss books speak directly to the subject of military strategy. For example, *I Can Lick 30 Tigers Today* (1969) offers a clear example of the fallacy of overestimation, where a military organization unrealistically assesses their own capabilities relative to their adversaries. In *30 Tigers*, after boasting unrealistically of his own capacity, the narrator discovers that a large number of tigers might prove more difficult to defeat than anticipated. So, he systematically reduces the number of tigers in the fight using a pretext of excusing them to clean their dirty fingernails, prevent heatstroke, eat lunch, and take a nap. "Well . . . You look sort of sleepy to me. Some of you chaps / Should go home and take naps. I only intend to lick three." Needless to say, reducing the initial estimate of opposing forces by suggesting they go home may solve the problem in the world of Dr. Seuss, but unfortunately there are not many examples of that approach working in real life.[5]

The Zax (1961) provides an excellent example of a strategic stalemate that both Clausewitz and Liddel Hart warned against.[6] Neither the North-Going Zax nor the South-Going Zax have the power to seize, hold, and exploit the initiative. They are mired in a stalemate, unable to do much except yell threats at each other. "I'll stay here, not budging! I can and I will / If it makes you and me and the whole world stand still!" Although the combatants in *The Zax* remain "standing un-budged in their tracks" while an overpass is built over their heads, in real life most stalemates eventually break, whether through development of new tactics or the employment of new technology.

Mountain warfare figures prominently in the Dr. Seuss book *I Had Trouble in Getting to Solla Sollew*. In *Solla Sollew*, General Genghis Kahn Schmitz foolishly orders his forces to march single file through a narrow mountain pass. Not only do they lack adequate weapons, cold weather gear, and logistical support, the route of the General's forces through a narrow defile makes them vulnerable to ambush. Having spotted only a single Poozer from the opposing Army, General Schmitz badly underestimates his adversary, and orders an advance. "Get that Poozer! Attack without fear! / The glorious moment of victory is near!" The narrator describes their mistake: "Then we went 'round a corner and found that, alas, / There was *more* than one Poozer in Pompelmoose Pass!" Consistent with recorded military history, the Poozers held an advantage during the battle of Pompelmoose Pass because of their strong defensive position, numerically superior forces, ability to leverage surprise, and lethality of their weapons (e.g., large claws).[7] General Genghis Kahn Schmitz should have known that local insurgent groups (such as the Poozers) operating in the mountains often have a better understanding of the high-elevation battlespace, possess superior maneuverability in the treacherous mountain terrain, and have advantageous physiological adaptations (e.g., fur) for combat at higher elevation levels. Dr. Seuss thus demonstrates that bravery cannot substitute for competence (nor for accurate intelligence).

SPECIAL OPERATIONS

While *Horton Hears a Who* has frequently (and mistakenly) been viewed as a story about compassion, the book actually concerns survival, evasion, resistance, and escape (SERE). At the beginning of the story, Horton fails to establish a secure perimeter while bathing in the Jungle of Nool. His adversary—the Sour Kangaroo—approaches unobserved and taunts Horton by calling him "the biggest blame fool in the Jungle of Nool!" Realizing that she is overmatched by Horton, the Sour Kangaroo engages in a tactical withdrawal by jumping into the pool. While the Sour Kangaroo mobilizes her reserve force before attacking, Horton flees into the jungle. Strange canopy-dwelling creatures track Horton "through the high jungle tree tops." He cannot evade the scouts from the main Wickersham force, who stage a cunning raid and abscond with Horton's highly valuable cargo (e.g., the clover with the dust speck).

Lacking the skills to dispose of the clover themselves, the Wickersham scouts approach the infamous mercenary, "a black-bottomed eagle named Vlad Vlad-i-koff." The eagle-for-hire accedes to their request for disposal, and "before the poor elephant even could speak, / That eagle flew off with the flower in his beak." At this point in the story, Horton's Tier One Special Mission Unit SERE training becomes apparent. Horton traverses steep, hazardous mountain terrain in the dark, pursuing Vlad Vlad-i-koff relentlessly. While running all "afternoon and far into the night" Horton endures significant physical pain, "with groans, over stones / That tattered his toenails and battered his bones." Despite his injuries, Horton's land-navigation skills enable him to move confidently through unknown territory at night. At dawn, Horton arrives at "a great patch of clovers a hundred miles wide!" Despite his extreme fatigue (e.g., he was "more dead than alive"), Horton retains the mission focus that characterizes Tier One operators.

After searching "clover, by clover, by clover," Horton locates and secures his stolen cargo. But the commander of the opposing force (e.g., the Sour Kangaroo) has apparently been monitoring his position via satellite.

She orders the Wickersham tribal faction to capture Horton and hold him as a prisoner of war (POW). Since the Sour Kangaroo is a marsupial while the forces under her command appear to be some type of simian, almost certainly the Sour Kangaroo is a foreign fighter from outside the Jungle of Nool. The Sour Kangaroo's mission appears to be training and equipping the indigenous military force, whose operations have been hindered by poor discipline and primitive technology (e.g., a large stew pot, a couple of spoons, and a very long rope). Like many insurgent groups, the Wickersham tribal militia appears to be genealogically organized, mobilizing forces based on extended clan relationships (e.g., "the Wickersham Brothers and dozens / Of Wickersham Uncles and Wickersham Cousins / And Wickersham In-laws)." Such forces mobilized according to a segmentary kinship system often prove difficult adversaries.[8]

When the Wickersham tribal faction captures Horton—in violation of Article 13 of the Geneva Convention, which requires that POWs be "humanely treated"—their methods of restraint intentionally humiliate and degrade Horton. "'Grab him!' they shouted. "And cage the big dope! / Lasso his stomach with ten miles of rope!" In violation of Article 17, which prohibits "physical or mental torture" and "any other form[s] of coercion," Horton is subjected to physical abuse and confinement, "They beat him! They mauled him! They started to haul / Him into his cage!" In addition, the Wickersham tribal faction subjects Horton to psychological abuse, including the threat of physical violence against him ("You're going to be roped! / And you're going to be caged!") and against his compatriots ("dunk that dumb speck in the Beezle-Nut juice!").

When the Wickersham tribal militia attempts to place Horton in a bamboo "tiger cage," he fights "back with great vigor and vim." Although Horton resists the Wickersham tribal militia, their numerically superior force cannot be overcome, "the Wickersham gang was too many for him." Despite the Sour Kangaroo's attempt to break him at the hands of the Wickersham militia, Horton exhorts his friends, "Don't give up!" At the end of the book, Horton saves the Whos from annihilation by presenting empirical proof of their existence to his captors. While Horton's heroism has been widely acknowledged, the animal denizens of Jungle of Nool ought to recognize that he endured captivity as a POW with great honor and never gave up hope.[9]

CORE VALUES

When he joined the Army, Ted Geisel swore an oath "to support and defend the Constitution of the United States" and to "bear true faith and allegiance" to the values enshrined in the Constitution and the other founding documents of the United States. Since 1789, all American military personnel have sworn the same oath. The oath is inculcated in soldiers even after separation from service.[10] So, we should not be surprised that Geisel honored that oath and continued to bear "true faith and allegiance" to the Constitution in his books. *Yertle the Turtle*, for example, sends a clear message of freedom from political tyranny. "And today the great Yertle, that Marvelous he, / is King of the Mud. That is all he can see. / And the turtles, of course . . . all the turtles are free. / As turtles and, maybe, all creatures should be." *Oh, the Places You'll Go!* carries a message about the rights of each individual to liberty and self-determination: "You're on your own. And you know what you know. / And *YOU* are the guy who'll decide where to go." Many other Dr. Seuss books also affirm this Constitutional philosophy.

In sum, in reading Dr. Seuss, children absorb the same values implicit in the Declaration of Independence and the US Constitution. Furthermore, the core values found in Dr. Seuss books are identical to the values that the US military has sworn to defend. Thus, the traces of Geisel's war-time experience found in Dr. Seuss books should come as no surprise. After all, "an elephant's faithful, 100 percent."

General David Petraeus
US Army (Ret.)

NOTES

1. Judith Morgan and Neil Morgan. *Dr. Seuss and Mr. Geisel: A Biography* (New York: Random House, 1995), 103.

2. Alan Derickson, "'No Such Thing as a Night's Sleep': The Embattled Sleep of American Fighting Men from World War II to the Present." *Journal of Social History* 47, no. 1 (2013): 1–26. http://www.jstor.org/stable/43306043.

3. Roger Morriss, "Colonization, Conquest, and the Supply of Food and Transport: The Reorganization of Logistics Management, 1780–1795." *War in History* 14, no. 3 (2007): 310–24. http://www.jstor.org/stable/26070709; Michael Owen Jones. "What's Disgusting, Why, and What Does It Matter?" *Journal of Folklore Research* 37, no. 1 (2000): 53–71. http://www.jstor.org/stable/3814665.

4. Howard E. McCurdy. *Space and the American Imagination*. Baltimore, MD: Johns Hopkins University Press, 2011.

5. Robert Jervis. "War and Misperception." *The Journal of Interdisciplinary History* 18, no. 4 (1988): 675–700. https://doi.org/10.2307/204820.

6. Hew Strachan, "Clausewitz and the Dialectics of War," in *Clausewitz in the Twenty-First Century*, edited by Hew Strachan and Andreas Herberg-Rothe (Oxford,

2007); B. H. Liddell Hart, *Strategy* (New York: Frederick A. Praeger, 1954; reprint 1967), 370.

7. Adrian Goldsworthy, *The Punic Wars* (London: Cassel & Co., 2001); Dexter Hoyos, *Mastering the West: Rome and Carthage at War* (Oxford: Oxford University Press, 2015).

8. Montgomery McFate, *Military Anthropology: Soldiers, Scholars and Subjects at the Margins of Empire* (New York: Oxford University Press, 2018).

9. Stuart L. Rochester and Frederick Kiley, *Honor Bound: The History of American Prisoners of War in Southeast Asia, 1961–1973* (Annapolis, MD: Naval Institute Press, 1998); John G. Hubbell, Andrew Jones, and Kenneth Y. Tomlinson, *P.O.W.: A Definitive History of the American Prisoner-of-War Experience in Vietnam, 1964–1973* (New York: Thomas Y. Crowell Company, 1976).

10. Thomas H. Reese, "The Oath of Allegiance," Proceedings 91/9/751, September 1965.

Acknowledgments

Katherine Blue Carroll: A big thank you to Spencer Hall.
John Lindsay: Thanks to my daughter Eleanor for her research assistance.
Kevin Eubanks: I would like to thank my family and our editor, Montgomery "Mitzy" McFate, for patiently bearing with me throughout the drafting stage.
Montgomery McFate: Thanks to Linda Claussen at the Seuss Archives at UC San Diego, who provided assistance, feedback, and advice. Many thanks to Danielle Burby for her advice and assistance. Thanks of course to Felicity McDonough for absolutely everything.
Gen Lester: Thanks to Andrea Zechman for her insights and creativity.
Becca Pincus: Thank you to Mike for love and support.
Saadia Pekannen: I thank the editor, Montgomery "Mitzy" McFate, and Peter Dombrowski for their thoughtful comments.
Chris Demchak: Thanks to Gabbi for her insights on Dr. Seuss.

Part I

INTRODUCTION

1

Learning the Art of War from Dr. Seuss

Montgomery McFate

I am subversive as hell.

—Ted Geisel

If you take an interest in military and national security affairs, you have probably read the works of Carl von Clausewitz, Sun Tzu, and Thucydides. But what about the books of the underappreciated military strategist Theodor Geisel, also known as Dr. Seuss? His books provide considerable insight into many military topics of contemporary importance. Just as *Hunches in Bunches* (1982) seems to invoke intelligence analysis, *The Cat in the Hat* (1957) appears directly relevant to cyber warfare.[1] Understanding the hidden wisdom of Dr. Seuss requires patience, imagination, and a little bit of codebreaking. Just as Sun Tzu's oracular pronouncements of archaic principles of warfare must be decoded to fully understand their meaning, Dr. Seuss books require cryptoanalysis to decode the hermetic military wisdom they contain. As Dr. Seuss observes in *On Beyond Zebra!* (1955): "You'll be sort of surprised what there is to be found / Once you go beyond Z and start poking around!"[2]

Once the secret code has been cracked, the extraordinary alignment of Dr. Seuss's prose with timeless axioms of classic texts of military theory becomes apparent. For example, on the subject of victory, Sun Tzu says, "If you fight with all your might, there is a chance of life; whereas death is certain if you cling to your corner."[3] Translated into the prose of Dr. Seuss in *Oh, the Places You'll Go!* (1990): "And will you succeed? / Yes! You will, indeed! / (98 and ¾ percent guaranteed.)"[4] On the subject of battle, Sun Tzu observes: "In battle, there are not more than two methods of attack—the direct and the indirect; yet these two in combination give rise to an endless series of maneuvers."[5] Dr.

Seuss reaches the same conclusion in his masterpiece *Fox in Socks* (1965): "When beetles / fight these battles / in a bottle / with their paddles / and the bottle's / on a poodle / and the poodle's / eating noodles . . . / . . . they call this / a muddle puddle / tweetle poodle / beetle noodle / bottle paddle battle."[6] On the subject of hopelessness, Sun Tzu notes the following: "So morning energy is keen, midday energy slumps, evening energy recedes—therefore those skilled in use of arms avoid the keen energy and strike the slumping and receding. These are those who master energy."[7] In the words of Dr. Seuss, this passage can be translated as: "when you're in a Slump, / you're not in for much fun. / Un-slumping yourself / is not easily done."[8]

The correspondence of Dr. Seuss and Sun Tzu should not be viewed as a mere accident. Although Dr. Seuss has been neglected in the field of military theory and security studies, Geisel's knowledge of contemporary history, political theory, and military strategy (which he acquired as both a patriotic citizen and committed soldier) run like a powerful undercurrent throughout his books. This volume therefore aims to reassess the contribution of Dr. Seuss to strategic thought and to ensure his place as one of the luminaries in the field.

To that end, this introductory chapter notes Dr. Seuss's current obscurity in the field of military art and science and provides a brief overview of Geisel's service in the US Army during World War II. These wartime experiences had a deep and significant influence on his books, apparent in imagery and themes. By denying that his books contained any symbolic meaning, Geisel deflected attention from the clues hiding in plain sight. Over time, the conventional interpretations of Dr. Seuss—by generations of parents and children, defenders and detractors, journalists and biographers—have become cemented in the mind of the audience. By relinquishing our assumptions about Dr. Seuss, many secrets pertaining to the art of war—long hidden in plain sight—are revealed.

DR. SEUSS IN MILITARY ART AND SCIENCE

How did this book come about? Without knowing anything about Ted Geisel's military background and experience, I had assigned his books for many years in my elective course at the US Naval War College. Dr. Seuss books provided some levity in a professional military education curriculum that frequently tended to be wearisome and technical. More importantly, assigning Dr. Seuss books to my students—who are mostly senior military officers—helped them to understand the psychosocial processes of transcultural warfare. Without the distraction of historical details typically found in a scholarly case study, the books of Dr. Seuss clarified various complex, abstract concepts about human violence that we discussed in class. But instead of long explanations of the origin, ideology, and strategy of the Khmer Rouge or Bosnian Serbs, the *Sneetches* (1961) offered a distilled example of the isolation, denial, and identity hardening experienced by oppressed groups that are often early precursors of genocide.[9] The sheer surprise of being assigned books by Dr. Seuss at a US military war college encouraged these officers to think more creatively about strategy, operations, and tactics.

Then, to my surprise, I discovered that a colleague at the US Naval War College, Capt. Thomas Culora (US Navy, ret.), had been assigning *Green Eggs and Ham* (1960) in his leadership seminar for many years. Sam-I-Am's unflagging perseverance in achieving his objective (e.g., voluntary agreement to eat unfamiliar and potentially dangerous food) provided the officers in his class with an example of how "unwanted crisis can become an unplanned opportunity for change and growth."[10] Then in 2022 I met Capt. Robert "Gus" Gusentine (a retired Navy SEAL), who directs and delivers the US Special Operations Command's Strategic Leaders International Course. He uses *The Cat in the Hat* as the primary textbook for this four-week professional development curriculum designed for senior government officials from partner nations (Colombia, Costa Rica, the Republic of Cyprus, Panama, Poland, Trinidad and Tobago, to name a few). "Nobody expects to be reading a children's book in a senior military course run by a Navy Seal," notes Gusentine, but *The Cat in the Hat* resonates as a "better example of complexity, wicked problems, and the non-linear nature of war," than any other book. At graduation, students receive an inscribed copy to bring home and share with their colleagues and family, thereby spreading the wisdom of Dr. Seuss across the globe. "That's the real magic."[11]

Since my colleagues and I had discovered Dr. Seuss's value in a war college classroom, naturally I began to wonder: How deeply has the US defense establishment been penetrated by the covert military fraternity of Dr. Seuss disciples? Do the Joint Chiefs of Staff gather to discuss *One Fish, Two Fish, Red Fish, Blue Fish* (1960) on the E Ring of the Pentagon? Does the Joint Special Operations Command consult *Bartholomew and the Oobleck* (1949) for historic lessons learned on effective counterproliferation of airborne green goop, applicable to other weapons of mass destruction (WMD)? Do Navy SEALs rely on *McElligot's Pool* (1947) as a technical manual for underwater demolition? Regretfully, I discovered after poking around in the dark corners of the national security establishment that the military genius of Dr. Seuss remains almost completely unrecognized even among the cognoscenti with TS/SCI clearances and a "need to know." Likewise, exhaustive scrutiny of obscure academic journals indicated that Dr. Seuss had been sadly neglected by scholars of national security.

Only *The Butter Battle Book* (1984), which Ted Geisel wrote as a commentary on the contentious topic of nuclear deterrence and escalation in the 1980s, had received any attention from scholars or practitioners of war. The book relates how the Yooks and the Zooks differentiate themselves based on whether they eat bread with the butter side up or down. Their philosophical conflict becomes an arms race, in which weapons become increasingly lethal until they reach the level of mass destruction. In a law review article, John Hursh (who is also a contributor to this volume) examined how the law of war would apply to the military escalation between the Yooks and the Zooks in *The Butter Battle Book*. In Hursh's view, "one could argue that Vanltch began an international armed conflict by firing his slingshot at Grandfather's Snick-Berry Switch," since firing a weapon into a foreign territory constitutes an act of international armed conflict.[12] Similarly, Roger S. Clark's 2013 law review essay "Is *The Butter Battle Book's* Bitsy Big-Boy Boomeroo Banned? What Has International Law to Say about Weapons of Mass Destruction," discusses the prohibition of explosive, expanding, or poisoned projectiles in customary international law with analogous weapons of the Yooks and the Zooks. "If using . . . a poison gas container to kill scores [of people] is a war crime, why is it not a war crime to use . . . a Bitsy Big-Boy Boomeroo which can kill hundreds of thousands? . . . [T]he mysterious Moo-Lacka-Moo with which the Boomeroo is filled does not poison or asphyxiate the adversary-it merely blows the enemy clear to Sala-ma-goo."[13]

Aside from *The Butter Battle Book*, the only other interpretation of Dr. Seuss from a military perspective appeared to be Chris Heatherly's 2020 blog entry for *Armchair General*! about *On Beyond Zebra* (1955). In the Dr. Seuss book, a boy draws on the blackboard imaginary letters that come after Z in the alphabet. The lesson of the story, according to Dr. Seuss, is that *"there's no end / To the things you might know, / Depending how far beyond Zebra you go!"* When translated into a military context, Heatherly argues that US armed services should "go beyond Zebra" and focus their training and doctrine on future conflicts. "Dr. Seuss," Heatherly writes, "was a genius whose wisdom extends beyond that of life's early lessons, and whose thoughts have application far beyond the kindergarten classroom."[14]

As this short review demonstrates, Dr. Seuss has been wrongly neglected in security studies, military history, and other academic disciplines concerned with war. Perhaps the reason for this unfortunate omission lies in the popular contemporary view of children's literature as nothing more than harmless, cheerful, entertaining bedtime stories designed to sooth weary toddlers. In the view of most parents in the United States and elsewhere, nothing dangerous, nothing frightening, nothing painful should be found between the pages of books intended for children. If one accepts this viewpoint, interpretation of Dr. Seuss from a military perspective would appear perplexing, incongruent, and perhaps even inappropriate. Yet across cultures and throughout history, children's stories do more than entertain; they teach children physical, emotional, and psychological survival skills to prepare them for a dangerous world. Almost all the classic European fairy tales portray the darkness of the human heart.[15] Consider, for example, the parental abandonment, witchcraft, and cannibalism in "Hansel and Gretel"; the jealousy, physical mutilation, and emotional abuse in "Cinderella"; and the secrecy, mass murder, and hoarding of victims' corpses in "Bluebeard," to name just a few examples.

Like these classic fairy tales, Dr. Seuss books teach survival skills for a dangerous world. In fact, many of the survival skills found in Dr. Seuss are actually *military* in nature. Consider, for example, the military importance of deception (in *Horton and the Kwuggerbug*), unconventional warfare (in *Yertle the Turtle*), resisting torture (in *Horton Hears a Who*), the fog of war (in *I Had Trouble Getting to Solla Sollew*), escape and evasion (in *Thidwick the Big Hearted Moose*), and so forth. Approaching Dr. Seuss from a military perspective will demonstrate that he deserves a place among the great strategists such as Carl von Clausewitz, Sun Tzu, and Thucydides.

LIFE IN UNIFORM

Not surprisingly, Geisel's biography includes military service. Born in Springfield, Massachusetts, in 1904, Geisel grew up in a close-knit German American community. He left to study at Dartmouth College in 1921, where he provided text and illustrations for a campus humor magazine. A professor encouraged Geisel to pursue a PhD at Oxford University, but he left after a single semester. Instead, Geisel returned to the United States and took an advertising position with Standard Oil. His advertising work led to a career drawing political cartoons for the liberal *PM Magazine*.[16]

On June 14, 1940, Ted Geisel heard a report on the radio that Nazi tanks had rolled into Paris. "While Paris was being occupied by the clanking tanks of the Nazis and I was listening on my radio," he wrote, "I found I could no longer keep my mind on drawing pictures of Horton the Elephant."[17] Instead, Geisel began drawing cartoons of fascist leaders, American politicians, and current events in order to call attention to the dangers of totalitarianism and American isolationism. In March 1942, for example, he produced a cartoon for *PM* with the caption "WHAT HAVE YOU DONE TODAY TO HELP SAVE YOUR COUNTRY FROM THEM?"[18] The cartoon communicates the idea that all citizens share a personal responsibility for the defense of the nation. But at that time, Geisel's only contribution to saving his country from totalitarianism had been drawing political cartoons. "I started receiving a lot of letters saying I was a dirty old man who had helped get us into the war," Geisel later recalled. The letters branded him as a hypocrite. Thus, at age thirty-eight, "I was too old to fight. So I enlisted."[19] On January 7, 1943, in New York City, Ted Geisel was sworn in as a captain in the US Army.

What drove Ted Geisel to enlist at age thirty-eight? Unlike many Americans, Geisel was aware of the danger posed by the Nazi regime and the threat of totalitarianism as a political philosophy more broadly. His family had suffered because of Germany's history of militaristic authoritarianism. During World War I, anti-German sentiment damaged his family's brewery business economically and provoked taunts from local boys.[20] He had seen the frenzied nationalism percolating in Germany in 1936 during a holiday in Bavaria. The Nazi spectacle of the Olympic Summer Games filled him with dread, and he found it impossible "to keep my mind off the storm that was going on."[21] Moreover, drawing cartoons concerning current events in Nazi Germany required deep and extensive knowledge of both Germany's past and present.[22] More than most Americans, whose imagination did not extend further than Mulberry Street, Geisel understood that becoming a soldier meant risking life and limb.

Assigned to the US Army's Information and Education division, Geisel found himself under the command of Major Frank Capra at Fox Studios in California. His new job was to make films in support of the Allied war effort, including the *Private Snafu* cartoon series. The twenty-six animated shorts of *Private Snafu* often appeared as the final segment of the *Army-Navy Screen Magazine*. Instead of the sleep-inducing approach typical of US military training programs, *Private Snafu* subverted the conventions. By adopting common Army slang (Situation Normal, All Fucked Up) as the title; inserting semi-nude pinup posters on the walls of characters' barracks that would elude the censor; and by featuring a clumsy, bewildered character totally unfit for military service, *Private Snafu* fully captured the attention of young recruits.[23] Every week Snafu made a different mistake: "telling army secrets to civilians while on leave; writing army secrets to his girlfriend; spreading rumors at his base; failing to take precautions against getting malaria."[24] As they watched the films, young soldiers laughed at Snafu's blunders, but the cascading consequences of Snafu's actions communicated the message that US Army rules and procedures (no matter how trivial, aggravating, or incomprehensible) existed for a legitimate reason and ought to be followed.

After the liberation of Paris in 1944, the Allies became increasingly optimistic about the defeat of Nazi Germany and the Axis powers. Anticipating the end of the war, Major Capra assigned Geisel to create a film to prepare young soldiers (and the American public) for the occupation of Germany.

When the film was finished, so the story goes, Captain Geisel had to obtain approval from all the US generals in Europe before the film was shown to US armed forces. After traveling to Paris in November 1944, Geisel spent the next few weeks visiting military headquarters and showing his film to senior commanders. (According to Geisel, "When someone else showed Gen. George Patton the film, he allegedly said 'Bullshit!' and walked out of the room.")[25]

While in Luxembourg in 1944, Ted Geisel serendipitously encountered an old friend from New York named Ralph Ingersoll. Before the war, Ingersoll had been managing editor of the *New Yorker* and *Fortune*. Subsequently, Ingersoll created *PM Magazine* to better express his leftist, liberal views. He hired Geisel (along with other talented, highly regarded cartoonists, essayists, and literati) to draw creative political cartoons, and they became close friends. Once the war began, Ingersoll volunteered for military service. His primary occupational specialty was deception planning, including crafting battlefield illusions to fool the Germans about the size and location of American forces.[26] When he and Geisel bumped into each other in Luxembourg, Ingersoll was a lieutenant colonel in Army intelligence working at Gen. Omar Bradley's 12th US Army Group headquarters in the Special Plans Branch.[27]

At Geisel's request, Ingersoll made arrangements for him to visit the front lines. Based on intelligence reports about the movement of German forces, Ingersoll "concluded that the 'safest' place would be the 106th Division front—a very calm sector."[28] On December 16, Geisel and his driver set off in a jeep north from Luxembourg City, arriving in the morning at a small town called Bastogne. Later that afternoon, they resumed their journey toward the front, where Geisel planned to link up with the 106th Division. Trundling through the French countryside, Geisel remained completely unaware of the Germans' surprise counteroffensive engulfing Bastogne, the "boring" little town in which he had eaten breakfast. According to Geisel, "the thing that probably saved my life" during the Battle of the Bulge (as it later became known), "was that I got there in the early morning and the Germans didn't arrive until that night."[29]

With Bastogne now under the control of the Wehrmacht, Geisel's route toward the front brought him deeper and deeper behind enemy lines. Disoriented by the rain, neither Ted Geisel nor his driver could make sense of the battered Michelin road map. They crossed paths with an American military police officer running in the opposite direction, who told them that they were ten miles behind enemy lines. "I thought the fact that we didn't seem to be able to find any friendly troops in any direction was just one of the normal occurrences of combat," Geisel recalled. "Day of semi-exhaustion semi-battle fatigue," he wrote in his journal. Having survived behind enemy lines during one of the bloodiest battles of World War II, this was certainly an understatement.

In the years that followed Geisel rarely discussed his survival at Bastogne or his rescue by a British unit.[30] But these wartime experiences left numerous traces in his children's books. Tucked in between the bright and vivid drawings of imaginary animals in Dr. Seuss books, the reader will sometimes encounter foreboding dark forests, ruthless authority figures, advanced weaponry, and other evocations of military life. In *Bartholomew and the Oobleck* (1949), for example, the green goo that falls from the sky (destroying livestock, ripping windows off their hinges, and terrifying civilians) resembles aerial bombardment of European cities.[31] In *Oh, the Places You'll Go!*, the protagonist comes "to a place where the streets are not marked. / Some windows are lighted. But mostly they're darked." Precariously balanced skeletons of destroyed buildings, a looming black church, and partially buried bunkers suggest the ravaged, decimated towns of Europe. In *Did I Ever Tell You How Lucky You Are?*, the poor, lonely protagonist wanders in a dark forest, full of menacing trees with menacing, prehensile branches. "Be grateful you're not in the forest in France," Dr. Seuss reminds us (or perhaps himself). "Where the average young person just hasn't a chance / To escape from the perilous pants eating plants." Having survived the Battle of the Bulge in a forest in France, Geisel certainly understood the role of chance in avoiding peril.

HIDDEN IN PLAIN SIGHT

If we could ask Geisel whether the books of Dr. Seuss contain important wisdom about the art of war, he would almost certainly respond with a flat and firm denial. Of course, disavowal is a strategy frequently used by artists to maintain "plausible deniability," as it is called in national security circles. For example, Picasso often refused to answer questions about his artwork, to the point where his denial of obvious facts became comical. Once when asked about the extensive collection of African sculpture in his studio, Picasso said, "African sculpture? . . . I have never heard of it!"[32] Likewise, when Picasso was asked about his most famous painting, *Guernica*, he intentionally obfuscated the truth: "If you give a meaning to certain things in my paintings, it may be very true, but it is not my idea to give this meaning. What ideas and conclusions you have got I obtained too, but instinctively, unconsciously, I make the painting for the painting. I paint the objects for what they are."[33]

American art critics accepted Picasso's statement at face value and "more or less completely ignored" the painting's obvious connection to the Spanish Civil War.[34] With *Guernica* decoupled from its historical context, new meanings could be attributed to Picasso's painting. For almost one hundred years, the meaning of *Guernica* has been debated in classrooms and in the pages of obscure, peer-reviewed journals. Does the bull in the painting represent the continuation of the Spanish nation, and is it therefore a symbol of hope?[35] Or does the bull signify the cruelty of fascism, and should it be viewed as a symbol of despair?[36] Or maybe the bull actually represents Picasso's ego?[37]

Despite the profusion of theories about the meaning of the painting and the confusion intentionally created by Picasso himself, the facts cannot be disputed. *Guernica* was painted on commission from the Spanish Republic for the International Exhibition held in Paris in 1937.[38] Picasso began working on the canvas following an aerial bombardment of Guernica by the German Luftwaffe, notably the first case of saturation bombing in military history. Without concern for collateral damage, German bombs decimated the Spanish Republican forces entrenched there.[39] Given that specific historical context, the allegorical meaning of *Guernica* cannot convincingly be denied.

So why did Picasso continue to insist that "the bull is a bull and the horse is a horse"?[40] Simply put, by denying that his artwork had any inherent meaning, Picasso deflected attention from the political content and cultural criticism inherent in the artwork. Although perhaps a bit duplicitous, this tactic spares the artist from political disagreements with either their audience or critics. Nobody could accuse Picasso of political opposition to Franco's regime if *Guernica* had no symbolic meaning.

Like Picasso, Geisel rejected the assertion that his books had any symbolic meaning. As he noted in an interview in 1986, "Sometimes people find morals where there are none. People have read all kinds of things into [my books]. . . . I'm getting blamed for a lot of stuff I haven't done."[41] If *The Cat in the Hat* had no symbolic meaning, then Geisel could not be accused of opposing traditional gender roles in the nuclear family, advocating destruction of middle-class materialism, or suborning anarchy as a political philosophy. Denial of symbolic meaning deflects political disagreement, allowing the reader to bumble along with their unexamined assumptions about the artwork, never recognizing the content hidden in plain sight.

Over time unexamined assumptions about the artwork's meaning become cemented in the minds of the audience, eventually acquiring so much power over perception as to render invisible the clues hidden in plain sight. Consider Edward Munch's painting *The Scream* (1893), which for more than one hundred years was misinterpreted as a self-portrait of existential despair.[42] But in 2019 the European news media reported an exciting new discovery: while preparing for a retrospective exhibition of Munch's work at the British Museum, curator Giulia Bartrum noticed an inscription on the back of an early lithograph of the painting, which said, "I felt the great scream throughout nature."[43] She concluded that "this rare version of *The Scream* that we're displaying makes clear that Munch's most famous artwork depicts a person hearing a 'scream' and not, as many people continue to assume and debate, a person screaming."[44] The meaning of *The Scream* had not been intentionally kept secret, locked away in a dark vault. The truth had just gone unnoticed.

In the case of Dr. Seuss, our deeply entrenched, unexamined assumptions about his books are rarely questioned. Readers generally assume that Dr. Seuss books adhere to the conventions of the genre (e.g., a friendly protagonist, plot resolution, simple language), contain no frightening or inappropriate content (e.g., no depictions of sex or death), do not have a hidden political agenda, and so forth. The general consensus is that Dr. Seuss books support the development of a democratic "civic identity,"[45] affirm the agency of children,[46] promote acceptance and tolerance,[47] and so on.

Inversion of such deeply entrenched assumptions can come as quite a shock. In 2019, Katie Ishizuka and Ramón Stephens's article "The Cat Is Out of the Bag: Orientalism, Anti-Blackness, and White Supremacy in Dr. Seuss's Children's Books" did just that. Using frequency counts of white and nonwhite characters and coding for themes (such as dominance, subservience, exotification, and so on) in fifty-nine Dr. Seuss books, they concluded that the "presence of anti-Blackness, Orientalism, and White supremacy span across Seuss's entire literary collection and career."[48] They claimed that far from promoting tolerance (e.g., "a person's a person, no matter how small"), Dr. Seuss books contained derogatory, offensive stereotypes of women and minorities. Apparently these demeaning portrayals had been hiding in plain sight, unnoticed by critics, teachers, and parents alike, right there on the pages of the books.

Reacting in shock at Ishizuka and Stephens's revelations, the National Education Association (NEA) removed a number of Dr. Seuss books from the Read Across America program.[49] Subsequently, the Seuss Foundation withdrew six books from publication in 2019: *And to Think That I Saw It on Mulberry Street* (1937), *If I Ran the Zoo* (1950), *McElligot's Pool* (1947), *On Beyond Zebra!* (1955), *Scrambled Eggs Super!* (1953), and the *Cat's Quizzer* (1976).[50] Defenders of Dr. Seuss argued that there "wasn't a racist bone in that man's body,"[51] that removing books from circulation was "extreme,"[52] and so on. Conservatives pointed to the decision of the Seuss Foundation as an example of "cancel culture,"[53] and the House minority leader, Kevin McCarthy, shared a video of himself reading from *Green Eggs and Ham* just to stir things up.[54]

Criticism of Dr. Seuss is nothing new. Over the years, detractors have claimed that *The Cat in the Hat* includes racist references to blackface,[55] that the "silent, secondary roles" of females in Dr. Seuss books indicate Geisel's sexism,[56] and so on. Like earlier critics, Iskuza and Stephens object to stereotyped imagery (such as turbans, chopsticks, and exotic birds) and textual themes that disempower minorities. For example, in their view *Horton Hears a Who* "positions the Whos in a deficit-based framework as the dominant, paternalistic Horton enacts the White Savior Industrial Complex."[57] Similarly, they argue that "Plain-Belly Sneetches never challenge their oppressor or the oppression itself. They never resist. . . . The Plain-Bellied Sneetches play out an unrealistic scenario of overcoming the intentional discrimination of individual Star-Bellied Sneetches through conformity and assimilation."[58]

The fundamental flaw in Ishizuka and Stephens's argument, however, involves unprovable assertions regarding unmeasurable effects (an unfortunate tendency found in most "critical theory"). Their explicitly stated research question is "How and to what extent are non-White characters depicted in Dr. Seuss's children's books?" However, their actual argument concerns *the effect of these depictions* on social groups and society in general. For example, the *depiction* of the Sneetches (according to Ishizuka and Stephens) as "moping and doping" refers to the "self-hatred" of oppressed groups. On the other hand, *the effect of that depiction* (according to Ishizuka and Stephens) is that it "actually *reinforces* White supremacy by upholding deficit-based, disempowered narratives of oppressed groups and promoting colorblindness."[59] By using the word "reinforce," Ishizuka and Stephens imply that Dr. Seuss books are somehow responsible for racism in the United States. Other words used in their essay (such as "promote," "shape," and "ingrain") also attribute causation to Dr. Seuss books. In short, the authors assert that Dr. Seuss books have *infected the minds of American children* with prejudice and intolerance (despite the explicit intentions of Geisel himself).

Proving causation in the realm of ideas is notoriously difficult, but very easy to satirize.[60] In 1977, for example, an essay entitled "Literary Rape in America: A Post-Coital Study of the Writings of Dr. Seuss" argued that his books encouraged promiscuous sexuality through "pubescent, florid and burlesque" drawings. Because some Dr. Seuss characters were "covered from head to foot in tremendous growths of hair that can only be described as pubic,"[61] Dr. Seuss should be "held responsible for the exponentially rising promiscuity rates of our post-beat, early post-hippie generation."[62] Needless to say, drawings of fuzzy animals in children's books do not cause promiscuity. Neither do illustrations of Inuit hunters in *Scrambled Eggs Super* (1953) *cause* racism (although they might reflect the cultural stereotypes of the time and place in which they were created).

LESSONS FROM THE MASTER

Like Ishizuka and Stephens's essay, this edited volume challenges commonly held assumptions about Dr. Seuss, specifically rejecting the idea that his books contain nothing violent, gruesome, or bleak. On the contrary, Dr. Seuss books reflect Geisel's World War II experiences and therefore include imagery of dark forests, lonely children, and bombed cities. His books also contain technical military knowledge that Geisel gained from his Army service, such as deception, propaganda, and covert operations. Warfare, in fact, permeates the books of Dr. Seuss.

What military lessons do the chapters in this edited volume extract from Dr. Seuss? Some of the chapters focus on the political conditions that justify warfare (such as abuse of power by authoritarian governments) and why individuals in those political systems chose to fight. In chapter 8, on *Yertle the Turtle*, Dr. Katherine Blue Carroll (an associate professor at Vanderbilt University) interprets Dr. Seuss's story about the rise and fall of the turtle king as a parable about totalitarianism. Using examples from the Middle East to illustrate theories of universal validity, Carroll asks how authoritarian rulers stay in power and why some individuals take great personal risks to resist dictatorship. Dr. Carroll brings tremendous knowledge and experience to her analysis, having served as a US Army civilian adviser in Iraq and written extensively about Middle Eastern politics, the US military, and political Islam.

Some of the chapters in this volume address the beliefs and experiences common to soldiers in every society. In chapter 11, about *Did I Ever Tell You How Lucky You Are?*, Erich Henry Warner and Montgomery McFate argue that soldiers have attributed their survival in combat to luck since the battles of antiquity. Geisel not only understood the role of luck in human affairs but also attributed his own good outcome to luck. Colonel Wagner brings deep knowledge and experience as an officer in the Marine Corps Reserve (and as the Marine Corps's senior uniformed historian) to his analysis of Geisel's work.

A number of chapters in this volume focus on military strategy. Antulio J. Echevarria II (chapter 2) makes the case that *I Had Trouble in Getting to Solla Sollew* illuminates many challenges in the development and execution of military strategy such as resource hijacking, poorly defined end-states, and (in Carl von Clausewitz's terminology) "friction." Dr. Echevarria brings deep practical and theoretical knowledge of military strategy to his analysis of this seminal Dr. Seuss text, having served as a US Army officer and currently holding the MacArthur Chair of Research at the US Army War College. With a similar focus on strategy, Saadia M. Pekkanen (chapter 3) uses *Oh, the Places You'll Go!* to illuminate three processes happening simultaneously in the space race: democratization, commercialization, and militarization. As the world's foremost expert in grand strategy in outer space, Dr. Pekkanen (who holds a PhD from Harvard and serves as a tenured faculty member at the University of Washington in Seattle) argues convincingly that this Dr. Seuss book provides the baseline grand strategy for the future of US outer space affairs. Similarly, in his chapter on *The Butter Battle Book*, Sam J. Tangredi (chapter 12) explores how contemporary theories of deterrence and escalation align with the implicit theories established in this classic of the Cold War era. Dr. Tangredi brings a deep knowledge of military strategy to his analysis of Dr. Seuss, having served as an officer in the US Navy and currently as Leidos Chair at the US Naval War College.

This volume also considers the consequences of warfare for individual military personnel and civil society in general. In her chapter on *How the Grinch Stole Christmas*, Montgomery McFate (chapter 13) argues that the Grinch suffers from post-traumatic stress disorder (PTSD) and that his symptoms serve as an allegorical representation of Geisel's traumatic wartime experiences.

A few chapters in this book concern neglected aspects of international military conflict, such as deception and propaganda. In chapter 10, on *Horton and the Kwuggerbug and other Lost Stories* (2014), Dr. Chris C. Demchak argues that western democracies have long abhorred or ignored deception as a practice in international relations. As a result of systemic changes caused by cyber warfare, deception must be reconsidered, and Dr. Seuss has much to tell us on this subject. As a full professor at the US Naval War College, Dr. Demchak brings powerful insights to her evaluation of cyber warfare, deception, and grand strategy. Propaganda, another neglected aspect of warfare, is the focus of Dr. Kevin P. Eubanks's chapter (7) on Geisel's World War II–era film series *Private Snafu*. Dr. Eubanks makes the case that although they were crafted to intentionally persuade the intended audience of young, male US Army recruits to obey military orders and regulations, the *Snafu* films reinforce the same set of core liberal democratic principles advocated for in Dr. Seuss's better known postwar children's literature. As a professor at the US Naval War College with a PhD from the University of North Carolina at Chapel Hill, Dr. Eubanks brings deep knowledge about the power of cinema in the context of war. In their chapter on *Thidwick the Big-Hearted Moose* (chapter 5), Dr. Rebecca Pincus and Montgomery McFate argue that this Dr. Seuss book contains a powerful warning about the security consequences of exceeding planet Earth's carrying capacity. In addition to a PhD from the University of Vermont, Dr. Pincus also brings extensive governmental experience on climate policy and strategy to her analysis.

Some of the chapters in this book touch on fundamental principles of warfare that directly affect contemporary operations. In their chapter about *The Cat in the Hat*, Dr. Jon R. Lindsay and Dr. Michael Poznansky (chapter 6) argue that processes and principles in the Dr. Seuss book pertain directly to cyber operations, including the target's cooperation—witting or unwitting—as a permissive condition, unintended consequences for the perpetrators, the challenge of establishing attribution, and broader cultural dimensions of hacking. As a former US Navy officer and a professor at Georgia Institute of Technology, Jon Lindsay brings deep knowledge and expertise to his analysis of this seminal text. As a professor at the US Naval War College with a PhD from the University of Virginia, Michael Poznansky contributes clear insights into our understanding of Dr. Seuss. Also concerned with fundamental principles of warfare with direct pertinence to contemporary operations, John Hursh's chapter on *Horton Hears a Who* (chapter 4) demonstrates how human rights law establishes international norms applicable to armed conflict, such as tolerance, protection of civilians, prohibition on torture, and collective action. As an attorney specializing in the laws of war and a former professor at the US Naval War College, Hursh brings considerable expertise to his analysis of Dr. Seuss. In their chapter on *Hunches in Bunches* (chapter 9), Dr. Genevieve Lester, John Nagl, and Dr. Montgomery McFate argue that this Dr. Seuss book illuminates many classic challenges pertaining to the use of intelligence in national security decision-making, such as bureaucratic pressure, cognitive fallacies, and untrustworthy sources, among other topics. With a PhD from University of California at Berkeley and extensive experience working in the US government, Dr. Lester brings both academic and practical knowledge to the subject of intelligence and national security. As a professor at the US Army War College with many years of service in the US Army, John Nagl brings deep insight on the pragmatic aspects of government policy.

CONCLUSION

The chapters in this edited volume interpret Dr. Seuss from a military perspective, revealing new insights about the art of war. Because this approach inverts the paradigmatic view of Dr. Seuss books as stories for children, the question arises: How can this newly revealed military dimension of Dr. Seuss be rectified with the messages of tolerance, equality, and protection of the weak in his books?

Political philosophers have long debated whether a contradiction exists between democratic ends and military means. In 1796, President George Washington urged Americans to "avoid the necessity of those overgrown military establishments which, under any form of government, are inauspicious to liberty, and which are to be regarded as particularly hostile to republican liberty."[63] Washington, like many of the Founding Fathers, believed that the concentration of power in a military institution poses a threat to democratic norms and practices. Yet given the bellicose nature of the international system (and perhaps the nature of humanity itself), all governments require military force to defend their states and ensure the continuity of their society.

The foundational ideas in Dr. Seuss (such as equality, individualism, skepticism of absolute authority) derive from the Constitution and the Bill of Rights. At a very tender age, children in the United States first encounter many of these concepts (such as liberty, free speech, and so on) in *Horton Hears a Who* and other Dr. Seuss books. The story of a kind-hearted elephant named Horton teaches children the principle of equality, that "a person's a person, no matter how small." But *Horton* also teaches children that recognizing the humanity of others means more than passive acceptance of their existence. Individuals must *act on their beliefs*. After hearing the voices of the Whos, Horton voluntarily assumes responsibility for their welfare, refuses to be bullied, acts according to his conscience, survives torture and humiliation, does not surrender when faced with impossible odds, does not complain, does not bear a grudge, and makes his community in the Jungle of Nool a better place.

Without question, Geisel believed in the democratic ideas found in his books. But he also subscribed to the idea of the citizen-soldier, that all members of a society shared a personal responsibility and civic duty to defend and protect the nation. Despite his fame and fortune, Geisel volunteered for military service during World War II, thereby bringing his actions into alignment with his ideals. Just as Horton the Elephant risked his life to defend his values, Geisel believed so strongly in these democratic ideals that he was willing to die for them. Far from contradicting the ethos of Dr. Seuss, this edited volume actually aligns with Geisel's life experiences, political beliefs, and civic values. In the words of Dr. Seuss in *I Can Read with My Eyes Shut* (1978): "The more that you read, / the more things you will know. / The more that you learn, / the more places you'll go."

NOTES

Epigraph quoted in Jonathan Cott, "The Good Dr. Seuss," in *Of Sneetches and Whos and the Good Dr. Seuss: Essays on the Writings and Life of Theodor Geisel*, ed. Thomas Fensch (Chesterfield, VA: New Century Books, 1997), 117.

1. Dr. Seuss, *Hunches in Bunches* (New York: Random House, 1982); and Dr. Seuss, *The Cat in the Hat* (New York: Random House, 1957).
2. Dr. Seuss, *On Beyond Zebra!* (New York: Random House, 1955).
3. Sun Tzŭ, *Art of War*, trans. Lionel Giles (1910), https://www.gutenberg.org/files/132/132-h/132-h.htm.
4. Dr. Seuss, *Oh, the Places You'll Go!* (New York: Random House, 1990).

5. Sun Tzŭ, *Art of War*.
6. Dr. Seuss, *Fox in Socks* (New York: Random House, 1965).
7. Sun Tzŭ, *Art of War*.
8. Dr. Seuss, *Oh, the Places*. On the subject of appropriate tactics for combat with a dispersed force, Sun Tzu says: "Now the shuai-jan is a snake that is found in the Ch'ang mountains. Strike at its head, and you will be attacked by its tail; strike at its tail, and you will be attacked by its head; strike at its middle, and you will be attacked by head and tail both." Dr. Seuss uses the same analogy of tail and head in the *Sleep Book* (New York: Random House, 1962): "Now the news has arrived / From the Valley of Vail / That a Chippendale Mupp has just bitten his tail, / Which he does every night before shutting his eyes. / Such nipping sounds silly. But, really, it's wise. He has no alarm clock. So this is the way / He makes sure that he'll wake at the right time of day. / His tail is so long, he won't feel any pain / 'Til the nip makes the trip and gets up to his brain./ In exactly eight hours, the Chippendale Mupp / Will, at last, feel the bite and yell 'Ouch'! and wake up." On the topic of preparedness, Sun Tzu notes: "So when the front is prepared, the rear is lacking, and when the rear is prepared the front is lacking. Preparedness on the left means lacks on the right, preparedness on the right means lack on the left. Preparedness everywhere means lack everywhere." Dr. Seuss comes to the same conclusion in *I Had Trouble Getting to Solla Sollew* (1965): "I learned there are troubles / Of more than one kind. /Some come from ahead and some come from behind." "But I've bought a big bat. / I'm all ready, you see. / Now my troubles are going / To have troubles with *me*!"
9. Karen Kovach, "Genocide and the Moral Agency of Ethnic Groups," *Metaphilosophy* 37, nos. 3/4 (July 2006): 618–38.
10. Thomas Culora, email message to author, October 1, 2023.
11. Captain Robert "Gus" Gusentine, email message to author, November 3, 2023.
12. John Hursh, "International Law, Armed Conflict, and the Construction of Otherness: A Critical Reading of Dr. Seuss's *The Butter Battle Book* and a Renewed Call for Global Citizenship," *New York Law School Law Review* 58, no. 617 (2014): 618–52.
13. Roger S. Clark,"Is *The Butter Battle Book's* Bitsy Big-Boy Boomeroo Banned? What Has International Law to Say about Weapons of Mass Destruction," *New York Law School Law Review* 58 (2013–14): 655–73.
14. Chris Heatherly, "Dr. Seuss and the Operational Art of War," *Armchair General*, April 27, 2010, http://armchairgeneral.com/dr-seuss-and-the-operational-art-of-war.htm.
15. Jack Zipes, *Irresistible Fairy Tale: The Cultural and Social History of a Genre* (Princeton, NJ: Princeton University Press, 2012), http://www.jstor.org/stable/j.ctt7sknm.
16. For basic biographical facts see Judith Morgan and Neil Morgan, *Dr. Seuss & Mr. Geisel: A Biography* (New York: Random House, 1995).
17. Ted Geisel, quoted in Philip Nel, *Dr. Seuss: American Icon* (New York: Continuum, 2003), 39.
18. Nel, *Dr. Seuss*, 45.
19. Nel, *Dr. Seuss*, 59.

20. Donald E. Pease, "Dr. Seuss in Ted Geisel's Never-Never Land," *PMLA* 126, no. 1 (January 2011): 198.

21. Mary Stofflet, *Dr. Seuss from Then to Now: A Catalogue of the Retrospective Exhibition* (San Diego, CA: San Diego Museum of Art, 1986), 29.

22. Richard H. Minear, *Dr. Seuss Goes to War: The World War II Editorial Cartoons of Theodor Seuss Geisel* (New York: New Press, 2001).

23. Benjamin L. Alpers, "This Is the Army: Imagining a Democratic Military in World War II," *Journal of American History* 85, no. 1 (June 1998): 148.

24. Alpers, "This Is the Army."

25. Rick Beyer, "Seuss on the Loose," *Dartmouth Alumni Magazine*, November–December 2004, https://dartmouthalumnimagazine.com/articles/seuss-loose.

26. Beyer, "Seuss on the Loose."

27. Beyer, "Seuss on the Loose."

28. Beyer, "Seuss on the Loose."

29. Morgan and Morgan, *Dr. Seuss & Mr. Geisel*, 114.

30. Ted Geisel, quoted in E. J. Kahn, "Children's Friend," *New Yorker*, December 17, 1960.

31. Pease notes that *Thidwick the Big-Hearted Moose* (1947) contains drawings that evoke World War II "territorial occupation." Donald E. Pease, *Theodor SEUSS Geisel* (New York: Oxford University Press, 2010), 81.

32. Andrea Giunta, "The Power of Interpretation (or How MoMA explained Guernica to its Audience)," trans. Jane Brodie, *Artelogie* (October 2017), https://doi.org/10.4000/artelogie.953 https://journals.openedition.org/artelogie/953?lang=en#ftn52.

33. Picasso, quoted in "Guernica," *Treasures of the World*, Public Broadcasting System, 1999, https://www.pbs.org/treasuresoftheworld/guernica/glevel_1/5_meaning.html

34. Jutta Held and Alex Potts, "How Do the Political Effects of Pictures Come about? The Case of Picasso's 'Guernica,'" *Oxford Art Journal* 11, no. 1 (1988): 33–39, http://www.jstor.org/stable/1360321.

35. Carla Gottlieb, "The Meaning of Bull and Horse in Guernica," *Art Journal* 24, no. 2 (1964): 106–12, https://doi.org/10.2307/774777.

36. Gottlieb, "Meaning of Bull and Horse in Guernica."

37. John Banville, "The Power and the Glory of Pablo Picasso," *New Republic*, November 16, 2021, https://newrepublic.com/article/164381/picasso-biography-richardson-review-power-glory.

38. Held and Potts. "Political Effects of Pictures,'" 33–39.

39. Ian Patterson, *Guernica and Total War* (Cambridge, MA: Harvard University Press, 2007).

40. Gottlieb, "Meaning of Bull and Horse in Guernica," 106–12.

41. Jonathan Cott, "'Somebody's Got to Win' in Kids' Books: An Interview with Dr. Seuss on His Books for Children, Young and Old," in *Of Sneetches and Whos and the Good Dr. Seuss: Essays on the Writings and Life of Theodor Geisel*, ed. Thomas Fensch (Chesterfield, VA: New Century Books, 1997), 125.

42. Vernon McCay and Marjie L. Baughman, "Art, Madness, and Human Interaction," *Art Journal* 31, no. 4 (1972): 413–20, https://doi.org/10.2307/775545.

43. Jack Guy, "Everything You Thought about 'The Scream' Is Wrong," *CNN*, March 21, 2019, https://www.cnn.com/style/article/munch-scream-british-museum-gbr-scli-intl/index.html.

44. Sami Quadri, "The Scream May Not Be Screaming," *Daily Mail*, 20 March 2019, https://www.dailymail.co.uk/news/article-6832823/The-Scream-not-screaming-Edward-Munchs-iconic-artwork-shows-no-thing.html.

45. Donald E. Pease, "Dr. Seuss's (Un)Civil Imaginaries," *New York Law School Law Review* 58, no. 509 (2013): 511.

46. Anne McGillivray, "Horton Hears a Twerp: Myth, Law, and Children's Rights in *Horton Hears a Who!*" *New York Law School Law Review* 58, no. 509 (2013): 597–629.

47. Kendall N. Lange, "Oh, The Things You Can Find (If Only You Analyze): A Close Textual Analysis of Dr. Seuss' Rhetoric for Children" (MA thesis, Kansas State University, 2007).

48. Katie Ishizuka and Ramón Stephens, "The Cat Is Out of the Bag: Orientalism, Anti-Blackness, and White Supremacy in Dr. Seuss's Children's Books," *Research on Diversity in Youth Literature* 1, no. 2 (2019): 28.

49. Barajas, "How Teachers Turned Away."

50. Bill Chappell, "Dr. Seuss Enterprises Will Shelve 6 Books, Citing 'Hurtful' Portrayals," *NPR*, March 2, 2021.

51. Quoted in Joshua Rhett Miller, "Chicago Public Library to Yank Six Dr. Seuss Books from Shelves," *New York Post*, March 9, 2021.

52. Ray Kelly, "Dr. Seuss' Great-Nephew Calls Museum Mural Removal 'Extreme,' Criticism 'a Lot of Hot Air Over Nothing,'" *Mass Live*, October 11, 2017, www.masslive.com/news/index.ssf/2017/10/dr_seuss_great_nephew_calls_mu.html.

53. Alexandra Alter and Elizabeth A. Harris, "Dr. Seuss Books Are Pulled, and a 'Cancel Culture' Controversy Erupts," *New York Times*, March 4, 2021.

54. Li Cohen, "'I Still Like Dr. Seuss': Kevin McCarthy Releases Video of Himself Reading 'Green Eggs and Ham,'" *CBS News*, March 6, 2021.

55. Philip Nel, "Was the Cat in the Hat Black? Exploring Dr. Seuss's Racial Imagination," *Children's Literature* 42, no. 1 (2014): 71–98.

56. Alison Lurie, "The Cabinet of Dr. Seuss," *New York Review of Books*, December 20, 1990.

57. Ishizuka and Stephens, "Cat Is Out of the Bag," 18.

58. Ishizuka and Stephens, "Cat Is Out of the Bag," 26.

59. Similarly, they believe *Horton Hears a Who* "reinforces themes of White supremacy, Orientalism, and White saviorism."

60. Gary King, Robert O. Keohane, and Sidney Verba, *Designing Social Inquiry: Scientific Inference in Qualitative Research* (Princeton, NJ: Princeton University Press, 1994).

61. E. Strong, "Literary Rape in America: A Post-Coital Study of the Writings of Dr. Seuss," *Journal of Contemporary Satire* 4 (1977): 35.

62. Strong, "Literary Rape in America."

63. "The Philadelphia Centennial," *American Advocate of Peace and Arbitration* 49, no. 5 (1887): 125, http://www.jstor.org/stable/45404791.

Part II

BOUNDING THE SUBJECT

2

I Had Trouble in Getting to Solla Sollew and Strategy

Antulio J. Echevarria II

Although somewhat obscure, Dr. Seuss's *I Had Trouble in Getting to Solla Sollew* (1965) contains significant lessons and advice for strategists everywhere. In fact, the book seems almost to have been constructed with this intention. With the (possible) exception of the palace guards in *Bartholomew and the Oobleck*, *I Had Trouble in Getting to Solla Sollew* contains the only representation of a soldier in the entire Dr. Seuss canon. Moreover, this is the only Dr. Seuss book with a narrative involving a military campaign, which in this instance includes deployment of forces, the commander's leadership, the battle, and the aftermath. For understanding Ted Geisel's views on warfare and strategy, *I Had Trouble in Getting to Solla Sollew* proves indispensable.

The plot is simple. At the beginning of *Solla Sollew*, a furry yellow protagonist walks through the Valley of Vung, where "nothing, not anything ever went wrong." After tripping over a rock in his path, our friend's troubles begin. He increases his vigilance in looking for more rocks in the path, but doing so makes him vulnerable to a rear echelon attack from "a very fresh green-headed Quilligan Quail." Now crouched in a defensive posture, the protagonist monitors both front and back, but a troublesome Skritz attacks from above, while an annoying Skrink attacks from below.[1] Just in the nick of time, a chap riding a camel-propelled cart invites our furry protagonist to join him on a trip "to the City of Solla Sollew / On the banks of the beautiful River Wah-Hoo, / Where they *never* have troubles! At least, very few." So begins the quest for the mythical utopia of Solla Sollew, involving a variety of improbable and yet cascading setbacks. When he arrives at his destination, our protagonist cannot enter the city gates and must return home to face his problems directly, with the aid of a baseball bat (a potentially lethal weapon).

Most readers interpret the message of *Solla Sollew* to mean running away from one's troubles solves nothing; it is much better to confront them directly. Indeed, dreaming of escape only deprives one of the energy and determination to overcome threats and challenges at home. One reviewer at the time likened Geisel's book to John Bunyan's *Pilgrim's Progress* (1678), with the City of Solla Sollew representing Bunyan's Celestial City.[2] The parallel only extends so far, however, as Geisel's narrative presents Solla Sollew as an escape rather than a heavenly reward. Geisel saw Solla Sollew as a dream destination and not even a good one. The book itself is an engaging anti-utopian allegory with some fundamental lessons for strategists.

Theodor Geisel (Dr. Seuss) did not consider *I Had Trouble in Getting to Solla Sollew*, published in 1965, to be one of his "more successful books."[3] It received some acclaim from reviewers, but its sales were unremarkable. One reason for the lackluster sales might have been the name Solla Sollew, which could well have come across as too alien for American adults, who may have doubted whether the story would interest their children. As this chapter will show, the story should interest strategists. Geisel served in the US Army during World War II as a propagandist (or a specialist in information operations, as the function might be called today), which gave him some exposure to the fundamentals of military strategy.[4]

What is strategy, one might ask? In its most essential form, strategy is the "practice of reducing an adversary's physical capacity and willingness to fight and continuing to do so until one's aim is achieved."[5] This is the competitive dynamic that lies at the heart of both military strategy and grand strategy, each of which consists of any combination of national power—military, political, economic, or cultural. Military strategists and grand strategists both use national power to out-position adversaries, though the former will typically craft military operations or campaigns to do so, whereas the latter will usually leverage alliances, coalitions, treaties, and other multilateral agreements to do so.

Using insights from Solla Sollew, this chapter argues that all strategies require the resolution of troubles or problems. However, strategic pitfalls, such as having one's interests and resources hijacked by other parties, may derail even the best strategy. Intervening in the affairs of others can develop into another pitfall because it can distract us from our goal or put an unwelcome drain on our resources. Geisel suggests that rather than becoming sidetracked by trying to solve troubles abroad, we should put more effort into solving those we have at home. An important theme running through the story is a phenomenon the famous Prussian theorist Carl von Clausewitz called "friction." Friction represents the sum of impediments, large and small, that can make any military action, any operation, or indeed any overarching military or grand strategy fail. Even a well-crafted strategy that accounts for all the aforementioned pitfalls, as well as others, stands to fail if the goal, or end state, we are pursuing is unrealistic or poorly chosen. Strategists should also note the warning in the book's conclusion, where Geisel raises the question of whether using force to solve our troubles will only make them worse or create more of them.

TROUBLES

The narrative of *I Had Trouble in Getting to Solla Sollew* rests on an element common to all types of strategies, namely, troubles or problems. The protagonist's efforts to solve problems as they arise, rather than dealing with their root causes, only results in more of them. After the furry protagonist accepts a lift from the camel driver, for instance, he complains that "the road got more bumpy, more rocky, more tricky. By midnight, I tell you, my stomach felt icky." When dawn arrives, both our protagonist and the camel driver have still not reached Solla Sollew but instead find themselves traversing a treacherous, steep mountain road.

> Instead of the city, we ran into trouble.
> Our camel was sick and he started to bubble.
> We had to pull *him* in the One-Wheeler Wubble!
> So there, there we were in a dreadful position.
> Our camel sure needed a camel physician.
> Now, doctors for camels are not often seen.
> Especially on mountains. They're far, far between.

Next, in the Vale of Vung, the camel driver offers a solution to the protagonist's troubles. But this solution becomes just another problem. The camel driver continually underestimates (or perhaps misrepresents) the duration and difficulty of the trip they have undertaken, and his cheerful demeanor perplexes and annoys the protagonist. Even the most optimistic readers are left wondering what terrible events will happen next. How can our poor yellow friend solve these never-ending problems?

All strategies emerge to solve troubles or problems. Problems occupy a spectrum running from threats (negative potentialities) to opportunities (positive possibilities). Naturally we want to reduce the former and increase the latter, in which case we need a strategy. Strategies are indispensable in competitive environments, in which our rivals actively work to diminish our influence and to gain advantages over us. (Whether the camel driver is an adversary remains unclear.) Absent a competitive environment, we can get by with a mere plan rather than a strategy. Building an automobile, for instance, does not require a strategy. But a strategy can be helpful if we are building an automobile in a competitive market in which consumers may choose from any number of other manufacturers.

Strategies and their components are easy to grasp in theory but difficult to put into practice. Executing a strategy can require considerable energy to overcome institutional inertia; agree to expend political capital; coordinate across organizations; allocate resources; build consensus concerning ends, ways, and means; and obtain public support. A twenty-first-century example is the challenge of global warming; it has taken years and considerable resources to convince international leaders to develop and implement a strategy to deal with this problem. Not all problems, moreover, are of sufficient gravity to galvanize the support necessary to overcome the aforementioned impediments. Instead, a more likely course of action is deciding simply to "muddle through," dealing with each problem only when it becomes a crisis and thus must be acknowledged as a priority. Muddling through, or treating troubles on a case-by-case basis rather than according to a grand vision, can be a viable form of strategy when we are in an uncertain environment and our actions may do more harm than good.[6] Our furry protagonist has a vague vision of Solla Sollew, but his efforts to get there constantly put him in jeopardy.

Solving troubles or problems is the basis of four major disciplines concerned with the study of strategy.[7] The first of these is the historical approach, which analyzes strategy through past events and interprets it largely inductively, as the product of political, cultural, and social influences.[8] The second approach befits the social sciences, which examine strategy largely deductively, as a body of generalizable theories wherein political actors behave rationally, according to their own interests.[9] Policymakers and practitioners make up the third approach, which attempts to apply accepted strategic theories and principles to specific situations to achieve desired outcomes.[10] Military strategists make up the last approach, which looks at strategy chiefly as a function of military power, with other forms of power (diplomatic, informational, and economic) serving in crucial but supporting roles.[11] Problem-solving, or an emphasis on dealing with troubles, underpins all of these strategic disciplines.

Our furry protagonist, muddling through sticky situations, most resembles the social scientist's and practitioner's approaches to strategy, since we see him make several deductions and attempt to apply theories and principles to new situations. We see little in the way of inductive reasoning or the application of military power by the protagonist. Moreover, Geisel's furry friend does learn from his experiences. At one point, for instance, the camel driver tricks him into pulling the cart, while the driver himself lounges on the seat with the camel. Our furry friend protests that "this is rather unfair!" But the camel driver replies by justifying his actions based on his own laziness and selfishness.

> But he said, "Don't you fuss. I am doing my share.
> "This is called teamwork. I furnish the brains.
> "You furnish the muscles, the aches and the pains.
> "I'll pick the best roads, tell you just where to go
> "And we'll find a good doctor more quickly, you know."

Our furry yellow friend complies by pulling the cart. But he no longer trusts the camel driver, who has used "his brain and his tongue" to manipulate the situation to suit himself.

Dr. Seuss has captured the spirit of adaptation in this passage; however, in so doing he portrays the protagonist as more reactive than proactive throughout the story. Our furry friend unwittingly follows or is drawn along an unfamiliar path, while demonstrating little foresight or willingness to prepare himself for what might lie ahead. Too many troubles pop up along the way for him to have stayed with any single strategy for very long. Perhaps only open-ended and flexible goals are possible in complex and rapidly changing situations, Geisel seems to say. Adaptation and resilience thus emerge as essential principles for guiding our strategies.

PSYCHOLOGICAL RESILIENCE

Traditionalists have interpreted *I Had Trouble Getting to Solla Sollew* as providing encouragement to young people to face their problems rather than running away from them. On some level, the book is about psychological resilience, or what in Geisel's day would have been referred to as mental toughness. Strategic thinkers from Sun Tzu to Clausewitz have praised military commanders who possessed the psychological resilience necessary to overcome the disappointments and setbacks of war and strategic competition.

At the midpoint of the protagonist's journey, we see his psychological resilience put to the test in a series of hardships and mishaps. One such hardship was pulling the cart over a very large mountain. After that, our furry friend takes his leave from the camel driver and his sick camel and eagerly hopes to continue the remainder of his journey by bus. But upon arriving at the bus stop, he learns the bus has punctured tires, and "until further notice, the 4:42 / Cannot possibly take you to Solla Sollew." No other buses are due, so the only option left is walking the very great distance to his destination. But other hardships and mishaps befall him along the way: "A hundred miles later /

My feet were so sore!" he cries. "THEN, wouldn't you know it! / It started to pour!" Miserable and wet, the protagonist finds shelter in a dark, abandoned house. He cowers under the covers on an uncomfortable cot, watching anxiously as the rain lashes the windows and lightning strikes a shadowy tree.

> I listened all night to the growls and the yowls
> And the chattering teeth of those mice and those owls,
> While the Midwinter Jicker howled horrible howls.

This moment is the nadir of our protagonist's morale, when the furry protagonist has fallen into despair after a constant barrage of troubles. Given the range of hardships and mishaps he faced, if he had quit it would have been understandable.

Geisel himself may well have considered quitting just before, or while, writing *Solla Sollew*. He had returned to civilian life after his Army service, and his success in writing children's books had steadily increased. The publishing house of Houghton Mifflin had approached him in the mid-1950s to start a series of Beginner Books for children to combat the growing problem of illiteracy among juveniles. The popularity of television, comic books, and other forms of pulp fiction was believed to be pulling down the literacy rates of young people.[12] His most popular work during this period, *The Cat in the Hat* (1957), achieved record sales, averaging some twelve thousand copies per month, making him a small fortune and turning Dr. Seuss into a household name.[13]

Although Geisel enjoyed considerable professional success during this period, he faced numerous personal troubles. From a psychological standpoint, *Solla Sollew* could reflect the author's efforts to convince himself to see his difficulties through, particularly those concerning his wife and creative partner, Helen. She had been diagnosed with Guillain-Barré syndrome in 1954. While she was thought to have recovered the following year, the symptoms—loss of balance, loss of vision, and paralysis below the waist—returned in 1964.[14] According to some accounts, she became fearfully possessive of Geisel during this period, which in turn impaired his creativity and caused him to want to distance himself from her. Friends later reported the two were "driving each other crazy," and at least one biographer suggested that he found a substitute family during this time.[15] The years 1964 and 1965, as he worked on *Solla Sollew*, were thus trouble-filled ones for Geisel, a time he later described as one of frustration for all involved. Tragically, this period of trouble and frustration culminated with Helen's suicide in October 1967. As Helen revealed in her suicide note, she felt a profound sense of failure and estrangement: "Loud in my ears from every side I hear, 'failure, failure, failure. . . .' I love you so much. . . . I am too old and enmeshed in everything you do and are, that I cannot conceive of life without you. My going will leave quite a rumor but you can say I was overworked and overwrought."[16]

I Had Trouble in Getting to Solla Sollew certainly expresses some of the darkness of Geisel's wife's suicide and his own attitude toward the event. Given his personal situation, a paradise "where they *never* have troubles! At least, very few" must have seemed quite alluring.

The book's conclusion shows that Geisel realized there were no Solla Sollews, no paradises where one might find refuge. The symbolism of the text makes this apparent when the little furry protagonist falls asleep in the house where he has taken refuge from the storm, dreaming he has arrived in Solla Sollew, slumbering comfortably on "soft silk and satin marshmallow-stuffed pillows." However, he suddenly awakes to discover the house has been swept off a cliff by raging floodwaters!

> I was crashing downhill in a flubbulous flood
> With suds in my eyes and my mouth full of mud
> And my nose full of water, my ears full of shrieks
> Of the owls flying off with mice on their beaks!
> And I said to myself, "Now I really don't see
> "Why troubles like this have to happen to *me*!"

Dreaming of utopias can be dangerous, indeed. It's better to be aware of one's surroundings.

MISADVENTURE

Having "floated twelve days without toothpaste or soap" due to the violent flood, our protagonist is thrown a rope by an obscure character, enabling him to escape the water and scramble up a cliff. When the protagonist arrives at the top, he encounters a man on horseback bedecked in full armor, extravagantly decorated with pink feathers and turquoise blue trim.

> I got to the top. But it *wasn't* a friend!
> I saw that my troubles were *not* at an end.
> A big man on a horse scared me out of my wits.
> He bellowed, "I'm General Genghis Kahn Schmitz."

The intimidating general officer immediately conscripts our poor protagonist into his army, which is already marching to battle. Schmitz plans to attack the Perilous Poozer of Pompelmoose Pass, and he promptly orders our protagonist to "Get into line! You're a Private, First Class!"

With this episode, Geisel's *Solla Sollew* warns of another strategic pitfall, namely a misadventure wherein one's interests and resources can be hijacked by an external party. Misadventures can occur with any type or at any level of strategy. They are a risk intrinsic to any alliance, coalition, or strategic partnership. Even without a preexisting strategic relationship, we can sometimes be pulled into others' conflicts. The misadventure caused by Schmitz illustrates how our resources can be appropriated by others. Hence, we must guard against being coerced, cajoled, or co-opted into solving another party's problems, which may only distract us from our own goals and place an unnecessary drain on our resources. In addition to his own troubles, Geisel might have had America's involvement in Vietnam on his mind, drawing parallels between the nation's problems and his own. By January 1964, the year Geisel drafted *Solla Sollew*, a military dictatorship under General Nguyen Khanh had taken over the government in Saigon. The similarities between the names Genghis Kahn Schmitz and General Nguyen Khanh are simply too great to be written off as a coincidence. Just as the forces of Schmitz were routed, so too were those of Nguyen Khanh throughout the summer of 1964, until the Gulf of Tonkin incident in August 1964 provided a pretext for further American escalation. The history of the US misadventure in Vietnam is well documented and will not be recounted here. It clearly serves as an example of how strategists must remain on their guard against incompetent and overzealous leaders, who can entice others to follow a flawed military plan.

ISOLATIONISM/ANTI-ISOLATIONISM

Although Geisel began his career committed to anti-isolationism, or the belief that America must not isolate itself from the rest of the world, he took the opposite position in *Solla Sollew*. A third-generation German immigrant and Evangelical Lutheran, Geisel felt some affinity for the finer elements of German culture and thus actively propagandized for American intervention to stop the spread of Nazism and fascism in the years preceding the Japanese attack on Pearl Harbor and the entry of the United States into World War II. He mobilized his ample talent as an illustrator for the cause of democracy and routinely lampooned the despotic and militaristic leaders of the Axis powers—Adolf Hitler, Benito Mussolini, and Hedeki Tojo—who repeatedly disparaged or threatened democratic values. Geisel applied the same skill and techniques years later to mock the militarism personified in the character of General Schmitz in *Solla Sollew*.[17] By the mid-1960s, however, Geisel, now in his early sixties, had matured; his outlook regarding US military interventions abroad had become more conservative, and he began emphasizing restraint. This stance put him at odds with those who believed America's policy of containment could only be carried out through military interventions, that is, blocking the spread of Soviet influence with military force.

Geisel transmitted his anti-interventionist message in *Solla Sollew* on two levels: individuals and governments. The first he accomplished by showing how the protagonist's attempt to flee his troubles only led to more troubles. Hence, the message was that *individuals* should mind their own gardens rather than meddling in those of their neighbors. In this sense, Geisel's message had much in common with Voltaire's *Candide* or modern film classics, such as *High Noon* (1952), starring Gary Cooper, who plays a town marshal who must stand up to an intimidating gang of outlaws.

In *Solla Sollew* Geisel also advocated a policy of anti-interventionism for *governments*. Here again, the parallels are strong with the misadventure in Vietnam and the buffoonery of General Schmitz. After a hasty initial advance, Schmitz's forces beat a swift retreat in the face of "more than one Poozer."

> Genghis Kahn Schmitz shouted out to his men.
> "This happens in war every now and again.
> "Some times you are winners. Some times you are loosers [*sic*].
> "We *never* can win against so many Poozers
> "And so I suggest it's time to retreat!"
> And the army raced off on its tin-plated feet.

By stating "some times you are winners. Some times you are loosers," Schmitz shifts blame for the failure of his attack to the vagaries of chance rather than to his own strategic planning and leadership, or lack thereof. Certainly we must always account for chance in strategic planning. But we must also account for the possibility that our estimates of enemy strength and disposition are wrong, which sometimes happens, and plan accordingly. Victory or defeat do not result merely from chance, but rather from human actions and reactions within the realm of chance. America's failed Bay of Pigs invasion of Cuba and attempted overthrow of Fidel Castro in April 1961, an example of faulty estimates of the enemy situation, may well have been on Geisel's mind. However, the series of defeats experienced by South Vietnamese forces before America's full entry into the war could also have influenced him. Whatever the source of his inspiration, Geisel clearly believed the US government's interventions abroad should receive a lower priority than its domestic troubles, which were indeed escalating by the early 1960s. In August 1965, the year in which *Solla Sollew* was published, severe riots broke out in the Watts neighborhood of Los Angeles, not far from Geisel's home in La Jolla, California.[18] Violence raged for six days, with smoke visible for hundreds of miles; more than a thousand people were injured, thirty-four of whom died, and at least $40 million in property damage occurred.[19] *Solla Sollew* thus demonstrates how Geisel's position on interventionism during the years leading up to World War II had completely reversed itself by 1965. Indeed, he had come to agree with the popular adage "foreign policy begins at home."[20]

MILITARY LEADERSHIP

In addition to his lack of skill as a strategist, the Kahn also exhibits poor leadership qualities in combat. Successful military commanders need to display personal bravery, presence of mind, diligence in preparation for battle, and enough wisdom not to underestimate their adversaries. Unlike Audie Murphy or other American heroes, Ted Geisel did not make a name for himself on the battlefield in World War II. In fact, the closest he came to actual combat was in December 1944, when he was sent to a position near Bastogne, Belgium, just as the Germans were launching a counteroffensive that came to be called the Battle of the Bulge. He and his military police escort found themselves behind enemy lines, where they wandered about for three days until rescued by British forces.[21] Nonetheless, Geisel knew enough about military leadership to satirize the character of General Schmitz for having none of the qualities necessary for successful command and for being, at root, a tin soldier:

> And the glorious general led the advance
> With a glorious swish of his sword and his lance
> And a glorious clank of his tin-plated pants.

Schmitz's vainglory and vanity are more than personality flaws. As Geisel shows, they also lead to negative strategic consequences: the vainglory of the "glorious general" causes him to forgo preparing properly for the battle and leads him to underestimate his Poozer adversaries. Such leadership failures can contribute overwhelmingly to strategic defeat. Interestingly, the Poozers' excellent use of cover and concealment among the rocks of Pomplemoose Pass echoes that of the Vietcong, who were consistently misunderstood and underestimated by the US government. When confronted by a small group of Poozers, Schmitz hastily flees the battle, thereby demonstrating his cowardice. His lack of bravery sets a poor example for his troops, who immediately drop their weapons and join their commander in flight. Our furry protagonist suddenly finds himself alone and unarmed, and he is quickly surrounded by menacing yellow Poozers. This abandonment scenario reveals a great deal about Schmitz's despicable character, since a long-standing principle of military honor and leadership is to put the health and welfare of one's troops first.

FRICTION

Solla Sollew also illustrates the problem of friction: the sum of impediments, large and small, that contribute to making any strategy fail. As the Prussian military theorist Carl von Clausewitz once explained, friction can come in two types: general and incidental. General friction includes those elements that make up the atmosphere of war—such as danger, physical exertion, and uncertainty—and essentially turn it into a resistant medium, like water, wherein movements are slowed or require more energy than normal. Incidental friction refers largely to chance, the random occurrence of things that can impede one's efforts in war, such as bad weather, lack of maps, worn-out muskets, and ammunition that fails to arrive. The best lubricant against friction, Clausewitz believed, was combat experience across military organizations in conjunction with the genius of the commander.[22]

Our poor protagonist clearly has no lubricant at his disposal. Everything goes wrong for him, until one thing goes right. He escapes the Poozers by a stroke of good luck (chance):

> I had terrible trouble in staying alive.
> Then I saw an old pipe that said, "Vent Number Five."
> I didn't have time to find out what *that* meant,
> But the vent had a hole. And the hole's where I went.

While friction falls into the two aforementioned categories, we can also think of it in terms of foreseeable and unforeseeable events or influences. Foreseeable events, such as sore feet, illness, rocks in the road, flat tires, and storms, can be anticipated and thus "hedged" against. Indeed, not hedging against them when we know they might occur is either laziness or negligence. Unforeseeable types of friction, by comparison, fall outside what is reasonable to prepare for, such as being conscripted by a delusional general. It is reasonable for us to anticipate our bus not arriving and to prepare ourselves for the alternative of walking a long distance; however, it is unreasonable to think we can hedge against the unforeseeable without driving up the costs of our enterprise beyond its benefits, except through the intuitive insight that comes from genius. While we should not condemn our protagonist for not preparing for the events he could not foresee, we must criticize him for failing to anticipate and hedge against the common frictions he reasonably ought to have foreseen.

END STATES

Having escaped from the Poozers by jumping down Vent Number Five, our protagonist finds himself trapped in a "frightful black tunnel" with "billions of birds / Going the wrong way." Of all the troubles described in *Solla Sollew*, the bird-filled tunnel is the scariest, even for those who do not suffer from claustrophobia or ornithophobia. It is the longest period our furry friend is exposed to danger, and it is where and when he sustains his most serious physical injuries. Readers would also do well to note the obvious parallels between Geisel's bird tunnel, Alfred Hitchcock's blockbuster thriller *The Birds* (1963), and the dangerous tunnel warfare of Vietnam. As our protagonist complains:

> I was down there three days in that bird-filled-up place.
> At least eight thousand times, I fell smack on my face.
> I injured three fingers, both thumbs and both lips,
> My shinbone, my backbone, my wishbone and hips!
> What's more, I was starved, I had nothing to eat.
> And damp! Was it damp! I grew moss on my feet.

Our furry protagonist finally emerges from the tunnel, joyously, after having surreptitiously discovered a trapdoor. He is surprised to find himself at "the banks of the beautiful River Wah-Hoo!" and not far from his final objective, the mystical city of Solla Sollew. He stands in awe before the city "with its glittering towers in the air" and approaches "a door way that shimmered and shined." The doorman, wearing a blue caftan decorated with flowers and an elaborate hat, beckons our protagonist to enter. Unfortunately, our protagonist is denied. A tiny green creature with a malevolent grin is perched near the keyhole and slaps the key away from the lock when anyone attempts to enter. The door man complains:

> Since then, I can't open this door any more!
> And I can't kill the Slippard. It's very bad luck
> To kill any Slippard, and that's why we're stuck
> And why no one gets in and the town's gone to pot.
> It's a terrible state of affairs, is it not!

Geisel's protagonist has seemingly arrived at the end of his journey. Ironically, however, he cannot complete it. To be sure, throughout his journey, our furry friend has had a strategic end state in mind: finding happiness in the magical city of Solla Sollew. But his end state proves to be an illusion, an unrealizable dream, and (as noted before) not even a good one. The protagonist ultimately abandons his objective, "a place where there are no troubles or only very few." Unlike classic dystopian literature, such as George Orwell's *Animal Farm* (1945) and William Golding's *Lord of the Flies* (1954), which critique the fundamental premises of specific utopias, anti-utopian tales such as *Solla Sollew* portray utopias as nonextant or unreachable. Orwell's *Animal Farm*, for instance, exploded the premise that power can be shared equally, while Golding's *Lord of the Flies* attacked the proposition that human nature is innately good.[23] Both portrayed utopias as the opposite of paradises because happiness required sacrificing something vital to the human spirit, such as free will. Geisel took a more pragmatic approach, rejecting the search for paradise itself as a fool's errand.

The lessons of *Solla Sollew* might help strategists avoid pursuing similar errands. This Dr. Seuss book, for example, contains an important paradox concerning the controversial concept of end states, the conditions necessary for us to consider our political goals accomplished. Strategy, whatever the type or level, must have objectives or goals toward which we may direct our efforts and resources. Polities naturally consist of varying interests, some at odds with one another, which cannot be managed without a goal the majority find acceptable. Hence, the process of defining end states, even if laden with pitfalls, can prove beneficial if it leads us to a consensus and compels our policymakers to think through what they wish to accomplish and what it might cost. Too often, that essential first step is ignored or foreshortened. Even if our concept of end states remains problematic in application, our discussion of what it would take to achieve them remains essential.

End states may be necessary to provide a sense of purpose and direction to a strategy; however, they typically represent ideal conditions, which are unachievable, and even achievable conditions often require more control over the situation than one has. And therein lies the proverbial rub. Wars can sometimes radically change the parties involved, altering their political and social structures, cultural values, and economic relationships. Consequently, many or most of the end states that provided the original sense of purpose or direction for the conflict can become unattainable or even undesirable, leaving one to doubt the wisdom of the enterprise overall. "Virtually every war in Europe has been the prelude to the next," as experts have written, "regardless of the peace settlements involved and the desired end state. . . . The U.S.-led victory in the first Gulf War did not bring a stable end state any more than the U.S. 'victories' in Afghanistan . . . or the U.S. invasion of Iraq."[24]

End states are also paradoxical in another sense. As most strategists would agree, many forms of conflict or competition do not truly end; they merely shift into other forms or dimensions or come to involve other means. Feelings of animosity or revanchism are not always easily assuaged and may even worsen over time. For those reasons, end states can be as illusory as utopias, or as disagreeable as dystopias, and the pursuit of them can lead to escalation or to war weariness. Determining desirable end states even in relatively stable situations may require considerable resolve.

Was our protagonist in *Solla Sollew* successful or unsuccessful as a strategist in this regard? Obviously, our yellow friend failed to reach his objective. On the other hand, that objective was never achievable from the outset, which he finally realized. Our furry friend had a vision of a fantastical destination, but he had no roadmap, no strategy, no plan for getting there. Even the best strategy, of course, will avail little if the goal itself is unrealistic or poorly chosen.

President Harry Truman is credited with having said, "The best ideas in the world are of no benefit unless they are carried out."[25] The story of *Solla Sollew* turns that saying on its head because the search for paradise could not be carried out in the first place. Geisel's protagonist eventually learned that lesson. Unfortunately, forsaking bad ideas can be enormously difficult, again partly because of inertia. Once we recognize our ideas as bad, as Geisel did with America's intervention in Vietnam and likely would have done with respect to Afghanistan and Iraq, we will still face troubles with the process of extraction. In short, even a good strategy cannot rescue a bad idea.

UTILITY OF FORCE

At the conclusion of *Solla Sollew*, our furry protagonist wisely declines an invitation from the city's doorman to accompany him on a journey to "the city of Boola Boo Ball / On the banks of the beautiful River Woo Wall, / Where they never have troubles! *No troubles at all!*" Instead, our little friend has learned an important lesson. He returns to the Valley of Vung. But this time he is mentally and physically prepared, with newfound determination and a handy weapon for dealing with his troubles.

> I know I'll have troubles.
> I'll, maybe, get stung.
> I'll always have troubles.
> I'll, maybe, get bit
> By that Green-Headed Quail
> On the place where I sit.

His determination is perhaps the most important change in the story. But also of note is the fact that he has armed himself with a large, pink baseball bat. Our furry friend confidently approaches the Green-Headed Quail, the Skritz, and the Skrink, and turns the tables on them, asserting:

> I've bought a big bat.
> I'm all ready, you see.
> Now my troubles are going to have troubles with *me*!

On the one hand, our friend's decision to confront his problems directly seems like an excellent message for children, and adults, to internalize. On the other hand, he is ready to resort to physical force to settle his problems, which seems uncharacteristic of Dr. Seuss. Moreover, it represents a contradiction in the story's plotline, since it runs counter to the negative depiction of the use of force in the middle of the book, namely, Schmitz's military failure during the battle with the Poozers.

An understanding of the political and social contexts in which the book was written may shed some light on the contradiction. Geisel, a demonstrably gifted writer, was clearly influenced by the sights, sounds, and events of the physical world around him. As we have seen, *Solla Sollew* reflects many of those influences. Yet he was also cloistered from the external world in important ways. He was partially insulated by his growing wealth, for instance, which made him a member of America's upper middle class. In 1964, his annual income from royalties was estimated at $200,000 ($1.7 million in 2020).[26] Such royalties gave him and his wife the means to live more than comfortably, frequently flying from coast to coast for meetings with his publishers and editors as members of America's elite "jet set." He was also a highly introverted personality, who delighted in living in the fantasy world of the intriguing characters he had created and allowing them to tell their own stories. Even in his splendid isolation, however, Ted Geisel could not have failed to notice the rebelliousness and mounting violence that marked America from the mid-1950s through the 1960s.

The period during which *Solla Sollew* was written was indeed one of significant troubles. In 1955, a decade before the book was published, the first live detonation of an atomic bomb was televised, bringing into vivid focus the potentially suicidal consequences of a nuclear war.[27] Three years before the book's publication, the Cuban missile crisis (October 1962) had taken place, a trouble of strategic proportions that brought the Soviet Union and the United States perilously close to a nuclear exchange. Then, in 1965, the year *Solla Sollew* appeared, President Lyndon Johnson announced his intention to increase the numbers of US troops in Vietnam, thereby beginning an escalatory spiral for Washington and Hanoi that ended in failure.[28] Also in that year, several young Americans openly protested Johnson's interventionist policies by publicly burning their draft cards. It was not the first time such an act had occurred, but it was the first time a large audience of Americans had witnessed it, due to a popular and rapidly expanding visual communications medium. This event also foreshadowed more trouble to come in terms of America's domestic turbulence.[29] Hollywood films, such as *The Wild One* (1953) and *Rebel without a Cause* (1955), had lionized social rebels prior to 1965, of course. In fact, the title of *Rebel without a Cause* derived from a groundbreaking sociological study that defined criminal psychopaths as would-be rebels with no cause but their own impulses.[30] As was becoming clear, however, the US government's use of military force in the 1950s and 1960s, whether domestically or internationally, had not fully extinguished America's troubles. To be sure, that force helped to break up destructive riots and to enforce policies of racial integration. But the troubles underlying those incidents were not truly resolved until political and cultural changes had taken root. The same holds true for Washington's use of force internationally, as Geisel duly warned; military force alone usually proves insufficient to close out an armed conflict, particularly when applied within the context of a limited war.

Yet curiously, Geisel's message in *Solla Sollew* inclined toward the use of force to solve one's problems at home. Perhaps his message would have been different after the slaying of four students at Kent State on May 4, 1970.[31] But the question remains: Why would Ted Geisel, who demonstrated utter incompetence with his military-issued .45 caliber pistol, so uncharacteristically encourage the use of force at the end of *Solla Sollew*? There are at least two possible answers. The first lies in the phenomenon of "character agency," wherein a writer's characters take over the storyline. Most authors enjoy this phenomenon, and we may assume Geisel did as well. But character agency may come at the cost of coherent messaging in the storyline. Second, the older, more mature Geisel may simply have been writing for himself in this case, a form of self-therapy to force himself to face his own troubles, which in turn overrode any messages to his readership. If this answer is less satisfying to us, it may be because we wish to see this famous author of instructional children's books as a wise and enlightened personality, rather than as a complex individual who at times struggled.

IN CONCLUSION

Although many readers may consider *I Had Trouble Getting to Solla Sollew* to be little more than a children's story meant to encourage young people to face their problems, as we have seen military strategists, too, have much to learn from *Solla Sollew*. Among the major lessons are, first, that all strategy centers on the need to resolve troubles or solve problems. Second, even the best strategy can be subtly or overtly derailed by various pitfalls. These pitfalls include having one's interests and resources hijacked by an external party, as illustrated by the camel driver; becoming entrapped in another party's military adventure, as in Schmitz's battle; and failing to anticipate friction, as with the sundry things that went wrong throughout the story. However, while as strategists we should always do our best to account for these and other pitfalls, we must admit that even a well-crafted strategy cannot account for every one of them. In short, we must accept some risk, somewhere in our planning. Third, even the best practical strategy will gain us little if our end state is unrealistic or poorly chosen. In the end, our furry protagonist rejected his unrealistic search for the utopia of Solla Sollew. He came to realize it was a fool's errand. So instead of heading out on a similar errand, he returned to the Valley of Vung, intent on achieving something more realistic, namely, more mental toughness, a better defensive posture, and with it perhaps some concrete deterrence capability. Finally, the most important lesson of all, especially from Geisel's point of view, is that our protagonist ought to have focused on those more realistic strategic objectives in the first place.

NOTES

1. Dr. Seuss, *I Had Trouble in Getting to Solla Sollew* (New York: Random House, 1965), 1–11.
2. Judith Morgan and Neil Morgan, *Dr. Seuss & Mr. Geisel: A Biography* (New York: Random House, 1995), 155.
3. Morgan and Morgan, *Dr. Seuss & Mr. Geisel*, 187.
4. Geisel received a direct commission into the US Army as a captain on January 7, 1943, and was assigned to Frank Capra's signal unit, which was responsible for biweekly newsreels and propaganda films such as *Why We Fight*. Richard H. Minear, *Dr. Seuss Goes to War: The World War II Editorial Cartoons of Theodor Suess Geisel* (New York: New Press, 1999), 7–8, 259, 264; and Morgan and Morgan, *Dr. Seuss & Mr. Geisel*, 106–9.
5. Antulio J. Echevarria II, *Military Strategy: A Very Short Introduction* (New York: Oxford University Press, 2017), 1.
6. Luke Johnson, "Muddling through Is a Strategy that Works," *Financial Times*, May 8, 2012.
7. These approaches, which cover more than US grand strategy, are detailed in William C. Martel, *Grand Strategy in Theory and Practice: The Need for an Effective American Foreign Policy* (Cambridge: Cambridge University Press, 2015), 7–19.
8. Compare Paul Kennedy, *The Rise and Fall of the Great Powers* (New York: Vintage Books, 1989); John Lewis Gaddis, *Surprise, Security, and the American Experience* (Cambridge, MA: Harvard University Press, 2005); Walter Russell Mead, *Special Providence: American Foreign Policy and How It Changed the World* (London: Routledge, 2002); Walter Russell Mead, *Power, Terror, Peace, and War: America's Grand Strategy in a World at Risk* (New York: Vintage Books, 2005); Williamson Murray, *War, Strategy, and Military Effectiveness* (New York: Cambridge University Press, 2011); Williamson Murray, Richard Hart Heinrich, and James

Lacey, eds., *The Shaping of Grand Strategy: Policy, Diplomacy, and War* (New York: Cambridge University Press, 2011); and MacGregor Knox and Alvin Bernstein, eds., *The Making of Strategy: Rulers, States, and War* (New York: Cambridge University Press, 1994).

9. Compare Joseph S. Nye, *Soft Power: The Means to Success in World Politics* (New York: Public Affairs, 2004); Joseph S. Nye, *The Future of Power* (New York: Public Affairs, 2011); John J. Mearsheimer, *The Tragedy of Great Power Politics* (New York: Norton, 2001); Robert Gilpin, *War and Change in World Politics* (Cambridge: Cambridge University Press, 1981); and Robert Gilpin and Jean M. Gilpin, *Global Political Economy: Understanding the International Economic Order* (Princeton: Princeton University Press, 2001).

10. Compare Henry Kissinger, *Nuclear Weapons and Foreign Policy* (New York: W.W. Norton, 1957); Henry Kissinger, *The World Restored* (London: V. Gollancz, 1974); Henry Kissinger, *Does America Need a Foreign Policy? Toward a Diplomacy for the 21st Century* (New York: Simon & Schuster, 2002); Zbigniew Brzezinski, *The Grand Chessboard: American Primacy and Its Geostrategic Imperatives* (New York: Basic Books, 1997); Zbigniew Brzezinski, *Second Chance: Three Presidents and the Crisis of American Superpower* (New York: Basic Books, 2007); Zbigniew Brzezinski, *Strategic Vision: America and the Crisis of Global Power* (New York: Basic Books, 2012); Richard Haass, *War of Necessity, War of Choice: A Memoir of Two Iraq Wars* (New York: Simon & Schuster, 2010); and Richard Haass, *Foreign Policy Begins at Home: The Case for Putting America's House in Order* (New York: Basic Books, 2013).

11. Compare Colin S. Gray, *Modern Strategy* (Oxford: Oxford University Press, 1999); Colin S. Gray, *The Sheriff: America's Defense of the New World Order* (Lexington: University Press of Kentucky, 2009); Colin S. Gray, *War, Peace and International Relations: An Introduction to Strategic History* (New York: Routledge, 2011); and Stephen Peter Rosen, *Winning the Next War: Innovation and the Modern Military* (Ithaca, NY: Cornell University Press, 1991).

12. Morgan and Morgan, *Dr. Seuss & Mr. Geisel*, 153; David Hajdu, *Ten-Cent Plague: The Great Comic-Book Scare and How It Changed America* (New York: Farrar, Straus, and Giroux, 2008); and Frank Gruber, *Pulp Jungle* (Los Angeles: Sherbourne Press, 1967).

13. Donald E. Pease, *Theodor Geisel: A Portrait of the Man Who Became Dr. Seuss* (New York: Oxford University Press, 2010), 102.

14. Pease, *Theodor Geisel*, 129–30.

15. Pease, *Theodor Geisel*, 130.

16. Pease, *Theodor Geisel*, 132.

17. Much to Geisel's later regret, however, he readily employed derogatory racial and ethnic caricatures in his drawings, especially of the Japanese. He attempted to atone for that practice by portraying other racial or ethnic groups more favorably in his postwar books.

18. Paul Gilje, *Rioting in America* (Bloomington: Indiana University Press, 1996).

19. John H. Barnhill, "Watts Riots (1965)," in *Revolts, Protests, Demonstrations, and Rebellions in American History*, ed. Steven L. Danver (Santa Barbara, CA: ABC-CLIO, 2011).

20. Haass, *Foreign Policy Begins at Home*.

21. Morgan and Morgan, *Dr. Seuss & Mr. Geisel*, 114.

22. Carl von Clausewitz, *On War*, ed. and trans. Michael Howard and Peter Paret (Princeton, NJ: Princeton University Press, 1976), Book I, ch. 3; ch. 8; and pp. 104, 122.

23. George Orwell, *Animal Farm* (New York: Signet, 1946); and William Golding, *Lord of the Flies* (London: Faber & Faber, 1954). These books contrasted with utopian narratives, such as B. F. Skinner's *Walden Two* (New York: Macmillan, 1948) and Aldous Huxley's *Island* (New York: Harper, 1962), which argued society could be perfected precisely by sharing power and curbing humanity's destructive tendencies.

24. Anthony H. Cordesman, "The 'End State' Fallacy: Setting the Wrong Goals for Warfighting," Center for Strategic and International Studies, September 26, 2016, https://www.csis.org/analysis/end-state-fallacy-setting-wrong-goals-war-fighting.

25. Cited from Hal Brands, *What Good Is Grand Strategy? Power and Purpose in American Statecraft from Harry S. Truman to George Bush* (Ithaca, NY: Cornell University Press, 2014), 11.

26. Morgan and Morgan, *Dr. Seuss & Mr. Geisel*, 182.

27. Bill Warren, *Keep Watching the Skies: American Science Fiction Films of the Fifties*, 2 vols. (Jefferson, NC: McFarland, 2009).

28. John Lewis Gaddis, *Strategies of Containment: A Critical Appraisal of American National Security Policy during the Cold War* (New York: Oxford University Press, 1981)197–217.

29. George Q. Flynn, *The Draft, 1940–1973 (Lawrence: University Press of Kansas, 1993);* and Adam Garfinkle, *Telltale Hearts: The Origins and Impact of the Vietnam Antiwar Movement* (New York: St. Martin's, 1995).

30. Robert Lindner, *Rebel without a Cause: The Story of a Criminal Psychopath* (New York: Grune & Stratton, 1944), 2.

31. H. W. Brands, *American Dreams: The United States Since 1945* (New York: Penguin, 2010), 201–4.

3

Oh, the Places You'll Go! and Grand Strategy in Outer Space

Saadia M. Pekkanen

In Dr. Seuss's final publication, *Oh, the Places You'll Go!* (1990), the protagonist begins his journey to the unknown future. "Congratulations!" says the narrator. "Today is your day. / You're off to Great Places! / You're off and away!" Since the book's publication, almost every American child has received it at high school or college graduation. *Oh, the Places You'll Go!* has been described as a "valedictory" book,[1] because this gift marks the passage from one phase of life to the beginning of another. As Ted Geisel's last book before his death, *Oh, the Places You'll Go!* also represents the author's farewell wave as he ended one journey and began another. Audrey Geisel (Dr. Seuss's wife) once wrote that his final book was not only a summation of his art and ideas, but also in some ways of his life.[2]

The book offers readers a message of autonomy, confidence, and hope as they begin their own journey through life:

> You have brains in your head.
> You have feet in your shoes.
> You can steer yourself
> any direction you choose.
> You're on your own. And you know what you know.
> And YOU are the guy who'll decide where to go.

But *Oh, the Places You'll Go!* also warns the reader that the journey has reasonably foreseeable difficulties. On the journey, as Dr. Seuss reminds us:

> I'm sorry to say so
> but, sadly, it's true
> that Bang-ups
> and Hang-ups
> *can* happen to you.
> You can get all hung up
> in a prickle-ly perch.
> And your gang will fly on.
> You'll be left in a Lurch.

Most actors—states and nonstates—find themselves in this ambivalent "prickle-ly perch" in the new space race today. Prosperity and progress beckon in outer-space affairs, but so does peril. Many dead ends and wrong turns naturally occur with technology that can serve the market as well as the military. Although many opportunities exist in outer space (after all, "it's opener / there in the wide open air"), the best path to reach the goal remains unknown. There is, in short, no grand strategy that serves as the "intellectual architecture" to guide the way.[3]

Luckily, the experiential journey of Dr. Seuss's narrator in *Oh, the Places You'll Go!* can provide the baseline grand strategy for the future direction of the United States in outer-space affairs. This chapter begins with a discussion of grand strategy and its utility in thinking about the present space race. Next, it discusses three processes happening simultaneously in the space race: democratization, commercialization, and militarization. The conjunction of all three processes can lead to weaponization. Their simultaneous presence means that space technology has the potential to do good but also harm if left unchecked. Finally, this chapter argues that Dr. Seuss's grand strategy of the "Great Balancing Act" represents the only realistic way to have peace in outer space.

GRAND STRATEGY

In *Oh, the Places You'll Go!* the protagonist bravely marches along through an ambiguous and uncertain landscape. Along the way, he must make choices about which direction to take without any signposts or guidance. As Dr. Seuss tells the reader:

> You'll look up and down streets. Look 'em over with care.
> About some you will say, "I don't choose to go there."
> With your head full of brains and your shoes full of feet,
> you're too smart to go down any not-so-good street.

Although the protagonist encounters dragons lurking in sewer pipes, blue elephants marching in formation, and Hakken-Kraks howling in the sea, he never encounters another person on a similar journey. Nevertheless, the narrator must still compete with these absent peers with the hope that he "won't lag behind," that he will "pass the whole gang," and that he will "soon take the lead." Like a nation-state in the international system, the protagonist has few allies, must follow his own counsel, and must compete with others in the same environment. His journey suggests the fragility of grand strategy given the unpredictable nature of the new space race.

The world of *Oh, the Places You'll Go!* tracks with the view of scholars such as Edward Mead Earle, who believed that statecraft often depends on nonmilitary factors, whether economic, political, psychological, technological, or even moral. At the base of the pyramid, tactics, Edward Mead Earle believed, involved handling forces in battle.[4] Strategy referred narrowly to dealing with, preparing for, and waging war. In practice, as Earle urged, the concept of strategy was an inherent element of statecraft, specifically an art that requires controlling and utilizing all given resources of a nation to secure it against known and potential enemies. Mead's broad view of strategy can be distinguished from others who had an "obsession with the battlefield."[5] Carl von Clausewitz, for example, believed that tactics referred to the use of armed forces in an engagement such as combat, while strategy referred to using those engagements cumulatively to obtain the object of war.[6] In Clausewitz's view, the object or aim of war is to impose one's will on the enemy such that they are disarmed and rendered powerless.[7]

In *Oh the Places You'll Go!*, the only illustration with any military overtone portrays blue elephants marching in loose formation with yellow and black flag poles balanced on their heads. However, the elephants do not appear to be threatened or frightened by an adversary. In fact, the elephants are smiling and cheerful. This nonviolent but competitive world imagined by Dr. Seuss corresponds to Edward Mead Earle's concept of grand strategy. Grand strategy, according to Earle, combined policies and armaments in a way that rendered war itself unnecessary or victory highly likely. Grand strategy looks beyond war, according to B. H. Liddell Hart, aiming to win the subsequent peace.[8] Put another way, grand strategy refers not just to coordination or marshalling all the resources of a nation (or even a band of nations), but to positioning resources so as not to be too exhausted to profit by the peace.

In the language of Dr. Seuss, grand strategy is the Great Balancing Act. Grand strategy requires foresight to regulate all the instruments at one's disposal to avoid damaging the security and prosperity of that future state of peace. To succeed, the essential end must be calculated and coordinated with regard to the limited means at one's disposal. As Dr. Seuss reminds us:

> You'll get mixed up
> with many strange birds as you go.
> So be sure when you step.
> Step with care and great tact
> and remember that Life's
> a Great Balancing Act.

Whether a state rises or falls, wins or loses, may well come down to the art of juggling all these scarce resources, long the "constant preoccupation" of monarchs and statesmen and ever the guiding imperative, as Paul Kennedy notes.[9] Whether in ancient empires or modern democracies, all decision-makers seek to survive and flourish by bringing together military and nonmilitary elements in a world that is always changing, forever threatening.

As we see in *Oh, the Places You'll Go!*, a grand strategy may be executed using many different means. Military means are only one among many to obtain the objective of one's grand strategy. The means mentioned by Dr. Seuss include "fun to be done" (e.g., charismatic statesmanship), "games to be won" (e.g., ruthless competition), "magical things you can do with that ball" (e.g., diplomatic trickery), being as "famous as famous can be" (e.g., strategic communications), and so on. As he makes his way across the terrain, the narrator encounters difficult choices. "Simple it's not," as Dr. Seuss notes. "I'm afraid you will find, / for a mind-maker-upper to make up his mind." Because strategy is "always a human project," designed and executed by enculturated people and organizations,[10] decision-makers (and their decisions) may well be the most important factor in both developing and executing a grand strategy. National leaders determine the present conditions (where are we?), the objective (where are we going?), and the means (how do we get there?). As Dr. Seuss reminds us, leadership means decision-making despite ambiguity, danger, unknown stakes, and confusing options.

> *IF* you go in, should you turn left or right . . .
> or right-and-three-quarters? Or, maybe, not quite?
> Or go around back and sneak in from behind?

Even those who single out the primacy of military power for attention in statecraft do so with the recognition that it is not effective by itself and is likely to set off unintended consequences.[11] For some, such as Williamson Murray, grand strategy forces all great states to face the perils of overstretch.[12] The theme of states overextending themselves, meaning a failure to synchronize material (economic) resources and military capabilities, is famously attached to the relative rise and fall of great powers by Paul Kennedy.[13] In *Oh the Places You'll Go!*, Dr. Seuss also considers the problem of overstretch and its consequences:

> You can get so confused
> that you'll start in to race
> down long wiggled roads at a break-necking pace
> and grind on for miles cross weirdish wild space,
> headed, I fear, toward a most useless place.
> The Waiting Place . . .

From this passage, we can conclude that Dr. Seuss understood how human choices—including not to act or even to blunder—push their countries down some paths and not others, which may affect their relative standing vis-à-vis their peers over time.

In sum, debates continue regarding the definition and applicability of grand strategy.[14] Some say grand strategy is defined as the "theory, or logic, that binds a country's highest interests to its daily interactions with the world."[15] Others opine that it is the intermingling of military and nonmilitary elements to preserve and enhance a nation's long-term interests,[16] done in such a way that the state manages to align its boundless aspirations with its limited capabilities.[17] Whatever the definition, grand strategy (as John Lewis Gaddis reminds us) requires aligning unlimited aspirations with limited capabilities across time, space, *and* scale.[18] Many of these key ideas about grand strategy can be found in *Oh, the Places You'll Go!*, including the inherent challenges of decision-making in the midst of ambiguity, balancing interests with constrained resources, and determining the correct means to employ.

THREE UNCERTAINTIES

In *Oh the Places You'll Go!*, Dr. Seuss actually anticipates the beginning of a new space race. "You'll be on your way up!" says the narrator. "You'll be seeing great sights! / You'll join the high fliers / who soar to high heights." These lines seem to refer to space flight in the stratosphere rather than only terrestrial transit. Later in the book, the narrator also warns about the consequences of failing to attain one's objectives in space, specifically spacecraft coming down "with an unpleasant bump," which will position the nation in a technological "Slump," which will then require additional financial and political resources. Because, of course, "Un-slumping yourself / is not easily done."

The pace of change and the unpredictability of events constitutes the central challenge in *Oh, the Places You'll Go!* and in the new space race. Rapid and dramatic technological advances in recent years have opened new frontiers in space observation, exploration, and resource extraction as humans increasingly direct attention and assets to moving beyond the confines of our globe. This emergent economy, often called "New Space,"[19] possesses revolutionary potential for the future and poses immediate challenges to national and international security today. While new space is developing technology with the potential to bolster military space power, the new space economy is fueled by the commercial sector. Governments are not the sole actors directing human activity in space. A new generation of private players, stakeholders, and transnational industrial complexes—both inside and outside the West—is democratizing the space industry, reigniting international competition for dominance in outer space. As Dr. Seuss notes: "Out there things can happen / and frequently do / to people as brainy / and footsy as you."

These trends hold the potential to drive informational, economic, and national security advantages in the years to come. Moreover, their fusion is likely to form a critical infrastructure for future societies. However, at this point no one quite knows what direction these technological innovations will tend. Nor can anyone say how the emerging architecture of new space and national security space endeavors will affect the relations of large and small powers in the world order, the balances between state and private actors, or the prospects for stability and security of people around the world. What we do know is that the world's powers have a deeply vested interest in harnessing these emerging trends and shaping the trajectory of the new space economy for security purposes. And sometimes what one country seeks in security threatens the security of others, leading them on the path to conflict. In this context, Dr. Suess advises caution and prudence:

> Just never forget to be dexterous and deft.
> And *never* mix up your right foot with your left.

The general context, the uncertainties, the chaotic paths that affect national interests make the new space race an ideal sandbox for thinking about grand strategy. Given the deeply intertwined trends in the new space race, national security decision-making cannot be isolated from the other elements of power. Among these trends, the most important are democratization, commercialization, and militarization.[20] To devise a realistic grand strategy for outer space, decision-makers will need to grasp and balance both the military and nonmilitary elements of these factors.

Democratization

Democratization means that there are many new actors, both state and non-state, that seek to advantage themselves in the new space race.[21] This is what distinguishes the present race from its more famous predecessor between the United States and the then Soviet Union. Today, the landscape of international space relations is "entirely different."[22] The United States and Russia remain important and established players. But today they are joined by a wide variety of established and emerging players spread across the world. Whether they will stand or fall on their own merits remains to be seen. The important thing for the construction of grand strategy is to recognize that there is no way, and no reason, to try to control the democratization process. It represents a glorious surge of human interest in the outer-space domain.

Beyond Europe and the United States, China, Japan, and India are among the more technically advanced and ambitious space powers.[23] China has landed on the far side of the moon, doing what no one has done before. Japan has sent two spacecraft hundreds of millions of kilometers away to asteroids and brought them back to tell the tale. India has made a name for itself in the commercial space business, launching hundreds of satellites at one go for paying customers. They are joined by a rising number of other states, which seek to capitalize on the promise of outer space in a wide variety of ways. Some have made a niche name for themselves, such as Luxembourg, which is interested in the business of regulating space resources. Pakistan will likely be an integral component of how China connects its Belt-and-Road infrastructure initiative, evident in road and port projects around the planet, upward into a digital space corridor. Israel has staked out a focused technology position, such as in the small satellites realm. The United Arab Emirates has become the first Arab country to send an interplanetary probe to Mars and continues to prioritize advancement of homegrown space capabilities. South Africa has taken on the mantle of space leadership on the African continent with an eye on the indigenous uses of satellite technologies for sustainable development purposes, such as weather forecasting and logistics tracking. Other states such as Australia, New Zealand, the Philippines, and Turkey have taken steps to create their own space agencies to position their countries in the new space race.

For the foreseeable future states will have a disproportionate impact on both the tenor and the trajectories of the unfolding international space saga. States will be the primary backers and consumers of space technologies, the prime architects of strategies and policies, and the designers of governance frameworks.[24] The rise of all these new states around the globe, with their different interests, means that for any one country, the construction of any kind of grand strategy is challenging. These states compete and conflict, each holding out for its uses of outer-space capabilities relative to others, making it difficult to reach consensus.[25] This makes it even more necessary to understand the rules of the road and how they will govern stakeholders' expectations and behaviors going forward.

While there is a foundational outer-space treaty from 1967, not everyone agrees on the interpretation of the underlying rules or principles about what to do or how to do it in the changing space domain. We will have to rethink what appropriation and property rights mean out there in space, given the complicated and often unsavory historical links between flags and trade down here on Earth. What does it mean, and how should newcomers in the space domain respond, when Russia, a principal architect of the 1967 Outer Space Treaty, proclaims that Venus is a Russian planet?[26] As David Baiocchi and William Wessler observe, the solutions to many problems are overdue, imminent, or merely emerging.[27] These include, for example, space resources, orbital debris, space traffic management, and planetary defense.

Commercialization

Commercialization has galvanized public interest and spurred dreams of a return to the moon and Mars and beyond. Private actors and individuals have long played a role in activities related to space, so it is not new to talk about the role of private actors relative to that played by government.[28] But today a distinct new generation of entrepreneurs and technologies has come to the fore, reshaping industrial prospects in and through space. Billionaire space entrepreneurs have unquestionably added glamour to the field, spurring others to follow suit from across the public, private, academic, and think-tank world. These private firms and entrepreneurs have unleashed swirls of shiny, unprecedented technologies.

As with any viable business, much depends on whether the private companies are ultimately competitive in their businesses. As Dr. Seuss reminds us, "There are points to be scored. There are games to be won. / And the magical things you can do with that ball / will make you the winning-est winner of all." Profits matter and are affected by the extent of the market and how it is projected to grow. There are challenges to defining and measuring just what the space economy is, and this is affected by how and what we include as goods and services and when we do so. This means that respectable estimates of the space economy range from about $167 billion to about $330 billion.[29] Bankers and investors, such as Morgan Stanley, Bank of America, and Goldman Sachs, project that the size of the space economy around 2040 will range between $1 trillion and $3 trillion. Forecasting is a difficult business, and future outcomes are always uncertain. Still, these financial prospects have fueled talk of a "trillion-dollar space economy" around us today.[30] Some technologies certainly hold the potential to shape the future of the space economy, on both the rocket and satellite fronts.

Reusable rocket technology promises faster and cheaper access to space. The most prominent companies paving the way are both American and led by billionaires. The startup vision for Elon Musk's SpaceX, incorporated in 2002, was not an easy sell to the establishment players, entrenched big contractors, members of Congress, and NASA skeptics.[31] But roughly two decades on, SpaceX has transformed expectations of the possible where rocketry is concerned, leading the public to assume that rockets should be going up but also coming back down in the interest of reusability. Reusability of rockets is less costly for the hard business of getting into space and better for the environment in the long run. SpaceX has also made inroads by being the first private company to take astronauts to the International Space Station atop its reusable Falcon 9 rocket—a historic occasion as it signified an astronaut launch from US soil after about nine years.[32] Jeff Bezos's Blue Origin launched and landed the first ever reusable rocket.[33] The company's mission is to build operational reusability into its rockets, which can enable a suborbital market for space tourism and long-term infrastructure; the ambition is to build a "road" for humanity's access to space. Outside the United States, there are potential entrants in the new commercial competition over rockets. Japan's Interstellar Technologies has been trying to develop rockets to lower cost barriers. Commercial or hybrid firms in China, such as Landspace, iSpace, Onespace, and Linkspace, are also interested in entering the launch market and are likely to affect competitive business prospects worldwide.[34]

Another significant commercialization trend involves satellites. As of January 2023, there were 6,718 satellites orbiting Earth.[35] Over the next decade, that number is going to increase dramatically with the explosive proliferation of small satellites (smallsats), raising concerns about the security, safety, and sustainability of outer-space activities.[36] Smallsats can serve a variety of purposes, such as Earth observation, telecommunications, and defense.[37] One estimate suggests that there may be well over one hundred thousand smallsats in low Earth orbit if the proposals by companies and governments around the world materialize.[38] For Earth observation satellites, Finland's Iceye, with the world's first radar satellite; the US Planet and Blacksky; and Japan's Axelspace showcase a range of business, environmental, and defense intelligence uses. Megaconstellations for beaming high-speed internet to any location on Earth promise a new era of global connectivity that may prove very lucrative.[39] This trajectory is being led by US firms, chief among them SpaceX, with its Starlink constellation that has already begun deployment. Amazon too has moved in this arena, with its Kuiper constellation. Outside the United States, UK-based Oneweb, which almost went bankrupt but was resurrected with a buyout from the UK Space Agency and an Indian telecommunications conglomerate, is also a player.[40] The rapid increase in these constellations is worrying for some scientific fields, such as astronomy. There is also widespread alarm about the expansion of orbital debris, which can be hazardous for space ventures.

Militarization

Militarization, stretched to the outright weaponization of outer space, would be deadly to the future of humanity in outer space and possibly Earth. "When you're alone," as Dr. Seuss reminds us, "there's a very good chance / you'll meet things that scare you right out of your pants," such as the weaponization of space. This situation becomes even more alarming when we recognize that commercialization and militarization are intertwined. Most of space technology is dual use, and this ambiguity makes it difficult to distinguish a space asset from a weapon, and offense from defense.[41] This fact poses challenges for peaceful prospects in outer space because of both economic and political realities.

The companies that make the shiny and exciting technologies need to profit to survive. Because relatively few customers exist especially for new space technology, the identity of the customers (and their requirements) exerts considerable force. Despite the emphasis on private enterprise and development, what companies make and sell depends also on government demand for goods and services from space, a result consistent in estimates across several organizations.[42] Depending on the underlying estimate of the total size and composition, government demand accounts for as much as half of the total space economy.

This state-centered facet of the space economy around the world has consequences, as it can affect the profitability, trajectory, and intent of the supposedly commercial technologies. Faced with the harsh realities of their bottom lines, virtually all the firms that started off with high talk of human settlements and science and exploration have come down to earth. Their space ambitions ("Back to the Moon to Stay!" "Occupy Mars!") are not curbed. But the imperative of making money disciplines as nothing else. SpaceX leadership has said it would launch a weapon to defend the United States, drawing applause for its show of staunch patriotism at one of the annual US Air Force Association meetings.[43] SpaceX has also doggedly pursued national security contracts and is now one of the two primary launchers for the US military, making it a member of the establishment.[44] Its Starlink internet satellite constellation that is being put into place is of interest to the military, a sizable and ready customer interested in the defense possibilities of the low earth orbit (LEO) broadband companies.[45] The commercial trends around us today are a fundamental part of the military space architecture and, leveraged well, can affect balances of power in space.[46]

Other space powers are no less motivated to gain military advantages in and through space, whether through commercial or civilian means. Japan has long played off the ambiguities of counterspace technologies, with the result that it can use the same cutting-edge capabilities to service friendly space assets and destroy adversarial ones.[47] China infamously used a missile for an anti-satellite test on one of its own satellites in 2007, generating alarming amounts of orbital debris.[48] China is also developing a wide range of counterspace capabilities to take on adversaries from the ground all the way to geosynchronous orbit.[49] Further, China has taken the lead in quantum satellites for secure military communications,[50] and it is also testing a reusable space plane that may mirror the US Air Force's secretive X-37b.[51] China, in short, is not sitting around on its laurels but is on a quest for space primacy. India too has carried out a direct-ascent, kinetic-kill test with a missile on its own satellite in 2019, joining the elite club of space powers (including the United States) who have demonstrated the capabilities to take out satellites and create debris.[52] Russia continues with the development and testing of both surface-to-space missiles and co-orbital satellite platforms.[53] Russia carried out a debris-generating anti-satellite test with a missile that targeted a defunct Soviet signals intelligence satellite in 2021.[54] France has announced that it is exploring the possibility of laser-carrying nanosatellites to "'dazzle'" hostile spacecraft.[55] These realities indicate that the balance of power is being reshaped in new and disturbing ways, causing a security dilemma and possibly an escalating arms race in outer space.[56]

For the foreseeable future, outer-space activities will be rooted in the geopolitics of Earth, where the structural flux is concerning. The world has returned to great power competition, which further exacerbates militarization.[57] From the viewpoint of the United States, this means its adversaries (principally China and Russia) are in a contest to overturn the international order and move it in their favor. The United States pinpoints their capabilities as the "greatest strategic threat" and their quest to weaponize space as a way to disrupt its own military effectiveness and freedom of operation in space.[58] The reverse is also true, seen from the eyes of these competitors. As Theresa Hitchens points out, the far more advanced US space capabilities, as well as the formation of the US Space Force and Space Command, are also factors in a military space race that could spur conflict.[59]

Any one of America's identified adversaries can exploit America's dependence on space assets, its Achilles heel.[60] Of the known operating satellites at present, the United States accounts for close to 70 percent of the total.[61] America's space dependence will also increase dramatically if the promised constellations of small satellites come about. Because of this vulnerability, the US military seeks the "Holy Grail" of space dominance, as Joan Johnson-Freese observes.[62] But the vulnerability remains an opportunity for America's adversaries, big or small, who may also suffer in a conflict but face comparatively less collateral damage.

As a result, the United States has found itself in the counterspace race, which is amplified in the return of great power competition today. The US military defines "counterspace" as "a mission that integrates offensive and defensive operations to attain and maintain the desired level of control and protection in and through space."[63] The United States and its allies know that militarization would threaten the desired level of control and protection on which our new space way of life depends. So in all that confusion, the United States attempts to hang onto those desirable outcomes, while its rivals use "wiggledy" ways to make that mission fail.

Without a grand strategy, it is not going to be simple for decision-makers to make up their minds. Dr. Seuss pinpointed the decisional dilemma well: we have "come to a place where the streets are not marked"; some things are lighted, "but mostly they are darked." We cannot stay out, we are all the way in. We have to ask how much can we lose, and what is it we win.

BRIGHT PLACES WITH BOOM BANDS

In *Oh, the Places You'll Go!*, the protagonist escapes from the Waiting Place, where "everyone is just waiting. / Waiting for the fish to bite / or waiting for wind to fly a kite." Rejecting the uselessness of waiting, the protagonist "find[s] the bright places / where Boom Bands are playing." In the new space race, a bright place with a boom band would be optimal: the musicians are playing their own instruments, seem even indifferent to each other, are spread far apart, but together balance each other's tunes. The boom band might be described as a "compeerment" or a comity of peers. The *Shorter Oxford English Dictionary* defines *compeer* as a person, companion, or associate of equal rank or standing.[64] *Black's Law Dictionary* defines "comity" as a principle or practice among political entities whereby their various acts are mutually recognized.[65]

Like the boom band playing together as equals, the only realistic grand strategy for space is compeerment, whereby a comity of peers recognize and balance national interests in space. For the foreseeable future, this state-centric grand strategy is "right enough."[66] A grand strategy of compeerment highlights the core nature of states, accounting for the fundamental importance of their core interests in and through space, and realistically reflects the strategic challenge in the outer-space domain. As Lukas Milevski puts it, "If the concept is meant to be practical, it should be a realistic reflection of what is actually possible."[67]

No state in any domain can push such frontiers without reactions from others. Compeerment is the recognition that all powers have such interests, and we need order among peers. Space dominance, primacy, and superiority are not viable options because they are not sustainable in practice. No one power, no matter how great, can prevail by pinning its hopes on fragile technologies, whether in orbits or on celestial surfaces. Further, for any one technology put out by a power there is an opposite one put out by a peer competitor. Compeerment means, therefore, that we must devise ways to balance the interests behind those technologies before they are all set to war against each other in outer-space activities. Whether this happens through some combination of the establishment of zones, or rules, or norms to keep the peace from Earth to orbits to celestial bodies is worth thinking very hard about—and in advance. The real test of compeerment as a grand strategy will be in how effectively it lets the comity of peers adapt and respond and adjust when, inevitably, things go awry.[68]

IN CONCLUSION: THE GREAT BALANCING ACT

American popular culture visuals of outer space often feature astronauts repairing equipment; starship troopers shooting aliens; billionaires' rockets launching into space; satellites whizzing around; habitable enclosures built on Mars; or Earth's very first interstellar visitor, the cigar-shaped visitor 'Oumuamua (Hawaiian for "'a messenger from afar arriving first'").[69] But the balance of power in space is rooted in the great power politics reshaping the world order down here on Earth every day. Weaponization is gaining ground because the balance of power in space depends on the geopolitical competition among powers interested in staking their own visions, claims, and agendas in space as on Earth. Because space is inhospitable and autonomous robotic systems can substitute for humans,[70] humanity appears unlikely to establish a peaceful permanent presence on a celestial body any time soon. Competition over space resources continues,[71] and recommendations have been made to put Space Force bases on the moon.[72]

This chapter has made the case that the United States ought to develop a grand strategy for outer space, given the current competition among peers and nonstate actors in the domain. A grand strategy is necessary now to address the deeply intertwined processes of democratization, commercialization, and militarization. Without a grand strategy to contain these forces, the weaponization of space becomes more likely, a prospect "down the road between hither and yon, / that can scare you so much you won't want to go on."

To ensure peace in space, as this chapter has argued, the United States and other states ought to consider themselves members of a compeerment mutually bound by diplomacy and alliances. Indeed, Peter Mansoor and Williamson Murray claim that alliances have been essential to grand strategies throughout history, more specifically the conduct of strategy in war and peace across centuries.[73] While the protagonist in *Oh, the Places You'll Go!* builds no coalitions and seeks no allies, many of his actions speak to the hard, tedious, and painstaking business of a fused civil-military diplomacy that will be critical as space powers aim for the boom band and encourage others to come along.

In the tumult of the new space race, where nothing is straightforward, developing a grand strategy and the alliances to support it will be challenging. State leaders executing grand strategy must be sufficiently competent to triumph over serial setbacks, which is a tall order if history is a guide.[74] Effective leadership in the era of grand strategy, as Hal Brands notes, "requires a supple mind that can reconcile multiple, often competing demands; it also requires a farseeing mind that can deal with the crisis or contingency at hand while simultaneously looking beyond it. Grand strategy is thus bound to be an exacting task, one that is full of potential pitfalls."[75]

In the face of all this, as Dr. Seuss reminds us:

> But on you will go
> though the weather be foul.
> On you will go
> though your enemies prowl.
> On you will go
> though the Hakken-Kraks howl.
> Onward up many
> a frightening creek,
> though your arms may get sore
> and your sneakers may leak.

In a world order in geopolitical flux, where great power competition is once again center stage and extended to outer space, formulating a coherent strategy in an "environment of struggle" poses one of the central challenges.[76] Williamson Murray captures the dilemma neatly: "The goals may be clear, but the means available and the paths are uncertain."[77] In *Oh, the Places You'll Go!*, the narrator traverses a complex terrain, in which he discovers cities with whimsical architecture, a malevolent forest populated by black trees, a toxic slump full of blue ooze, and so on. No explanation is offered regarding how, by whom, and for what purpose these environments were created. Despite his uncertainty, the protagonist must not quit: "On and on you will hike, And I know you'll hike far and face up to your problems whatever they are."

NOTES

1. Brian Jay Jones, *Becoming Dr. Seuss: Theodor Geisel and the Making of an American Imagination* (New York: Dutton, 2019), 423–26.
2. Audrey Geisel, "Living with the Cat," in *Your Favorite Seuss: Thirteen Best Loved Stories,* comp. Janet Schulman and Cathy Goldsmith (New York: Random House, 2004), 338.
3. Hal Brands, *What Good Is Grand Strategy? Power and Purpose in American Statecraft from Harry S. Truman to George W. Bush* (Ithaca, NY: Cornell University Press, 2014), 1.
4. Edward Mead Earle, introduction to *Makers of Modern Strategy: Military Thought from Machiavelli to Hitler*, ed. Edward Mead Earle (Princeton, NJ: Princeton University Press, 1943), viii.
5. Williamson Murray and Mark Grimsley, "Introduction: On Strategy," in *The Making of Strategy: Rulers, States, and War*, ed. Williamson Murray, MacGregor Knox, and Alvin Bernstein (New York: Cambridge University Press, 1994), 4.
6. Carl von Clausewitz, *On War*, ed. and trans. Michael Howard and Peter Paret (Princeton, NJ: Princeton University Press, 1989), 128.
7. Von Clausewitz, *On War*, 75, 77.
8. B. H. Liddell Hart, *Strategy* (New York: Meridian, 1954), 321–26, 353, 357–58.
9. Paul Kennedy, "Grand Strategy in War and Peace: Toward a Broader Definition," in *Grand Strategies in War and Peace*, ed. Paul Kennedy (New Haven, CT: Yale University Press, 1991), 4–6; and Paul Kennedy, "American Grand Strategy, Today and Tomorrow: Learning from the European Experience," in *Grand Strategies in War and Peace*, ed. Paul Kennedy (New Haven, CT: Yale University Press, 1991), 171.
10. Colin S. Gray, "Strategic Culture as Context: The First Generation of Theory Strikes Back," *Review of International Studies* 25, no. 1 (January 1999), 55–56n23, https://doi.org/10.1017/S0260210599000492; Colin S. Gray, *The Strategy Bridge: Theory for Practice* (Oxford: Oxford University, 2010), 36–38; and Colin S. Gray, *Theory of Strategy* (Oxford: Oxford University Press, 2018), 58–61.
11. Robert J. Art, *A Grand Strategy for America* (Ithaca, NY: Cornell University Press, 2003), 6.
12. Williamson Murray, "Thoughts on Grand Strategy," in *The Shaping of Grand Strategy: Policy, Diplomacy, and War*, ed. Williamson Murray, Richard Hart Heinrich, and James Lacey (New York: Cambridge University Press, 2011), 1–2.
13. Paul Kennedy, *Rise and Fall of the Great Powers: Economic Change and Military Conflict from 1500 to 2000* (New York: Vintage, 1987).
14. Thierry Balzacq, Peter Dombrowski, and Simon Reich, "Introduction: Comparing Grand Strategies in the Modern World," in *Comparative Grand Strategy: A Framework and Cases*, ed. Thierry Balzacq, Peter Dombrowski, and Simon Reich (New York: Oxford University Press, 2019), 5–6.
15. Hal Brands, *Promise and Pitfalls of Grand Strategy* (Carlisle Barracks, PA: US Army War College), 3.
16. Kennedy, "Grand Strategy in War and Peace," 5.

17. John Lewis Gaddis, *On Grand Strategy* (New York: Penguin Press, 2018), 21.
18. Gaddis, *On Grand Strategy*, 21 (emphasis in original).
19. Matthew Weinzierl and Mehak Sarang, "The Commercial Space Age Is Here," *Harvard Business Review*, February 28, 2021, https://hbr.org/2021/02/the-commercial-space-age-is-here.
20. Saadia M. Pekkanen, "Governing the New Space Race," *American Journal of International Law Unbound*, 113 (2019), https://doi.org/10.1017/aju.2019.16.
21. David Baiocchi and William Wesler IV, "The Democratization of Space: New Actors Need New Rules," *Foreign Affairs* 94, no. 3 (May/June 2015): 98–104.
22. Tanja Masson-Zwaan, "New States in Space," *American Journal of International Law Unbound*, 113 (2019): 98, https://doi.org/10.1017/aju.2019.13.
23. Saadia M. Pekkanen, "China, Japan, and the Governance of Space: Prospects for Competition and Cooperation," *International Relations of the Asia-Pacific* 21, no. 1 (January 2021), https://doi.org/10.1093/irap/lcaa007.
24. Pekkanen, "Governing the New Space Race," 92.
25. Paul Larsen, "Outer Space: How Shall the World's Governments Establish Order among Competing Interests?," *Washington International Law Journal* 29, no. 1 (2019): 3, https://digitalcommons.law.uw.edu/wilj/vol29/iss1/3; and Masson-Zwaan, "New States in Space," 98–99.
26. Lianne Kolirin, "Venus Is a Russian Planet—Say the Russians," *CNN*, September 18, 2020, https://www.cnn.com/2020/09/18/world/venus-russian-planet-scn-scli-intl/index.html.
27. Baiocchi and Wesler, "Democratization of Space," 103.
28. Alexander MacDonald, *Long Space Age: The Economic Origins of Space Exploration* (New Haven, CT: Yale University Press, 2017).
29. Keith W. Crane et al., *Measuring the Space Economy: Estimating the Value of Economic Activities in and for Space* (Alexandria, VA: Institute for Defense Analyses, 2020), iv–viii, 27–41.
30. Jeff Foust, "Commerce Department to Develop New Estimate of the Size of the Space Economy," *SpaceNews*, January 2, 2020, https://spacenews.com/commerce-department-to-develop-new-estimate-of-the-size-of-the-space-economy/.
31. Christian Davenport, *Space Barons: Elon Musk, Jeff Bezos, and the Quest to Colonize the Cosmos* (New York: Public Affairs, 2018).
32. Caleb Henry, "SpaceX Raises $1.9 Billion in Equity," *SpaceNews*, August 18, 2020, https://spacenews.com/spacex-raises-1-9-billion-in-equity/.
33. Davenport, *Space Barons*.
34. Andrew Jones, "Chinese Space Launch Firm iSpace Raises $173 Million in Series B Funding," *SpaceNews*, August 25, 2020, https://spacenews.com/chinese-space-launch-firm-ispace-raises-173-million-in-series-b-funding/.
35. Union of Concerned Scientists (UCS), *UCS Satellite Database*, January 1, 2023, https://www.ucsusa.org/resources/satellite-database.
36. Saadia M. Pekkanen, "Zooming in on the Promise and Peril of Satellite Imagery," *Seattle Times*, August 26, 2022, https://www.seattletimes.com/opinion/zooming-in-on-the-promise-and-peril-of-satellite-imagery/.

37. Saadia M. Pekkanen, Setsuko Aoki, and John Mittleman, "Small Satellites-Big Data: Uncovering the Invisible in Maritime Security," *International Security* 47, no. 2 (2022), https://doi.org/10.1162/isec_a_00445.

38. C. Walker Walker and J. Hall, eds., "Technical Document: Impact of Satellite Constellations on Optical Astronomy and Recommendations toward Mitigations," National Science Foundation NoirLab, accessed September 2, 2020, https://noirlab.edu/public/products/techdocs/techdoc003/.

39. Jonathan O'Callaghan, "The Risky Rush for Mega Constellations," *Scientific American*, October 31, 2019, https://www.scientificamerican.com/article/the-risky-rush-for-mega-constellations/.

40. Michael Sheetz, "How OneWeb's $1 Billion Bankruptcy Rescue Changes the Competitive Landscape for Elon Musk's Starlink," *CNBC*, July 10, 2020, https://www.cnbc.com/2020/07/10/onewebs-bankruptcy-rescue-changes-the-competition-for-elon-musks-spacex-starlink.html.

41. Joan Johnson-Freese, *Space as a Strategic Asset* (New York: Columbia University Press, 2007), 6–7; and James Clay Moltz, *The Politics of Space Security: Strategic Restraint and the Pursuit of National Interests* (Stanford, CA: Stanford University Press, 2008), 42–43.

42. Crane et al., *Measuring the Space Economy*, 29, 32, fig. 3.

43. Sandra Erwin, "SpaceX President Gwynne Shotwell: 'We Would Launch a Weapon to Defend the U.S.,'" *SpaceNews*, September 17, 2018, https://spacenews.com/spacex-president-gwynne-shotwell-we-would-launch-a-weapon-to-defend-the-u-s/.

44. Eric Berger, "In a Consequential Decision, Air Force Picks Its Rockets for Mid-2020s Launches," *ArsTechnica*, August 7, 2020, https://arstechnica.com/science/2020/08/the-air-force-selects-ula-and-spacex-for-mid-2020s-launches/; and Jeff Foust, "With Pentagon Award, SpaceX Joins the Establishment," *SpaceNews*, August 7, 2020, https://spacenews.com/news-analysis-with-pentagon-award-spacex-joins-the-establishment/.

45. Sandra Erwin, "Air Force Laying Groundwork for Future Military Use of Commercial Megaconstellations," *SpaceNews*, February 28, 2019, https://spacenews.com/air-force-laying-groundwork-for-future-military-use-of-commercial-megaconstellations/.

46. US Department of Defense, *Defense Space Strategy: Summary*, June 2020, 4, https://media.defense.gov/2020/Jun/17/2002317391/-1/-1/1/2020_DEFENSE_SPACE_STRATEGY_SUMMARY.PDF?fbclid=IwAR2TYXPZVQm0o-ybSXM3UHOYK-sRsdlvaETKChf1raiGaFMYR64QkMxu_9o (accessed 2020/07/26).

47. Saadia M. Pekkanen, "Thank You for Your Service: The Security Implications of Japan's Counterspace Capabilities," *Texas National Security Review*, Policy Roundtable, ed. Jonathan D. Caverley and Peter Dombrowski, October 1, 2020, 70–89, https://tnsr.org/roundtable/policy-roundtable-the-future-of-japanese-security-and-defense/.

48. Ashley Tellis, "China's Military Space Strategy," *Survival* 49, no. 3 (2007), https://doi.org/10.1080/00396330701564752.

49. Kevin Pollpeter, "China's Space Program: Making China Strong, Rich, and Respected," *Asia Policy* 15, no. 2 (2020): 13–14, https://www.nbr.org/publication/asia-in-space-the-race-to-the-final-frontier/.

50. Saadia M. Pekkanen, "China Leads the Quantum Race While the West Plays Catch Up," *Forbes*, September 30, 2016, https://www.forbes.com/sites/saadiampekkanen/2016/09/30/china-leads-the-quantum-race-while-the-west-plays-catch-up/#1f98d8eb5928.

51. Andrew Jones, "China Carries Out Secretive Launch of 'Reusable Experimental Spacecraft,'" *SpaceNews*, September 4, 2020, https://spacenews.com/china-carries-out-secretive-launch-of-reusable-experimental-spacecraft/.

52. Shounak Set, "India's Space Power: Revisiting the Anti-Satellite Test," Carnegie India, Carnegie Endowment for International Peace, September 6, 2019, https://carnegieindia.org/2019/09/06/india-s-space-power-revisiting-anti-satellite-test-pub-79797; and Daniel Oberhaus, "India's Anti-Satellite Test Wasn't Really about Satellites," *Wired*, March 27, 2019, https://www.wired.com/story/india-anti-satellite-test-space-debris/.

53. Theresa Hitchens, "Russia Builds New Co-Orbital Satellite: SWF, CSIS Say," *Breaking Defense,* April 4, 2019, https://breakingdefense.com/2019/04/russia-builds-new-co-orbital-satellite-swf-csis-say/; and Kyle Mizokami, "Meet Russia's Imposing New Satellite-Destroying Missile," *Popular Mechanics*, April 16, 2020, https://www.popularmechanics.com/military/weapons/a32173824/nudol-missile-anti-satellite/.

54. Andrew E. Kramer, "Russia Acknowledges Antisatellite Missile Test That Created a Mess in Space," *New York Times*, November 16, 2021, https://www.nytimes.com/2021/11/16/world/europe/russia-antisatellite-missile-test.html.

55. Theresa Hitchens, "Space Lasers for Satellite Defense Top New French Space Strategy," *Breaking Defense*, July 26, 2019, https://breakingdefense.com/2019/07/france-envisions-on-orbit-lasers-for-satellite-defense/.

56. Ben Marino, "China and the US: The Arms Race in Space," *Financial Times*, August 27, 2020, https://www.ft.com/video/80b1eb31-6cbc-422d-865b-29686dc9b235; and Sandra Erwin, "New Studies Provide Fresh Insights into the Escalating Space Arms Race," *SpaceNews*, April 4, 2019, https://spacenews.com/new-studies-provide-fresh-insights-into-the-escalating-space-arms-race/.

57. White House, *National Security Strategy of the United States*, December 2017, https://trumpwhitehouse.archives.gov/wp-content/uploads/2017/12/NSS-Final-12-18-2017-0905-2.pdf.

58. US Department of Defense, *Defense Space Strategy: Summary.*

59. Theresa Hitchens, "US, Allies Agree on Threats in Space but Struggle with Messaging," *Breaking Defense*, September 11, 2020, https://breakingdefense.com/2020/09/us-allies-agree-on-threats-in-space-but-struggle-with-messaging/.

60. Mary Beth Griggs, "Trump's Space Force Aims to Create 'American Dominance in Space' by 2020," *Popular Science,* August 10, 2018, https://www.popsci.com/space-force-2020/.

61. *US Satellite Database.*

62. Joan Johnson-Freese, *Space Warfare in the 21st Century: Arming the Heavens* (New York: Routledge, 2017), 15.

63. US Air Force, *Air Force Doctrine: Annex 3-14 Counterspace Operations* (Maxwell Air Force Base, AL: Curtis E. Lemay Center for Doctrine Development and Education, 2018), 2–3, https://www.doctrine.af.mil/Doctrine-Annexes/Annex-3-14-Counterspace-Ops/.

64. Angus Stevenson, ed., *Shorter Oxford English Dictionary on Historical Principles—Sixth Edition* (Oxford: Oxford University Press, 2007), 469.

65. Bryan A. Garner, ed., *Black's Law Dictionary*, 10th ed. (St. Paul, MN: Thompson Reuters, 2014), 324.

66. Brands, *What Good Is Grand Strategy?*, 3 (quoting Colin Gray).

67. Lukas Milevski, *The Evolution of Modern Grand Strategic Thought*, 143.

68. Brands, *What Good Is Grand Strategy?*, 199.

69. US National Air and Space Agency, "Oumuamua Overview," December 19, 2019, https://solarsystem.nasa.gov/asteroids-comets-and-meteors/comets/oumuamua/in-depth/.

70. Nathan Strout, "The Space Force Doesn't Want to Send a Human to Do a Robot's Job," C4ISRNET, accessed September 30, 2020, https://www.c4isrnet.com/battlefield-tech/space/2020/09/29/no-the-space-force-wont-be-sending-humans-into-space-anytime-soon/.

71. Sandra Erwin "U.S. Military Eyes a Role in the Great Power Competition for Lunar Resources," *SpaceNews*, August 20, 2020, https://spacenews.com/u-s-military-eyes-a-role-in-the-great-power-competition-for-lunar-resources/.

72. Sarah Scoles, "U.S. Air Force Cadets Study Idea of Space Force Bases on the Moon," *Science*, July 15, 2020, https://www.sciencemag.org/news/2020/07/us-air-force-cadets-study-idea-space-force-bases-moon.

73. Peter R. Mansoor and Williamson Murray, "Introduction: Grand Strategy and Alliances," in *Grand Strategy and Military Alliances*, ed. Peter R. Mansoor and Williamson Murray (New York: Cambridge University Press, 2016), 2–3.

74. Walter A. McDougall, "Can the United States Do Grand Strategy?," *Orbis* 54, no. 2 (2010): 167, 173, 176. https://doi.org/10.1016/j.orbis.2010.01.008.

75. Brands, *What Good Is Grand Strategy?*, 10.

76. Murray and Grimsley, "Introduction: On Strategy," 5.

77. Murray, "Thoughts on Grand Strategy," 11.

4

Horton Hears a Who and International Human Rights Law

John Hursh

*H*orton Hears a Who is one of Dr. Seuss's most appreciated and widely read books. Published to immediate acclaim, the book sold millions of copies and was made into a Hollywood film, assuring Horton's place in American literary history and influence on American culture. The book tells the story of Horton the Elephant's encounter with a foreign society called the Whos, who inhabit a dust speck. Horton rescues the Whos from imminent destruction, proclaiming his most famous line: "A person's a person no matter how small." The Whos are threatened by the Sour Kangaroo, who is unable to hear their cries and threatens to boil them in Beezlenut oil. Disaster is averted when the Whos exercise their voice by making enough noise to be heard by the animals in the Jungle of Nool.

Theodor Geisel (Dr. Seuss) originally conceived Horton's story as a parable for the importance of voting rights.[1] Scholars have since interpreted the text in a variety of ways, including illustrating the ethical importance of personhood and human dignity,[2] celebrating individualism within community life,[3] and underscoring the need to treat all people equally.[4] While these readings show the complexity of Geisel's work, *Horton Hears a Who* also demonstrates a number of concepts and principles in international human rights law.

This chapter begins with Horton's struggle to convince the other animals of the Whos' existence (and the Whos' concurrent struggle to make themselves heard) that illuminates the concept of "voice" in international human rights law. Recognition of human rights begins with listening to the voices of others, both individually and collectively. After Horton hears the voices of the Whos during his initial encounter, he responds by enacting fundamental human rights norms such as tolerance. When Horton tries to save Who-ville from destruction, he acts to protect the Whos as a people and to protect

Who-ville as a well-developed community. Horton's efforts on behalf of the Whos result in his detainment and torture, the latter an absolute prohibition in international law. The story's conclusion demonstrates how the Whos succeed through collective action, which lends force to the voices of individuals and collective groups demanding recognition of their claims to human rights.

What is the relationship between human rights and armed conflict? Some human rights, such as the right to life and the prohibition against torture, are recognized as such fundamental legal norms that may not be derogated from even during armed conflict. And while international humanitarian law serves as the primary source of law during armed conflict, numerous states and international organizations now apply international human rights law alongside international humanitarian law in armed conflicts without clearly defined battlefields or easily discernible distinctions between combatants and civilians. A lack of human rights may lead to civil unrest and violence, and eventually result in armed conflict. This is certainly true in *Horton Hears a Who*, as the Whos' claim for recognition of their rights leads to Horton's persecution and torture, the narrowly averted destruction of Who-ville, and the near death of all the Whos. Disconnected and excluded from the community of animals in the Jungle of Nool, the Whos are extremely vulnerable and lack the ability to exercise their voice and assert their rights until the pivotal moment of the story when the smallest Who adds her voice to the struggle to be heard, resulting, at last, in recognition from the rest of the animals.

THE MAKING OF HORTON

Between 1941 and 1943, Geisel drew political cartoons for *PM Magazine*, a liberal daily newspaper published in New York, which went out of business in 1948. The newspaper held an uncommon position during the 1940s, supporting US intervention in World War II at a time when most Americans did not, but also clinging to the social and economic commitments of the New Deal when most Americans thought primarily of the war.[5] In their biography of Geisel, Judith Morgan and Neil Morgan note that Geisel "admired its [*PM*'s] bluntness when the newspaper hammered the America First Committee, whose isolationism was permeated with racism."[6] Geisel appreciated that *PM* fought for the marginalized and the overlooked members of society. As Geisel remarked, "*PM* was against people who pushed other people around. . . . I liked that."[7]

Unlike many other political cartoons intended merely to mock political adversaries, Geisel's cartoons were designed to prompt the reader to act or change their attitudes. On domestic issues, Geisel urged readers to support labor unions, Franklin D. Roosevelt's New Deal, civil rights, and women's emancipation, and to reject racism, anti-Semitism, and segregation.[8] In his cartoons, Geisel portrayed the United States as Uncle Sam, rather than as President Roosevelt, to promote a sense of togetherness and downplay political differences.[9] His cartoons reminded readers that they should purchase war bonds to sustain the war effort,[10] collect scrap metal, and limit holiday travel.[11]

Recently, accusations have been made that Dr. Seuss books harbor racist views, that Geisel's wartime cartoons depicted minorities negatively, or even that Geisel himself was a racist. Certainly his depictions of Japanese people and Japanese culture in his *PM* cartoons and his US Army films were based on stereotypes commonly held by most Americans at the time.[12] But Geisel held progressive political views that were quite unusual in the United States in the early 1940s. In fact, Geisel drew many cartoons for *PM* that opposed racism and anti-Semitism.[13] Both before and after the war Geisel supported full integration of Black Americans in every domain of civil society. In particular, he hoped that Americans of different racial and ethnic backgrounds working and fighting side by side during World War II would improve race relations in the United States after the war.[14] In a 1944 US Army memo, Geisel wrote about his concern that this opportunity was being squandered: "Racial tension within our Army threatens to grow. . . . Disillusionment, cynicism, distrust, bitterness, are already souring the milk of human kindness; maggots are already eating the fruits of victory."[15]

At age thirty-eight, Geisel left *PM* to enter the US Army. In the Information and Education Division, Geisel worked alongside numerous accomplished writers, directors, and producers in Los Angeles, which became known as Fort Fox (Fox Studios) or the Hollywood Front.[16] Geisel worked on a variety of projects, including a series of training films called *Private Snafu*. In March 1944 Geisel was promoted to major and assigned to complete an instructional film for US soldiers who would occupy Germany after the war ended.[17] After the War Cabinet in Washington, DC gave the film its blessing, final approval had to come from US generals commanding troops in Europe.[18] Geisel traveled to Paris before setting out to meet US general officers in Belgium, Luxembourg, the Netherlands, and Germany. He even ended up trapped behind enemy lines for three days during the Battle of the Bulge before being rescued by British forces.[19] Returning to California in January 1945, Geisel left active duty a year later, retiring as a lieutenant colonel and a recipient of the Legion of Merit.[20]

One of Geisel's final projects for the US Army was a film titled *Your Job in Japan*, which mirrored the film Geisel had completed for US soldiers occupying Germany. Following his retirement from the army, Geisel and his wife Helen wrote and produced another film about Japan, called *Design for Death*, which won the Academy Award for Best Documentary in 1947.[21] This film explained how the Japanese government had psychologically manipulated the civilian population to establish the political and social preconditions for war.[22]

In 1953 *Life* magazine offered Geisel the opportunity to travel to Japan to complete an article examining the effects of the US military occupation on education and child-rearing and how changes to these practices had affected the ambitions of young Japanese children.[23] Geisel's article, "Japan's Young Dreams," details his meetings with hundreds of Japanese elementary schoolchildren and their teachers. The students drew pictures of what they wanted to be when they grew up. Most students drew themselves in modern attire performing a variety of jobs, ranging from teachers to astronauts. These drawings demonstrated a clear shift in preference from military to civilian professions. As one Japanese teacher remarked, "If we had given them this assignment ten years ago, every boy in Japan would have drawn himself as a general."[24]

Various scholars conclude that Geisel's trip to Japan, the *Life* magazine article, and his subsequent book *Horton Hears a Who* can be read as an atonement for his racist stereotyping of Japanese people in his prewar cartoons.[25] Geisel neither affirmed nor denied that this was the case, but following his trip to Japan he showed a deeper appreciation for the universal nature of humanity. His new humanistic understanding of the Japanese can be seen in his dedicating the book to Mitsugi Nakamura, a Japanese professor whom Geisel met and befriended in Kyoto.[26] His reaction to the final version of the *Life* magazine story also shows his shift in attitude. In his view, the edits made by *Life* magazine exoticized Japanese culture, whereas he had wanted to avoid sensationalism in the hopes that Americans would change their own attitudes toward the Japanese. His annoyance with *Life* magazine seems, at least in part, to have propelled him toward writing *Horton Hears a Who*.

LISTENING

Shortly after retiring from the army, Geisel witnessed the establishment of the United Nations and the founding of modern international human rights law in the latter half of the 1940s. Concepts, principles, and ideas about human rights were part of public discourse following World War II. Geisel would have followed the public debate regarding the 1948 *Universal Declaration of Human Rights*, which contains thirty articles that affirm the fundamental rights of all people. The concept of "human rights" in contemporary political discourse emerged from the *Universal Declaration of Human Rights*. Horton's assertion that "a person's a person no matter how small" invokes the substantive content and ethical imperative of several articles within the declaration, but perhaps none so much as Article 1, which states, "All human beings are born free and equal in dignity and rights. They are endowed with reason and conscience and should act towards one another in a spirit of brotherhood."[27]

Human rights exist legally because of international law, as well as regional and domestic laws that reinforce human rights norms. Whether governments recognize, respect, and enforce human rights depends on several factors, including how marginalized groups use their voice to assert a collective demand for recognition. Recognition, however, depends on how well a government listens to, and how willing it is to listen to, its people. Listening is central to *Horton Hears a Who*, which begins with Horton enjoying a moment of peaceful relaxation:

> On the fifteenth of May, in the Jungle of Nool,
> In the heat of the day, in the cool of the pool,
> He was splashing . . . enjoying the jungle's great joys. . . .
> When Horton the elephant heard a small noise.

In the accompanying illustration, Horton stops splashing and looks toward the noise, but concludes "there's no one around." Then he hears the very faint noise again, "as if some tiny person were calling for help." Horton still cannot see the person calling for help, and asks, "But *who* are you? *Where?*" Realizing that the voice is coming from the speck, Horton becomes engaged and attentive. Recognizing that Who society will be decimated by a natural disaster if the dust speck blows into the pool, Horton resolves to save the Whos: "Because, after all, / A person's a person, no matter how small." Acting on this ethical imperative, Horton "gently, and using the greatest of care" uses his trunk to move the speck to safety.

Horton's encounter with the individual on the speck exemplifies how listening is a frequently overlooked aspect of human rights. Hearing a call for help, Horton responds decisively and effectively, moving from comfortable relaxation to purposeful action, but such a response could not have occurred without Horton's willingness to listen.[28] Thus, one scholar characterizes Horton's encounter with the Whos as "hearing the Other." She states: "In standing firm against the animals of Nool, Horton exhibits the true generosity that is so rare among those not subject themselves to discrimination and persecution."[29]

THE ENCOUNTER

Horton can hear the Whos, but he cannot see them. "He looked and he looked. He could see nothing there / But a small speck of dust blowing past through the air." Regardless of whether he can see them or not, Horton has faith that "a creature of very small size" actually does exist. Moreover, Horton feels compassion for these people whom he cannot see, surmising that they must be "shaking with fear / That he'll blow in the pool!" Empathy for their suffering as fellow beings leads Horton to the conclusion that he must act. "I'll just have to save him."

In addition to illustrating the importance of listening, Horton's initial interaction with the person on the speck also invokes one of the most challenging aspects of human rights practice and human rights discourse: the encounter. Broadly stated, the encounter speaks to the duty one person owes another during a specific engagement. Various disciplines, including ethics, moral philosophy, and anthropology, have wrestled with this concept. Human rights law too has struggled to define and understand this concept but perhaps in a less satisfying manner than other disciplines, due to conflicting theoretical approaches to understanding the law.[30] While older understandings of legal responsibility, particularly within domestic common law traditions, typically impose only very limited duties on individuals to help others, more contemporary understandings are less rigid and often require individuals to at least make reasonable efforts to assist those in danger. This progressive development of the law is a favorite example of tort professors and legal historians alike and shows how social attitudes can reshape legal meaning.[31]

Within international law, one of the clearest and most dramatic examples of the encounter is that of the migrant or refugee trying to enter a different state. Legal scholar Itamar Mann deftly explores this topic in *Humanity at Sea*, in which he describes the "rights of encounter" as "those rights that stem not from inclusion in particular political communities but from the bare life of humans as such, as experienced by those of us who are bound by human rights."[32] Mann continues that these rights "arise when refugees make demands in the name of their own humanity and authorities are pressed to respond."[33] He concludes that human rights law provides a "thin but firm modicum of legal responsibility individuals may experience toward all other individuals upon encounter," but also notes the tension that exercising human rights may bring, as recognizing the rights of the refugee during the encounter can lead to allowing new members into a state's social contract, in turn implicating, and likely challenging, a state's sovereignty.[34]

Horton's encounter with the Whos reflects this dynamic. Like the migrant or refugee, the Who calls for help not as a recognized member of the jungle community, but through an appeal to Horton's humanity. Thus, one scholar reads Horton's recognition and protection of the Whos as an affirmation of the inherent dignity in all people,[35] while another reading compares Horton's actions to a Kantian ethical imperative, insisting "all people matter" and "[a]ll people possess an inherent, inviolable value beyond any price or measure; all people possess dignity."[36]

DO NO HARM

After Horton rescues the dust speck, the Sour Kangaroo (and the Young Kangaroo in her pouch) mock him, questioning his belief that people live on the dust speck. Horton implores the kangaroos to believe him, noting that "a family with children just starting to grow" might be living on the dust speck. In response the kangaroos insult Horton, calling him "the biggest blame fool in the Jungle of Nool!" The Kangaroos plunge into the pool, causing a "terrible splashing" (likely to harm or destroy the dust speck inhabitants), either through reckless negligence or malicious intent. "I've got to protect them," says Horton, picking up the clover and moving it away from the water.

Horton responds to this call for help not because of a well-defined legal obligation but because of a strong sense of moral responsibility and ethical duty. At the same time, his actions mirror fundamental human rights norms.[37] First, Horton acts in the "spirit of brotherhood." Under the *Universal Declaration of Human Rights*, people should "act towards one another in the spirit of brotherhood."[38] Second, Horton acts according to the "do no harm principle." While this principle is most often applied in international environmental law,[39] humanitarian practitioners also invoke it in situations of conflict and fragility.[40] The "do no harm" principle asks those who would intervene to consider first whether their intervention will have unintended negative consequences, and if so, whether these consequences will outweigh the benefits of intervention.[41] Horton reasons that there could be multiple people on the speck—even families. Horton urges caution, asking the Kangaroos to "just please let them be." In contrast, the Sour Kangaroo does not know with certainty that the dust speck is unpopulated, but the Kangaroos' reckless splashing endangers the people on the speck. Moreover, the Sour Kangaroo gains nothing by splashing the speck; it poses no threat, and she acts only out of hostility toward Horton—the opposite of brotherhood.

TOLERANCE

Horton's concern for the people on the dust speck exemplifies respect for others and tolerance for difference, key aspects of the modern human rights movement and progressive understandings of international law. Tolerance has a long history in human rights discourse and its precursors. For example, in the seventh century, Muslims recognized Christians and Jews as "people of the book" (*ahl al-kitāb*), which afforded protections and allowed them to practice their religion openly in Islamic lands, albeit while following special rules and paying a tax.[42]

Within modern human rights, showing tolerance and respect for others is recognized in numerous treaties, as well as in legal and political declarations. The preamble to the United Nations (UN) Charter calls for "we the peoples of the United Nations" to "practice tolerance and live together in peace with one another as good neighbors."[43] Article 26 of the *Universal Declaration of Human Rights* provides for the right to education, and within that right it requires that education "shall promote understanding, tolerance and friendship among all nations, racial or religious groups, and shall further the activities of the United Nations for the maintenance of peace."[44] More recently, in 1995 the United Nations Educational, Scientific, and Cultural Organization (UNESCO) put forth its *Declaration of Principles of Tolerance*, which referred to numerous international treaties that recognized the importance of tolerance before resolving to "take all positive measures necessary to promote tolerance in our societies, because tolerance is not only a cherished principle, but also a necessity for peace and for the economic and social advancement of all peoples."[45]

In contrast, the Sour Kangaroo shows a lack of tolerance for the people on the dust speck:

> "I think you're a fool!" laughed the sour kangaroo
> And the young kangaroo in her pouch said, "Me, too!
> "You're the biggest blame fool in the Jungle of Nool!"
> And the kangaroos plunged in the cool of the pool.

The Sour Kangaroo's belief that the Whos do not exist (and her mockery of Horton, who accepts their existence) shows a pronounced xenophobia and indifference to others. In response to the xenophobia endured by many migrants and refugees, but also as part of general UN efforts to "promote mutual understanding and global harmony,"[46] the Secretary-General has emphasized the importance of tolerance, particularly given the diversity present in today's world.[47] Similarly, the United Nations Human Rights Council has emphasized the promotion of tolerance, inclusion, and respect for diversity when confronting racial discrimination, attacks on minority rights, and systemic issues of exclusion and inequality.[48] Further, in 2016, the United Nations launched its Together campaign to promote tolerance, respect, and dignity across the world.[49]

But Horton's tolerance for the Whos is met with violence, not just from the Sour Kangaroo but from others living in the Jungle of Nool:

> Through the high jungle tree tops, the news quickly spread:
> "He talks to a dust speck! He's out of his head!"

There are abundant historical examples of vulnerable groups, including children, women, minorities, and the disabled, who endure harassment and physical and mental harm. Vulnerable groups face such abuse during times of peace and armed conflict. Conflict and civil unrest may heighten harassment and persecution, but systemic racism, sexism, and exclusion based on sexual preference or identity during peacetime are just as problematic and likely even more so given the normalizing effects of their occurring during "normal times."

DEVELOPMENT

Concerned that the Whos may come to great harm if he puts the clover down, Horton resolves to protect the inhabitants, repeating his earlier affirmation of the dignity of all people: "A person's a person. No matter how small." Then Horton again hears a very faint voice from the speck. Putting his ear next to the speck, he asks the voice to speak louder.

> "My friend," came the voice, "you're a *very* fine friend.
> "You've helped all us folks on this dust speck no end.
> "You've saved all our houses, our ceilings and floors.
> "You've saved all our churches and grocery stores."

The accompanying illustration shows a bustling town with people engaged in daily life, including a parent pushing children in a stroller, an individual walking home with their groceries, and a parent holding the hand of a child. The mayor of Who-ville tells Horton that their community takes pride in its buildings, which to Horton "would seem terribly small," but to the Whos "are wonderfully tall." The depiction of Who-ville serves as a reminder that human rights are not always a struggle against an authoritarian despot in a faraway place. Human rights are also profoundly everyday experiences. Rights such as the right to food, the right to shelter, the right to education, and the right to work are fundamental human rights because they are fundamental to being human.[50]

The depiction of Who-ville as a developed and welcoming city, and of the threat that the community of this city faces, speaks to the right to development. Following World War II, consensus opinion recognized the right to development, and many government officials and legal scholars held that individual freedom required this right. US president Franklin Roosevelt made this point in his 1944 State of the Union Address and characterized the right to development as "the freedom from want."[51] In the decades that followed, most government officials and many academics conceived of development simply as macroeconomic growth at the state level, and it was often neglected during the first half of the Cold War.[52]

But from the 1970s onward, the right to development took on a much different meaning and came to include environmental concerns and human rights. This conceptual evolution led to the 1986 *Declaration on the Right to Development*, which recognized the right to development as "an inalienable human right."[53] Article 5 of the declaration requires states to "take resolute steps to eliminate the massive and flagrant violations of the human rights of peoples and human beings" resulting from numerous situations including aggression, interference with and threats against national sovereignty and territorial integrity, and war.[54] Such actions obviously undermine development and frustrate the rights of people to enjoy this right. Still, throughout the 1990s development remained a contested concept, and the relationship between development and human rights was often unclear.[55]

The understanding of the right to development changed again with Amartya Sen's *Development as Freedom*.[56] Published in 1999, a year after Sen received the Nobel Prize for Economics, in many ways this work returns the concept to its original postwar meaning. In fact, Sen's opening sentence aligns squarely with Roosevelt's earlier understanding of this right: "Development can be seen, it is argued here, as a process of expanding the real freedoms that people enjoy."[57] Still, more than thirty years after the near unanimous adoption of the *Declaration on the Right to Development*, the concept remains contested, and the content and implications of this right are still debated.[58] Likewise, scholars and practitioners have long noted the gap between the rhetoric that states use when discussing the right and the neglect of this right in practice.[59]

Despite these challenges, the right to development continues to evolve, and the exercise of voice is central to this evolution.[60] The Whos have already achieved a well-functioning and orderly society. Thus, their concern is with keeping these gains and protecting the development already present within their community, even though other animals cannot witness this development. This passage also shows how voice is critical to asserting rights and claiming identity. The mayor's declaration of identity ("I am a Who") and of place and community ("My town is Who-ville") is a turning point in the story, exemplifying how "scenes of sociopolitical recognition" are central to human rights.[61] Indeed, the remainder of the story focuses on Horton and the Whos seeking recognition from the other animals in the Jungle of Nool.

TORTURE

After finding the clover carrying the Whos, Horton is accosted again by the Sour Kangaroo and the Young Kangaroo. In the illustration, they look down on Horton with contempt, their hands on their hips and their eyes closed. Nearby the Wickersham monkeys, who are carrying a rope, smile malevolently. The Sour Kangaroo assumes the role of enforcer:[62]

> "For almost two days you've run wild and insisted
> "On chatting with persons who've never existed.
> "Such carryings-on in our peaceable jungle!
> "We've had quite enough of your bellowing bungle!
> "And I'm here to state," snapped the big kangaroo,
> "That your silly nonsensical game is all through!"
> And the young kangaroo in her pouch said, "Me, too!"

Backed by dozens of Wickersham brothers, uncles, cousins, and in-laws carrying a rope, the Sour Kangaroo threatens Horton: "You're going to be roped! And you're going to be caged!" and threatens to boil the Who civilization "in a hot steaming kettle of Beezle-Nut oil!" Horton now faces unlawful detention and possibly torture,[63] while the Whos face not only torture but also an existential threat to their lives and their community. Within international law, the prohibition on torture is absolute, creating an international legal norm that forms part of customary international law and a duty owed by all states to the international community.[64]

The prohibition on torture is also an *erga omnes* obligation, unlike some other human rights, and may not be derogated from at any time under any circumstances.[65] Very few legal principles reach the level of an *erga omnes* obligation, and those that do are actions deemed so offensive to the international community, and to all humanity, that they could never be lawful.[66] The prohibition against torture is also present in treaty law. Expressed in numerous human rights and international humanitarian law treaties,[67] the *Convention against Torture and Other Cruel, Inhuman or Degrading Treatment or Punishment* most directly addresses this issue.[68] Opened for signature in 1984, the convention entered into force in 1987,[69] and 170 states are now parties to the treaty.[70]

Due to its status as customary international law, the prohibition on torture is a legal requirement that all states must follow regardless of whether they are parties to the convention. Nonetheless, the convention makes valuable contributions to public international law and to international human rights law more specifically. It provides the first legal definition of torture in a modern international treaty,[71] which reduces the ability of actors to perform coercive and harmful acts by removing ambiguity about what constitutes an offense. Functionally, the convention contains an optional protocol that establishes processes and mechanisms that aim to prevent torture, including regular visits to places of detention—where torture most frequently occurs—and the UN Subcommittee on Prevention and National Preventive Mechanisms.[72] While these processes will not prevent all instances of torture, they are likely to reduce the number of offenses because they help to denormalize torture and undermine its acceptance.

The prohibition of torture applies to all aspects of international law, but it resonates most in international human rights law. A primary objective of human rights law is to ensure that all people can access their voice and speak their truth. Torture is the opposite of giving voice, and it obliterates truth. As Joseph Slaughter observes, torture seeks to destroy the subject by silencing their voice: "As a human rights violation, torture is paradigmatic in its implementation as a tool to destroy a speaking subject. Human rights violations target the voice, and therefore, the voice should be the focus of international human rights instruments."[73] Understanding human rights abuses as an infringement of the modern subject's ability to narrate their story, Slaughter concludes that torture destroys voice and individual subjectivity.[74]

Slaughter calls the ability to tell one's own story "narratability," noting that the greater the abuse, the more the subject loses the ability to tell their story. In state-sanctioned torture connected to political violence, the goal of the torturer is to mute the subject's voice, thereby destroying their subjectivity as an individual. Primo Levi's account of surviving the Auschwitz concentration camp (one of only 20 survivors from his arriving group of 650 people) captures this idea:

> We became aware that our language lacks words to express this offence, the demolition of a man. In a moment, with almost prophetic intuition, the reality was revealed to us: we had reached the bottom. It is not possible to sink lower than this; no human condition is more miserable than this, nor could it conceivably be so. Nothing belongs to us any more; they have taken away our clothes, our shoes, even our hair; if we speak, they will not listen to us, and if they listen, they will not understand. They will even take away our name: and if we want to keep it, we will have to find ourselves the strength to do so, to manage somehow so that behind the name something of us, of us as we were, still remains.[75]

All human rights abuses infringe upon or disrupt the subject's ability to narrate their story. At the far end of the spectrum, narration is silenced with death. Violation of the right to life ends narration through the irreversible destruction of the subject.[76] In *Horton Hears a Who*, Horton faces torture to silence his story, while the Whos face annihilation to silence theirs.

COLLECTIVE ACTION

When the Whos are threatened with being boiled alive if they do not make themselves heard, Horton implores the mayor to have them make as much noise as possible. Only by exercising their voice can the Whos prove their existence: "We are here! We are here!" But Horton's plan for collective action fails when the Sour Kangaroo hears nothing. After the Wickershams trap Horton in a cage, he implores the Whos:

> Don't give up! I believe in you all!
> A person's a person, no matter how small!
> And you very small persons will *not* have to die
> If you make yourselves heard! *So come on, now, and TRY!*

Despite making noise that "rattled and shook the whole sky," the Whos are still not heard. The mayor desperately searches Who-ville for citizens to add their voice and discovers Jo-Jo, a very small Who who is seemingly unaware of the Whos' existential plight. They climb to the top of the Eiffelberg Tower, where the mayor (à la Churchill) implores Jo-Jo's help "in the town's darkest hour."[77] Jo-Jo shouts out "YOPP!," which pushes the noise into audible range for the Sour Kangaroo. "They've proved they ARE persons," says Horton. "No matter how small. / And their whole world was saved by the Smallest of All!" This evidence causes the Sour Kangaroo to cooperate with Horton in their protection.

The story's conclusion shows that the Whos succeed through collective action (uniting all their voices together to be heard), and that every citizen (even the smallest) must participate for the community to win political recognition. Collective action lends force to individual and group claims to human rights.[78] Social movements, civil rights campaigns, and other human rights mobilization efforts rely on the collective voices of many to achieve social change and recognition of rights, whether urban or rural, particularly local or sweepingly global.[79] Even in highly repressive and authoritarian states, where being a part of a group deemed unfriendly to or unsupportive of the government can result in terrible consequences, collective action can work. When an authoritarian government detains, tortures, and kills its opponents, this repression does not always result in a decline in collective action. Sometimes severe repression leads to the formation of human rights organizations and increases in collective action.[80] As just one example, in Chile after General Augusto Pinochet seized power through a military coup, numerous human rights organizations formed during the height of the regime's repression.[81]

Collective action is often associated with labor rights, in which labor organizers and union leaders rely on the collective strength of their group to advance the rights of all members.[82] However, collective action and human rights share the underlying principle that the voices of a marginalized group can lead to political inclusion internally and externally.[83] Examples include efforts of agrarian movements focusing on land rights,[84] coalitions confronting corruption, and alliances against human trafficking and modern-day slavery.[85]

Collective action undertaken by individuals to secure their rights as citizens exemplifies the importance of inclusion. Without citizenship, an individual will face incredible barriers in realizing their economic, social, and cultural rights.[86] As Susan Bibler Coutin has noted, "Citizenship has been defined as a legal status, a form of political membership, a marker of entitlement, a boundary that separates the included from the excluded."[87] In her research on the experiences of Salvadoran migrants fleeing civil war in El Salvador and their attempts to gain citizenship in the United States,[88] Coutin observed a group of Salvadorans assembled outside a Los Angeles County building, chanting, "*Estamos aquí! Y no nos vamos!* (We are here! And we're not leaving!). Struck by this moment of solidarity, she was reminded of the Whos' desperate attempt to gain recognition.[89]

> I was suddenly reminded of the Dr. Seuss story, "Horton Hears a Who." In order to prevent the destruction of their world by those who doubted its existence, all of the Whos in Whoville had to shout, "We are here! We are here!" The assertion, "*Aqui estamos*" ("We are here") began to seem more important than the defiant "*y no nos vamos*" ("and we're not leaving"). The lives that these immigrants had created in the interstices of law and illegality were threatened by those who doubted the legitimacy of their world. By asserting their presence, these unauthorized immigrants claimed both legitimacy and formal membership in the polity.[90]

CONCLUSION

Although sometimes characterized as a subversive influence, Geisel did not object to authority per se; rather, he was opposed to unchecked political power and unbounded self-interest. He did not expect faultless leaders, but he did demand that leaders prioritize the needs of their citizens and perform their tasks without gross self-aggrandizement or excessive personal gain. Numerous Dr. Seuss books illustrate the fallibility of leaders and the corrupting influence of political power. As Cook notes, "Seuss shows political authority as potentially selfish and exploitative, thirsting for more power, heedless of the best interests of the community."[91]

When faced with the wrongs of authoritarian leaders, Geisel's books show how citizens can confront and defeat such leaders. Geisel did not put the individual above all else, as he recognized the need for strong communities to support individual rights. Thus, in Horton (as in other Dr. Seuss books) individuals can and usually will cooperate to find a solution to the social problems confronting their communities.

Perhaps surprisingly, the use of force against authoritarian leaders to advance human rights remains an open question in the world of Dr. Seuss. Having witnessed firsthand the destruction brought by war as an army officer moving through Europe in 1944, Geisel accepted that war could be an unavoidable outcome. But in contrast to World War II, which Geisel defended as a "war that had to be fought," he showed less support for later armed conflicts, including the Korean and Vietnam Wars and US intervention in Cambodia.[92] Similarly, to save the Whos and Who-ville, Horton was forced to resist and eventually to fight. But he did so after attempting to settle the dispute peacefully. When pushed, Horton acted out of a sense of ethical duty and moral responsibility that highlights some of the most important tenets of human rights, including encountering the vulnerable with compassion and not as Other, treating difference with tolerance and respect, and refusing persecution while respecting the right of development. Further, Horton's actions underscore the absolute prohibition on torture and how torture irrevocably harms and even destroys its subjects, including their voice. In contrast, collective action can amplify voice and move the marginalized and excluded to a place of safety and inclusion. But above all, Horton shows the power of voice and how by listening to others and recognizing their rights, societies both strengthen human rights and make themselves more secure.

NOTES

1. Brian Boyd, "The Origin of Stories: Horton Hears a Who," *Philosophy and Literature* 29, no. 2 (October 2001): 201.

2. Dean A. Kowalski, "Horton Hears You, Too! Seuss and Kant on Respecting Persons," in *Dr. Seuss and Philosophy: Oh, the Things You Can Think!*, ed. Jacob M. Held (Lanham, MD: Rowman & Littlefield, 2011), 119.

3. Shira Wolosky, "Democracy in America: By Dr. Seuss," *Southwest Review* 85, no. 2 (Spring 2000): 173.

4. Philip Nel, "'Said a Bird in the Midst of a Blitz . . .': How World War II Created Dr. Seuss," *Mosaic: An Interdisciplinary Critical Journal* 34, no. 2 (June 2001): 78.

5. Richard H. Minear, *Dr. Seuss Goes to War: The World War II Editorial Cartoons of Theodor Seuss Geisel* (New York: New Press, 2001), 13.

6. Judith Morgan and Neil Morgan, *Dr. Seuss & Mr. Geisel: A Biography* (New York: Random House, 1995), 101. *PM* editor-in-chief Ralph Ingersoll called the "Professional Isolationists," which included North Dakota senator Gerald Nye and famed aviator Charles Lindbergh, the "American Enemies of Democracy." Minear, *Dr. Seuss Goes to War*, 17.

7. Morgan and Morgan, *Dr. Seuss & Mr. Geisel*, 101.

8. Donald E. Pease, *Theodor SEUSS Geisel* (New York: Oxford University Press, 2010), 65; and Minear, *Dr. Seuss Goes to War*, 6.

9. Minear, *Dr. Seuss Goes to War*, 191–92.

10. Minear, *Dr. Seuss Goes to War*, 192–93.

11. Minear, *Dr. Seuss Goes to War*, 193–94.

12. Pease, *Theodor SEUSS Geisel*, 92.

13. Minear, *Dr. Seuss Goes to War*, 23. "Dr. Seuss saved some of his most biting cartoons for issues of anti-black racism and anti-Semitism."

14. Pease, *Theodor SEUSS Geisel*, 73.

15. Pease, *Theodor SEUSS Geisel*, 73.

16. Pease, *Theodor SEUSS Geisel*, 68.

17. Morgan and Morgan, *Dr. Seuss & Mr. Geisel*, 110.

18. Morgan and Morgan, *Dr. Seuss & Mr. Geisel*, 111.

19. Morgan and Morgan, *Dr. Seuss & Mr. Geisel*, 114.

20. Morgan and Morgan, *Dr. Seuss & Mr. Geisel*, 115–16.

21. Morgan and Morgan, *Dr. Seuss & Mr. Geisel*, 119–20.

22. For reasons unknown, Geisel attempted to destroy all existing copies of the film.

23. Pease, *Theodor SEUSS Geisel*, 92–93.

24. Theodor Geisel, "Japan's Young Dreams," *Life*, March 29, 1954.

25. Sophie Gilbert, "The Complicated Relevance of Dr. Seuss's Political Cartoons," *Atlantic*, January 31, 2017, https://www.theatlantic.com/entertainment/archive/2017/01/dr-seuss-protest-icon/515031/. "They [the cartoons] also have their own flaws, most notably their racist portrayal of both Japanese citizens and Japanese Americans. Geisel's bigoted treatment of both only a few months before the forced internment of Japanese Americans was something many believe he tried to atone for in his later books."

26. Pease, *Theodor SEUSS Geisel*, 93.

27. United Nations General Assembly, Resolution 217 A (III), *Universal Declaration of Human Rights*, December 10, 1948, United Nations Treaty Series, https://www.un.org/sites/un2.un.org/files/2021/03/udhr.pdf.

28. Dr. Seuss's depiction of Horton as an elephant is telling, as Horton's large ears suggest a disposition toward listening, which allows him to hear the concerns of others. Likewise, Horton's physical strength suggests a moral responsibility to ensure the safety and well-being of vulnerable persons.

29. Tanya Jeffcoat, "From There to Here, from Here to There, Diversity Is Everywhere," in *Dr. Seuss and Philosophy: Oh, the Things You Can Think!*, ed. Jacob M. Held (Lanham, MD: Rowman & Littlefield, 2011), 98.

30. See, for example, Giselle Corradi, Eva Brems, and Mark Goodale, eds., *Human Rights Encounter Legal Pluralism: Normative and Empirical Approaches* (London: Hart Publishing, 2017).

31. Here, the best example is the legal progression from having almost no duty to help a person in distress, such as a drowning child or an injured person, to the emergence of Good Samaritan laws and the affirmative duty to rescue.

32. Itamar Mann, *Humanity at Sea: Maritime Migration and the Foundations of International Law* (New York: Cambridge University Press, 2016), 13. Mann's inquiry on the rights of the encounter relies on several historical examples of maritime migrants seeking access to a specific state. Legal scholars have increasingly applied the concept of the encounter to other settings. For example, see Sara Miellet, "Human Rights Encounters in Small Places: The Contestation of Human Rights Responsibilities in Three Dutch Municipalities," *Journal of Unofficial Law and Legal Pluralism* 51, no. 2 (2019): 213–32.

33. Mann, *Humanity at Sea*, 13.

34. Mann, *Humanity at Sea*, 13.

35. Kowalski, "Horton Hears You, Too!," 126.

36. Jacob M. Held and Eric N. Wilson, "What Would You Do if Your Mother Asked You? A Brief Introduction to Ethics," in *Dr. Seuss and Philosophy: Oh, the Things You Can Think!*, ed. Jacob M. Held (Lanham, MD: Rowman & Littlefield, 2011), 107.

37. Horton acts out of moral duty and ethical obligation: I can protect these vulnerable people, so I must. International human rights law makes similar legal obligations to protect the vulnerable. However, international law requires much less for the enforcement of these rights, particularly if the use of force is required and only allows other states to intervene within another state under authorization from the UN Security Council invoking its Chapter VII authority or by responding in self-defense to an armed attack under Article 51 of the UN Charter.

38. Legally the word "shall" creates a duty to act, whereas "should" does not. And as Wendy Hesford notes, the *Universal Declaration of Human Rights* "confers universal personhood upon every human being, but it implies that this recognition—the incorporation of the subject into the regime of rights—has yet to be attained." Wendy S. Hesford, "Human Rights Rhetoric of Recognition," *Rhetoric Society Quarterly* 41, no. 3 (2011): 284.

39. Like the precautionary principle, the do no harm principle encourages restraint and requires that individuals and organizations properly assess and understand complex situations before intervention to prevent or limit unintended negative consequences.

40. Mary B. Anderson, *Do No Harm: How Aid Can Support Peace—Or War* (Boulder: Lynne Rienner Publishers, 1999); and Organisation for Economic Cooperation and Development, *Do No Harm: International Support for Statebuilding* (Paris: OECD, 2009).

41. Rory Stewart and Gerald Knaus, *Can Intervention Work?* (New York: W. W. Norton, 2012).

42. John L. Esposito, ed., *The Oxford Dictionary of Islam* (Oxford: Oxford University Press, 2003), 10.

43. United Nations General Assembly, *Charter of the United Nations and Statute of the International Court of Justice*, 1945, United Nations Treaty Series 993, https://treaties.un.org/doc/publication/ctc/uncharter.pdf.

44. *Universal Declaration of Human Rights*, art. 26(2).

45. United Nations Educational, Scientific and Cultural Organization, *Declaration of Principles on Tolerance*, November 16, 1995, https://www.refworld.org/docid/453395954.html.

46. United Nations, "Our Aim," Together: Respect, Safety and Dignity for All campaign, accessed October 4, 2023, https://together.un.org/our-aim.

47. United Nations, "Tolerance Is a Commitment 'To Seek in Our Diversity the Bonds That Unite Humanity'—UN," *UN News*, November 16, 2016, https://news.un.org/en/story/2016/11/545462-tolerance-commitment-seek-our-diversity-bonds-unite-humanity-un. Further, the UN Secretary-General, along with the head of UNESCO, recalled the "we the peoples" clause within the UN Charter's preamble and called for seeing the world through this prism and "to collectively build societies that are more inclusive, more peaceful and more prosperous." United Nations, "Tolerance Is a Commitment."

48. United Nations Human Rights, Office of the High Commissioner, "Human Rights Council Holds Panel Discussion on Promoting Tolerance, Inclusion, Unity and Respect for Diversity," March 19, 2018, https://www.ohchr.org/EN/HRBodies/HRC/Pages/NewsDetail.aspx?NewsID=22845&LangID=E. For example, in 2016 the United Nations launched a campaign called Together that promotes tolerance, respect, and dignity across the world. The Together campaign is a specific response to the xenophobia endured by many migrants and refugees, but it is also part of general UN efforts to "promote mutual understanding and global harmony." See United Nations, "Our Aim."

49. United Nations, "Our Aim."

50. For example, Article 6 of the *International Covenant for Economic, Social, and Cultural Rights* establishes the right to work and the importance of this right for human dignity: "The right to work is essential for realizing other human rights and forms an inseparable and inherent part of human dignity." This connection between work and dignity can be especially true for the most marginalized within society. See International Rescue Committee, *Policy Brief: Overview of Right to Work for Refugees: Syria Crisis Response: Lebanon and Jordan*, IRC, January 1, 2016, 1, https://www.rescue.org/report/policy-brief-overview-right-work-refugees-syria-crisis-response-lebanon-and-jordan. "The right to work is a protected human right under international law, which recognizes that being able to access work is fundamental to human dignity and central to survival and development of the human personality."

51. Arjun Sengupta, "Right to Development as a Human Right," *Economic and Political Weekly* 26, no. 27 (July 7–13, 2001): 2527.

52. David P. Forsythe, "The United Nations, Human Rights, and Development," *Human Rights Quarterly* 19, no. 2 (May 1997): 334.

53. United Nations General Assembly, Resolution 41/128, *Declaration on the Right to Development*, December 4, 1986, art. 1(1), https://www.ohchr.org/sites

/default/files/rtd.pdf. Regionally, the *African Charter on Human and Peoples' Rights* first recognized this right explicitly in 1981.

54. *Declaration on the Right to Development*, art. 5.

55. Forsythe, "United Nations, Human Rights, and Development," 334–35.

56. Amartya Sen, *Development as Freedom* (New York: Anchor Books, 2000).

57. Sen, *Development as Freedom*, 1.

58. Karin Arts and Atabongawung Tamo, "The Right to Development in International Law: New Momentum Thirty Years Down the Line?," *Netherlands International Law Review* 63, no. 3 (2016): 230–36.

59. Stephen Marks, "The Human Right to Development: Between Rhetoric and Reality," *Harvard Human Rights Journal* 17 (2004): 137–68.

60. Debapriya Bhattacharya and Andrea Ordóñez Llanos, *Southern Perspectives on the Post-2015 International Development Agenda* (New York: Routledge, 2017), 4–5.

61. Hesford, "Human Rights Rhetoric of Recognition," 283.

62. The Kangaroos also take offense at Horton for effectively disturbing the peace: "For almost two days you've run wild and insisted / On chatting with persons who've never existed./Such carryings-on in our peaceable jungle!" As Kowalski notes, Horton's actions do cause temporary unrest in the jungle, but he argues that the blame belongs to other animals that did not properly investigate the facts of the situation. Kowalski, "Horton Hears You, Too!," 130.

63. Unlawful detention, particularly solitary confinement for a prolonged period, can constitute torture.

64. Malcolm Shaw, *International Law* (Cambridge: Cambridge University Press, 2017, 8th ed.), 93.

65. Shaw, *International Law*.

66. For example, the prohibitions on genocide, crimes against humanity, and slavery are also *erga omnes* obligations.

67. Article 5 of the *Universal Declaration of Human Rights* states: "No one shall be subjected to torture or to cruel, inhuman or degrading treatment or punishment." The 1984 *Convention against Torture and Other Cruel, Inhuman or Degrading Treatment or Punishment* followed the 1975 *Declaration on the Protection of All Persons from Being Subjected to Torture and Other Cruel, Inhuman or Degrading Treatment or Punishment*. Article 7 of the *International Covenant on Civil and Political Rights* (1976), the two Additional Protocols to the Geneva Conventions (1977), and key regional human treaties also prohibit torture. See Malcolm Shaw, *International Law*, 246 and n265–66.

68. United Nations General Assembly, *Multilateral Convention against Torture and Other Cruel, Inhuman or Degrading Treatment or Punishment*, December 10, 1984, United Nations Treaty Series 1465, https://treaties.un.org/doc/publication/unts/volume%201465/volume-1465-i-24841-english.pdf.

69. United Nations General Assembly, *Multilateral Convention against Torture*.

70. United Nations Human Rights, Office of the High Commissioner, "Status of Ratification: Interactive Dashboard," accessed October 4, 2023, https://indicators.ohchr.org/.

71. Manfred Nowak and Elizabeth McArthur, *The United Nations Convention against Torture: A Commentary* (Oxford: Oxford University Press, 2008), 28.

72. Nowak and McArthur, *United Nations Convention against Torture*, 895–924.

73. Joseph Slaughter, "A Question of Narration: The Voice in International Human Rights Law," *Human Rights Quarterly* 19, no. 2 (May 1997): 407.

74. Slaughter, "Question of Narration," 413.

75. Primo Levi, *Survival in Auschwitz: The Nazi Assault on Humanity*, trans. Stuart Woolf (New York: Simon and Schuster, 1996), 26–27.

76. Article 3 of the *Universal Declaration of Human Rights* states: "Everyone has the right to life, liberty and security of person."

77. Here, Seuss's use of "Eiffelberg Tower" and "darkest hour" is probably not coincidental. The darkest hour likely references Prime Minister Winston Churchill's famous speech imploring the British people to resist the Nazis during World War II, at a time when many government leaders sought to make peace with Hitler. The Eiffelberg Tower is likely a stand-in for the Eiffel Tower in Paris, the iconic French structure designed to commemorate the centennial of the French Revolution.

78. Alex Neve, "Amnesty International: 58 Years On, Collective Action for Human Rights Matters More Than Ever," Amnesty International Canada, May 28, 2019, https://www.csjcanada.org/blog/2019/5/29/amnesty-international-58-years-on-collective-action-for-huma.html.

79. United Nations, "Collective Action Now, the Only Way to Meet Global Challenges, Guterres Reaffirms in Annual Report," *UN News*, September 23, 2019, https://news.un.org/en/story/2019/09/1047042.

80. Mara Loveman, "High-Risk Collective Action: Defending Human Rights in Chile, Uruguay, and Argentina," *American Journal of Sociology* 104, no. 2 (September 1998): 517. "The generalized demobilization that is the expected outcome of dramatic increases in the scope and scale of state repression does not capture the entire picture; state repression may stimulate collective organization and opposition from certain sectors as a direct result of the severity and cruelty of its attempts to stifle it in others."

81. Loveman, "High-Risk Collective Action," 488.

82. For example, Article 28 of the *European Union Charter of Fundamental Rights* recognizes the right of collective bargaining and action. Adopted in 2000, this charter enshrines the political, social, and economic rights of European Union citizens and European Union residents into law.

83. Beniamino Cislaghi, Diane Gillespie, and Gerry Mackie, *Values Deliberation and Collective Action: Community Empowerment in Rural Senegal* (London: Palgrave Macmillan, 2016).

84. Priscilla Claeys, "The Right to Land and Territory: New Human Right and Collective Action Frame," *Interdisciplinary Journal of Legal Studies* 75, no. 2 (2015): 115–37.

85. Anthony J. Cooper, Chris N. Bayer, and Mark S. Winters, *It Takes a Community: Collective Action Initiatives Confronting Corruption and Forced Labour* (n.p.: Development International and Konung International, 2019).

86. For example, although the right to work is recognized in Article 6 of the *International Covenant for Economic, Social, and Cultural Rights*, governments may still regulate those individuals lacking citizenship. See, for example, International Rescue Committee, *Policy Brief: Overview of Right to Work for Refugees.*

87. Susan Bibler Coutin, "Citizenship and Clandestiny among Salvadoran Immigrants," *Political and Legal Anthropology Review* 22, no. 2 (November 1999): 53.

88. Coutin, "Citizenship and Clandestiny."

89. Coutin, "Citizenship and Clandestiny," 63.

90. Coutin, "Citizenship and Clandestiny," 63.

91. Timothy E. Cook, "Another Perspective on Political Authority in Children's Literature: The Fallible Leader in L. Frank Baum and Dr. Seuss," *Western Political Quarterly* 36, no. 2 (June 1983): 329.

92. Nel, "Said a Bird in the Midst of a Blitz," 66.

Part III

SPECIALIZED DOMAINS OF WARFARE

5

Thidwick the Big-Hearted Moose and Environmental Security

Rebecca Pincus and Montgomery McFate

In *Thidwick the Big-Hearted Moose* (1948), Dr. Seuss tells the story of a moose named Thidwick, whose generous nature leads him to accommodate a growing number of creatures who take up residence in his antlers. Eventually, the burden of carrying the load of residents endangers Thidwick's life, and he sheds his antlers in order to survive, jettisoning the community that had lived on his head.

Since its publication, *Thidwick* has been mined for lessons about property rights,[1] viewed as a commentary regarding "imperial aggrandizement and unjust occupation that precipitated World War II,"[2] and interpreted as a commentary on the limits of charity.[3] In this chapter we approach *Thidwick* as a metaphor for the relationship between humankind and the natural environment with consequences for security, whether at the individual, national, or planetary levels.

Reading the works of Dr. Seuss through an environmental lens is not new. In fact, many scholars have identified deep-rooted environmentalism in his work. However, virtually all scholarship has focused on Seuss's most explicitly environmentalist text, *The Lorax* (1971). *The Lorax* conveys a stark warning about the environmental impacts of unregulated capitalism and is widely used to unpack ideas about environmental ethics with both children and adults. While *The Lorax* is an unambiguous first-wave environmentalist parable that reflects the environmental concerns of the 1960s–70s, such as preservation of wilderness and prevention of pollution,[4] *Thidwick* offers a more useful environmental parable for the global and existential environmental concerns of the twenty-first century. It contains themes of deep ecology and environmental security and conveys a warning tale about selfishness exceeding the limits of tolerance. *Thidwick* also contains a powerful narrative about the limits of Earth's carrying capacity and the potential for conflict over environmental resources. Seen through an environmental security lens, *Thidwick* offers darkly humorous conclusions about the current trajectory of the age of the Anthropocene.[5]

CAPACITY AND COLLAPSE

The story begins when Thidwick the moose encounters a Bingle Bug who asks, "It's *such* a long road / And it's *such* a hot day, /Would you mind if I rode / On your horns for a way?" The Bingle Bug's request appears to be a simple favor, conditioned on the excessive heat, that will be of short duration. Thidwick graciously grants the Bingle Bug's request. But the Bingle Bug does not leave as expected. Instead, he invites an additional guest to join him on Thidwick's horns, who then invites another guest, and so on, until poor Thidwick's large antlers are loaded with creatures that Thidwick himself did not invite.

While the Bingle Bug couches his initial request in ostentatiously polite terms, that *politesse* vanishes as soon as he has gained a secure position in Thidwick's antlers. The Bingle Bug pivots to an attitude of blatant entitlement, showing no sign of respect for Thidwick's personal autonomy: "'There's plenty of room!,' laughed the bug. 'And it's free!'" Soon, all the residents of Thidwick's antlers begin to act with reckless entitlement and selfish greed, inviting additional residents without ever asking Thidwick's permission. Eventually a bear takes up residence in Thidwick's antlers. Soon the horde weighs five hundred pounds.

From an ecological perspective, the population living on Thidwick's head clearly exceeds the carrying capacity of his antlers. "Carrying capacity" is an ecological term that refers to the maximum number of individuals in a species that can be sustained indefinitely by the available resources in a specific area.[6] Carrying capacity varies, depending on the availability of resources like food, water, and shelter. For example, the carrying capacity of a forest would be altered after a forest fire because fewer resources would exist to sustain the populations of browsing species like deer, but additional resources would become available to sustain the insects that feed on dead wood. A healthy ecosystem contains many species living together in balance, supporting each other in a way that can go on indefinitely without disturbance.

If a population exceeds the carrying capacity of its environment, it will eventually crash, due to starvation or emigration. For example, feeding activities of grazing animals, like deer, often damage their food plants in the process of feeding on them. Usually this damage is minimal and sustainable and is part of a healthy food web. However, when the population of deer overshoots the carrying capacity (e.g., if humans eliminate predator species like wolves), the deer will begin to exert damaging pressure on their food plants. Since the deer graze down the plants to the root and strip outer layers of bark from trees and saplings, many of these plants die. The reduction of food supplies in the area will often lead to an even greater die-off of deer.

In some cases of ecosystem collapse the previous structure of relationships cannot be regained, and the ecosystem itself is rearranged into a new set of balances. The International Union for the Conservation of Nature (IUCN), which works to protect ecosystems and species around the globe, explains ecosystem collapse in this way: "Unlike species, ecosystems do not disappear; rather they transform into novel ecosystems with different characteristic biota and mechanisms of organization.... Collapse is a transformation of identity, a loss of defining features, and/or replacement by a different ecosystem.... Transitions to collapse may be gradual, sudden, linear, non-linear, deterministic or highly stochastic.... An ecosystem may thus be driven to collapse by different threatening processes and through multiple pathways."[7]

For example, the world's largest tropical lake, Lake Victoria, was once home to more than 350 species of vividly colored cichlid fish, which provided food for local communities. Today Lake Victoria is a depleted shadow zone. Deforestation around the lake and an increase in fertilizer use led to increased runoff, causing algal blooms. The colonial British government introduced Nile perch into the lake (which were larger than the cichlids), which proceeded to devour almost every prey species, including cichlids. Without small, algae-eating fish like cichlids, the unchecked algae blooms depleted the lake's oxygen in all but a thin surface layer. Now life in Lake Victoria is confined to this surface region, and even the voracious perch are in danger of extinction. Lake Victoria's sad fate is an example of humans exceeding the carrying capacity of the lake, through introduction of nonnative species and local habitat modification. While Lake Victoria perhaps could have survived one or the other, the combination of two serious pressures overloaded the capacity of the lake.[8]

To sum up thus far, *Thidwick the Moose* would seem at first blush to be a parable about the dire consequences of exceeding the carrying capacity of an ecosystem. The various forest animals making a home in Thidwick's antlers have a combined weight of five hundred pounds, which is literally being *carried* by Thidwick. These animals quickly exhaust the limited resources of the environment they inhabit, potentially causing great harm to the protagonist. Even small children reading the book will experience a sense of doom as they consider the implications of the pattern: eventually no more room will exist in Thidwick's antlers, and Thidwick himself might collapse.

SHORT-TERM INTERESTS, LONG-TERM CONSEQUENCES

Throughout the story, the inhabitants of the moose's antlers are concerned only with their own immediate needs, showing no consideration about how their burgeoning population might affect the future condition of their collective resource (i.e., Thidwick's antlers). Each of the forest animals residing in Thidwick's antlers seeks to fulfill a personal, short-term interest: the Bingle Bug wants a comfortable ride because "it's such a long road / And it's such a hot day"; the Tree-Spider wants a "fine place" to build a web; and the Zinn-a-zu Bird seeks architectural novelty, exclaiming, "I've been living on *trees* ever since I was born / But here's something *new*! Why not live on a *horn*!"

Far from "leaving no trace" like responsible campers, these forest creatures proactively modify their environment, eventually causing actual harm. The Zinn-a-zu bird, for example, pulls out Thidwick's hair to make a nest. Eventually the bird "plucked out exactly two hundred and four! 'Don't worry,' he laughed. 'You can always grow more!'" Other animals follow the bird's example, such as the woodpecker that drills several holes into Thidwick's antlers, which then become a cozy nest for a squirrel family.

Lacking any sense of the long-term consequences, the parasitic Zinn-a-zu Bird and the other the forest animals never suggest a restriction on additional antler-riders to protect the host upon whom they depend. Moreover, the forest animals fail to consider that Thidwick might be endangered by the loss of 204 hairs, which has certainly compromised his ability to endure the frigid winter cold. Instead of being concerned about the damage they have caused to the environment (i.e., Thidwick's body), they worry only about their short-term needs.

As winter approaches, the forest animals' selfish attitudes are revealed as more than just harmless bad manners. Now their selfishness poses a threat to the life of the host. Because of the seasonal shortage of moose-moss on the northern shore of Lake Winna-Bango, Thidwick prepares to depart on his annual migration to the south, where copious amounts of moose-moss grow.

> For moose-moss gets scarce when the weather gets freezy.
> The food was soon gone on the cold northern shore
> Of Lake Winna-Bango. There just was no more!
> And all Thidwick's friends swam away in a bunch
> To the south of the lake where there's moose-moss to munch.

The crowd on Thidwick's antlers interferes with the moose's migration, employing an argument based on the concept of fairness. "'STOP!' screamed his guests. 'You can't do this to us!'" In essence, the forest animals assert that their territorial occupation of the host's antlers entitles them to decide whether Thidwick migrates south in order to survive or remains on the frozen northern shore of Lake Winna-Bango for their benefit (but where he will die of starvation). "These horns are our home and you've no right to take / Our home to the far distant side of the lake!" The forest animals living in Thidwick's antlers have claimed squatter's rights to the territory they occupy.[9] In the language of property law jurisprudence, "squatter's rights" is known as the doctrine of adverse possession, which allows individuals to claim legal title to a vacant property after openly inhabiting it for a period of time.[10] Since the forest animals have been inhabiting the moose's antlers, they believe that they can claim legal title to the property.

Since the forest animals have claimed possession of the antlers, in their view Thidwick's attempt to swim to the southern shore of Lake Winna-Bango constitutes a "taking" (also known as a claim of "eminent domain"). Under the Fifth Amendment to the US Constitution, if the government seizes ("takes") private property for public use through eminent domain, it must provide "just compensation."[11] From the perspective of the forest animals, Thidwick has threatened to "take" their home using the power of eminent domain but has not compensated them in any way.

From Thidwick's perspective, on the other hand, the demands of the forest animals are unfair because his long-term interest in his own survival has been completely ignored. "'Be fair!' Thidwick begged, with a lump in his throat." In response to Thidwick's desperate plea, the forest animals again employ "fairness" as a counterargument. "We're fair," they proclaim. To prove their "fairness," the forest animals employ an ostensibly democratic process and call for a vote. Of course, since the population of forest animals constitutes an overwhelming majority (against the single vote of Thidwick), they "win" the election. The illustration on the following page shows the spiders, birds, turtles, and other creatures living in Thidwick's antlers jubilantly celebrating their political victory.

Although the decision to cross the lake has been determined by a fair, "democratic" vote (insofar as everyone voted), the result devalues long-term environmental considerations. *Thidwick* thus contains an implicit warning about the unintended consequences of majoritarian democracy. Although the forest creatures believe that a "vote" on whether the moose can migrate to the southern shore of Lake Winna-Bango constitutes a "fair" decision-making practice, this decision dooms Thidwick to starvation. Blindly ignoring the long-term consequences of denying Thidwick food (i.e., the death of the host that sustains them), they fail to recognize that this democratic decision also endangers their survival. By asserting their short-term, selfish interests, the forest animals have set themselves on a long-term path to ecological failure.

Thidwick thus offers an implicit criticism of western democracy, in which a majority group achieves their short-term objective by imposing a long-term cost on the minority without their knowledge or consent. Consider the hundreds of treaties concluded between the US government and Native American tribal peoples: the practice of state-to-state negotiations and the treaties themselves were predicated on a profoundly European conception of state, property, land, and rights.[12] The treaties largely formalized the land expropriation that had already occurred. Moreover, the US government abrogated most of the treaties when it was convenient to do so.[13] While democracy is the dominant mode of achieving fairness in Eurocentric political processes, many non-western indigenous societies have made fair decisions through processes that were not democratic (such as the consensus-based decision-making of the Igbo of Nigeria).[14] In this sense, Thidwick's tale is a condemnation of colonial expropriation masquerading as "democracy," in which newcomers impose a definition of fair process that benefits them to the detriment of the original residents.

The tale of *Thidwick* contains echoes of other myths and fables dramatizing the consequences of short-term thinking. For example, in the story of the Pied Piper of Hamelin the townspeople hire the piper to lure rats out of the town by playing his flute. But after he performs the task, the townspeople refuse to pay him. The piper then retaliates by luring the town's children away.[15] The story of the Pied Piper warns about the costs of shortsightedness and self-interest, which are inevitably far costlier in real terms than the small benefit of saving a few coins. While the townspeople thought they were saving money by reneging on their agreement with the Pied Piper, the long-term consequence of losing their children proves horrific. Likewise, the animals in *Thidwick* think they are gaining a free ride on Thidwick's antlers, but in the end their selfish shortsightedness costs them much more.

COMMON POOL RESOURCE DILEMMAS

Both *Thidwick* and *The Lorax* concern what environmental scholars call "common-pool resource dilemmas" (also known as "the tragedy of the commons"), in which exploitation of a resource held in common inevitably leads to short-sighted extinction of the resource.[16] In *The Lorax*, the Once-ler's greed destroys a whole ecosystem. Having become rich by manufacturing Sneeds out of the foliage of Truffula Trees, the Once-ler can only meet public demand through large-scale industrial production. To supply the factory manufacturing Sneeds, the Truffala Trees must be chopped down, thereby destroying the habitat of the Brown Bar-ba-loots, the Humming-Fish, and the Swomee-Swans. Industrial pollution eventually makes the environment completely uninhabitable, leaving a depopulated wasteland.

The Lorax is unambiguously an environmentalist story intended to raise ecological awareness, particularly about the effects of unregulated overharvesting of resources. Some Dr. Seuss scholars call it a "foundational ecopolitical text."[17] Although Geisel claimed that his inspiration for *The Lorax* came on a safari trip to Kenya in 1970 (where he saw patas monkeys and acacia trees), the clear-cut forests of Truffala Trees suggest deep concern about the effect of logging on the environment.[18] It is worth noting that Geisel wrote the book at a time when logging had surged in the Pacific Northwest and the United States had become a major exporter of timber.[19]

Whatever the inspiration for the book, Garrett Hardin's article "The Tragedy of the Commons," which was published in *Science Magazine* in 1968,[20] provided the scientific basis for the overall theme of *Thidwick*. Using grazing on common lands as an example, Hardin's essay succinctly explains the clash between the demands of a growing human population and the limits of natural ecosystems. Each individual rancher has a short-term incentive to graze as many animals as possible on common land, since feeding animals on free grass maximizes the rancher's profit. Any ranchers who decline to graze their animals on common land cede their advantage in the marketplace to their competitors. Like the Once-ler, each rancher seeking to maximize profit by exploiting a free resource will inevitably consume the entirety of the free resource, whether pasture or Truffala Trees. Moreover, individuals have an additional incentive to pollute the commons. Since dumping trash, wastewater, or smokestack emissions costs less than clean disposal, every economically rational individual will do so. When no restrictions limit consumption of common resources, behavior that appears to be logical leads to catastrophe and destruction.

While *The Lorax* conveys a blunt environmental message, *Thidwick* expresses a message about the "tragedy of the commons" more subtly. In *Thidwick,* each successive guest invites additional creatures to live in Thidwick's horns. "There's plenty of room and it's free!," as the Zinn-a-zu Bird correctly observes. In the view of the forest creatures, Thidwick's antlers are unregulated, lack any oversight, and belong to everyone. "'This big-hearted moose runs a public hotel! / Bring your nuts! Bring your wife! Bring your children as well!' / So the whole squirrel family all jumped on, pell mell." These free riders (literally and figuratively) treat Thidwick's horns as a commons, behaving in accordance with Hardin's observation that a lack of regulation incentivizes harm. For example, after the Zinn-a-zu bird invites his extended family to establish residence in a new abode, Uncle Woodpecker begins drilling holes in Thidwick's antlers. While Thidwick and his herd of moose appear shocked and horrified by this grave violation of his pristine antlers, the forest animals seem to approve of the drilling. The next illustration calls additional attention to the damage caused by Uncle Woodpecker: "Now the big friendless moose walked alone and forlorn, / With four great big woodpecker holes in his horn." In *Thidwick*, as in *The Lorax*, the tragedy of the commons plays out: each individual maximizes their own gain, inevitably destroying or consuming the resource.

SOCIAL NORMS AND VULNERABILITY

As noted thus far, *Thidwick* contains implicit warnings about the crash of ecosystems when their carrying capacity is exceeded, about the danger of prioritizing short-term interests at the expense of long-term concerns, and about the dilemmas created by unregulated common pool resources. How does any of this pertain to security per se?

While many types of security are at play in *Thidwick*, the theory of alliance politics, which focuses on how alliances form and function during peacetime and crisis,[21] dominates the story. As the security umbrella provided by Thidwick's antlers expands to include all sorts of new actors, the hegemon (Thidwick) providing the security guarantee finds himself increasingly at odds with his allies/dependents. When war comes (in the form of hunters), Thidwick ultimately abandons his useless allies because they give him no strategic or tactical benefit in the face of violent conflict. Astute readers might notice that Thidwick behaves as a political realist would, making a calculated decision to toss off his dependent antler-dwellers and thereby maximize his own position.[22] More specifically, Thidwick's rejection of his free riders seems to offer a pessimistic commentary on the North Atlantic Treaty Organization, since guaranteeing the security of member states potentially jeopardizes the security of the most powerful member of the alliance by drawing them into an armed conflict. (Indeed, in 1947–48, while Ted Geisel was writing *Thidwick*, the US Congress and the executive branch were debating whether a postwar collective security alliance ought to exist and if so, what form it should take.)[23] Conversely, even sheltering under the security umbrella provided by a hegemon is no guarantee of security, since the hegemon (e.g., Thidwick) might jettison allies when circumstances warrant.

While different types of political theory could be applied to *Thidwick*, constructivism offers the most insight into the security dynamics in the text. Generally, a constructivist perspective would view a nation's culture, norms, beliefs, and doctrine as a powerful force in decision-making.[24] Here, Thidwick's normative belief that hosts must show generosity to guests compels him to welcome an endless stream forest animals seeking shelter. "I'm happy my antlers can be of some use. / There's room to spare, and I'm happy to share! /

Be my guest and I hope that you're comfortable there!" The moose's generous nature leads the forest creatures to correctly conclude that he will never refuse to honor their requests. As the bug notes, "That moose won't object. He's the big-hearted kind." Instead of complaining, Thidwick merely accepts the situation and even rationalizes it as evidence of his own character: "I'm a good sport, so I'll just let him rest, / For a host, above all, must be nice to his guest."

Rather than being a source of strength, Thidwick's cultural norms and values eventually become a source of vulnerability. Thidwick's belief that "a host, above all, must be nice to his guest" provides him no recourse for managing abusive guests. He feels bound by his values to endure the burden of the animals in his horns.

> You couldn't say "skat!" 'cause that wouldn't be right.
> You couldn't shout "Scram!" 'cause that isn't polite.
> A host has to put up with all kinds of pests,
> For a host, above all, must be nice to his guests.
> So you'd try hard to smile, and you'd try to look sweet
> And you'd go right on looking for moose-moss to eat.

Just like Thidwick, states sometimes endanger (or even sacrifice) their own security in order to act in accordance with their cultural values.[25] Indeed, as Geisel was writing *Thidwick,* the United States potentially endangered its own security to uphold its commitment to democracy. Following World War II, Germany's infrastructure, economy, and political institutions had been almost completely devastated. In 1945 the US drafted the Morgenthau Plan to guide the postwar reconstruction of Germany. Under the Morgenthau Plan, German armed forces would be abolished and industrial production would be severely restricted, thereby preventing Germany from threatening Europe ever again. While the reduction of Germany to a "pastoral state" would have curtailed any future threat from Germany, the punitive effect of the Morgenthau Plan came to be seen by many US diplomats and members of the executive branch as contrary to the basic values of western democracy.[26] In the end, the Allies' belief in the abstract rights of self-determination, economic progress, and democratic governance trumped security concerns, leading to the adoption of the Marshall Plan.

NASTY, BRUTISH, AND SHORT

Thidwick stands apart from other Dr. Seuss books because of the realistic representation of the characters and their situations. Unlike many fantastical creatures in other Dr. Seuss books (such as the Biffer Baum Birds building their nest in the *Sleep Book* or the Nooth Grush standing on a toothbrush in *There's a Wocket in My Pocket*), the illustrations in *Thidwick* can be recognized as spiders, birds, bugs, and so on. Unlike many fantastical landscapes in Dr. Seuss (such as the empty desert stretching to infinity in *The Zax* or the vertiginous rock formations in *Happy Birthday to You!*), the landscape of Lake Winna-Bango has realistic craggy mountains, pine trees, lakeshore, flowers, rock formations, and river rocks. Even the interior of the Harvard Club is portrayed accurately.

What is the effect of such realism? Accurate portrayals signify to the reader that Lake Winna-Bango exists, that tender moose-moss grows upon its shores, that Thidwick is a real moose, and therefore that the violence of the hunt must also be real. The moose's terror certainly feels real. Weak from exhaustion and lack of food, Thidwick sinks "down, with a groan, to his knees. / And *then*, THEN came something that made his heart freeze." The next illustration shows Thidwick gasping in alarm as *"bullets came zinging right past Thidwick's face! / Guns were bang-binding all over the place!"* The hunters wear realistic caps, boots, and bow ties. From an elevated position on a hill they shoot ordinary rifles with ordinary bullets. "Fire again and again / And shoot straight, one and all! We *must* get his head / For the Harvard Club Wall!" Needless to say, very few Dr. Seuss books portray ammunition rounds being fired. Most contain only minimal danger, including parental disapproval (such as *And to Think That I Saw It on Mulberry Street* and *Cat in the Hat*), angry authority figures (such as *One Fish, Two Fish, Red Fish, Blue Fish* and *Yertle the Turtle*), trickery and lies (such as *Fox in Sock* and *the Sneetches*), lurking figures in dark landscapes (such as *What Was I Scared Of?* and *Oh, the Places You'll Go!*), territorial aggression (*The Zax* and *I Can Lick 30 Tigers Today!*), and environmental destruction (such as *The Lorax* and *Bartholomew and the Oobleck*).

Far more serious than any other Dr. Seuss book, the danger portrayed in *Thidwick* is death. He was unable to run fast enough to escape the hunters, and *"then finally they had him! / Because of those pests, he had run out of luck, / Because of those guests on his horns, he was stuck!"* At this moment in the book, when confronted with the moose's imminent demise, small children often experience powerful dread (often leading to tears and nightmares). Although children find the idea of the moose's death upsetting, quite often children also express dismay that the hunters want Thidwick's head as a *trophy*. The hunters see Thidwick as the quarry of the hunt, reduced to nothing but flesh and bones. But over the course of the story, children have stopped thinking of Thidwick as an animal; instead they perceive him as a *person*. The illustration of Thidwick fleeing in terror from human predators armed with deadly weapons gives children their first understanding that violence dehumanizes others by making them into *objects*.

This brutal competition for survival so realistically portrayed in *Thidwick* resembles the international political system, in which strong, well-armed nation-states prey on weaker states. This view of the international political system originates in Thomas Hobbes's (1588–1679) classic text, *Leviathan* (1651). Hobbes argued that human beings are warlike by nature, existing in a perpetual state of "war of every man, against every man."[27] In a hypothetical state of nature (specifically, without government), Hobbes proposed that life would be "solitary, poor, nasty, brutish, and short."[28] (Indeed, this appears to be an accurate description of Lake Winna-Bango.) In the realist political tradition, the constant expectation of warfare and pervasive feeling of insecurity in a Hobbesian world creates a balance of power among states. Thus, as states pursue their national interests and defend against external hostilities, a political equilibrium is reached in the international political system.[29]

Reflecting perhaps on his own experiences in World War II, Geisel's political philosophy expressed in *Thidwick* (written in the immediate postwar period) posits that weak states (whether because of famine, internal conflict, depopulation, etc.) become easy prey for strong states (whether hunting for resources, territory, or economic expansion). In *Thidwick*, violence is communal: a group of hunters rather than a single individual hunts the moose. Moreover, certain rituals accompany this bloodshed, such as firing a continuous volley of shots as a group ("shoot straight, one and all!), chasing the wounded victim, wearing traditional plaid coats and red hats, and so on. This ritualized communal violence seeks "deadly trophies," whether scalps, fingers, or ears stripped from the victim's body (or in this case, a whole moose head).[30] The victim (in this case, the moose) performs an act of sacrifice, absorbing the violence and thereby preventing the other community members from becoming needless victims.[31] During World War II, war deaths were often described as an individual soldier making a "sacrifice" for the nation and thereby ensuring security.[32] This pattern is ancient. During the Neolithic period, as Barbara Ehrenreich argues in *Blood Rites*, our human ancestors were not the predators found in the ethnological theory of "man the hunter," but rather the prey.[33] In the primordial struggle to survive in a nasty, brutish, and short world, Neolithic humans devised blood rituals to appease angry deities (such as human sacrifice). Such rituals can be found in the present day (and in *Thidwick*), except that sacrifice now takes the form of the sacred violence of killing and dying for community survival in war between the states.[34]

MASS EXTINCTION

Cornered by the hunters, Thidwick's life seems to be in grave peril, and he contemplates his imminent demise. "He gasped! He felt faint! And the whole world grew fuzzy! / Thidwick was finished, completely . . . / *or WAS he . . . ?*" In that desperate moment, Thidwick reaches beyond behavioral norms to biological imperatives: he remembers that he is a moose, a member of the Cervidae family. Cervids have deciduous horns, meaning that they are shed each fall.[35] "It's true, he was in a most terrible spot,/ *But NOW he remembered a thing he'd forgot!* A wonderful something that happens each year / To the horse of all moose and the hors of all deer." As the hunters close in, Thidwick shakes off his old antlers "so that NEW ones can grow!"

As biologists know, when moose shed their antlers each year no permanent harm occurs. But when Thidwick sheds his antlers, he destroys the ecosystem existing there: "And he called to the pests on his horns as he threw 'em, / 'You wanted my horns; now you're quite welcome to 'em!'" Thidwick's shedding of his antlers is analogous to the mass extinctions in Earth's history—but with one key twist. Throughout Earth's epochs of evolution and destruction, including five major cycles of extinction, its inhabitants did not cause the mass extinction event. The dinosaurs bore no responsibility for the asteroid that wiped them out sixty-six million years ago, at the end of the Mesozoic era.[36]

However, the present sixth cycle of extinction in the age of the Anthropocene bears the imprint of humanity. In the last century four hundred species have gone extinct. The UN Living Planet Report found that about 70 percent of all individuals of vertebrate species have disappeared since 1970.[37] The loss of species around the world is a product of habitat destruction, pollution, climate change, and the spread of invasive species. Human activity has created this "vale of biological impoverishment."[38] Humans have modified Earth to suit their needs, and as a result the planet is warming at an unprecedented rate. Similarly, the animals living on Thidwick's head modified his antlers to suit themselves. They drilled holes, extracted hair, spun webs, and made other permanent alterations but never considered whether the changes were sustainable. In *Thidwick* (just as on planet Earth), refusal to live within sustainable limits risks a total collapse of the environment.

While mass extinctions devastate the array of existing life, they also provide an opportunity to develop new forms of life and jump-start evolution. The theory of punctuated equilibrium in evolution posits that life is generally characterized by stasis, punctuated by episodes of intense evolution—including following the mass extinction episodes.[39] For example, during the Permian extinction 251 million years ago, roughly 95 percent of species were wiped out. Geologists believe the Permian extinction was driven by either an asteroid impact or massive volcanism, which triggered runaway greenhouse gas buildup and planetary warming of 6°C.[40]

What *Thidwick* proves (in the eloquent words of Dr. Malcolm in *Jurassic Park*) is that "life . . . finds a way." While particular species come and go through waves of extinction, life finds new expressions. In this Dr. Seuss book, life also finds a way. After disposing of the pests in his antlers, Thidwick "swam Winna-Bango and found his old bunch, / And arrived just in time for a wonderful lunch / At the south of the lake, where there's moose-moss to munch." Thidwick's lifecycle will continue in its enduring pattern, untouched by the ephemeral community that resided on his antlers one summer.

DEEP ECOLOGY

Scholars of environmental science generally overlook *Thidwick*. Instead, they focus on the environmentalism expressed in *The Lorax*, in which human greed, factory production, and unfettered capitalism have caused ecological collapse. *The Lorax* adheres to the dominant body of environmental thinking at the time of publication (now referred to as "shallow" or anthropocentric environmentalism), which posits that humans can control the natural world and engineer solutions to environmental problems.[41] The environmental laws enacted in the United States during the 1970s reflect this human-centric body of thought, showing great confidence that managing human activity (especially corporate activity) through regulations and permits will provide protection from the system pressure of unfettered capitalism toward self-destruction.[42] Anthropocentric environmentalism thus emphasizes technological innovations to reduce pollution, scientific resource management, and lifestyle changes like recycling. However, this type of environmentalism ignores the effects of economic and population growth, assuming that through proper management, human societies and natural environments can coexist.

Thidwick, on the other hand, emerges from a different stream of nature-centric environmentalism called "deep ecology" or ecocentrism. In 1973 Norwegian philosopher Arne Naess argued that humans exist *within* nature, neither apart nor above nature. "We may be said to be in, of and for Nature from our very beginning."[43] In *Thidwick*, deep ecological theory underlies the entire plot: when forest creatures hop aboard the moose's antlers, they fail to appreciate that their very survival depends on Thidwick. He is their home, as they are carried on his antlers. From a deep ecology perspective, they have abandoned their "ecological selves."[44]

Unlike other forms of environmentalism, deep ecology advocates for major transformation of human society. As Drengson and Inoue write, "the deep [ecology] approach aims to achieve a fundamental ecological transformation of our sociocultural systems, collective actions, and lifestyles."[45] Only by returning to our "ecological selves" can human life be reorganized in harmony with nature. Thidwick's epiphany that he is indeed a moose exemplifies the return of the ecological self to nature. "It's true, he was in a most terrible spot / *But NOW he remembered a thing he'd forgot!*" Having become a habitat for forest creatures, Thidwick has forgotten his moose nature and the natural processes of life as a cervid. But suddenly he remembers that cervids like him shed their antlers "so that NEW ones can grow!" His rediscovered ecological self returns to nature.

In addition to deep ecology, *Thidwick* can be seen as an example of the Gaia hypothesis, which is the scientific theory that the global ecosystem is a self-regulating superorganism that collectively maintains conditions conducive to life on Earth.[46] Through decades of writing, James Lovelock (who is widely recognized as the father of the Gaia hypothesis) has argued for a planetary perspective, warning that human interference in global atmospheric and biospheric processes risks destroying the long-standing balance on Earth. The tale of *Thidwick* reflects Lovelock's warnings, since the creatures riding on Thidwick's antlers see the moose as nothing more than a pair of antlers. In their selfish, short-term thinking, they fail to see Thidwick as a living being (just as Earth is a living being for believers in the Gaia hypothesis).

While the Gaia hypothesis and deep ecology might seem unrelated to security, new scholarship in the field of environmental security theory makes connections between environmental conditions (such as climate change or resource scarcity) and peace or conflict.[47] Some recent research explicitly focuses on climate security as a domain of security studies.[48] In the environmental security approach, the moose's antlers would be seen as an environment where inhabitants have radically changed the climatic conditions, causing conflict between the inhabitants.

IN CONCLUSION: "ALL STUFFED, AS THEY SHOULD BE"

Thidwick nominally concerns the struggle for supremacy and security between a moose and the creatures on his antlers. Like *The Lorax*, *Thidwick* has a clear environmental theme but a much bleaker vision. The final scenes of *The Lorax* inspire some hope for the future. When the Once-ler passes the last Truffala Tree seed to the boy, he says hopefully, "UNLESS someone like you / cares a whole awful lot, / nothing is going to get better. / It's not." Caring for the environment and regrowing the forest of Truffala Trees will presumably bring back the Lorax and the thriving ecosystem he inhabited. In *Thidwick*, after shedding his antlers the moose rejoins his herd at the south end of the lake and enjoys the delicious moose-moss. However, his former guests are killed and now exist only as trophies on the Harvard Club wall: "all stuffed, as they should be."

Ending the book with an image of dead trophies on a wall demonstrates Geisel's contempt for the perpetrators of environmental destruction, whom he holds accountable for their destruction of the antler environment. Geisel thereby conveys an implicit warning about the genocide of the whole human species, which might happen if we ignore the crash of ecosystems, prioritize short-term interests at the expense of long-term concerns, and fail to regulate common pool resources. Moreover, Thidwick's survival and concurrent destruction of the inhabitants on his antlers suggests that planetary security provides the foundation upon which all other types of security rest. Without planetary security, we doom ourselves to extinction, as *Thidwick* demonstrates.

NOTES

1. Aeon J. Skoble, "Thidwick the Big-Hearted Bearer of Property Rights," in *Dr. Seuss and Philosophy: Oh, the Thinks You Can Think!*, ed. Jacob M. Held (New York: Rowman & Littlefield, 2011), 159–66.

2. Donald E. Pease, *Theodor SEUSS Geisel* (New York: Oxford University Press, 2010), 84.

3. Alison Lurie, "The Cabinet of Dr. Seuss," *New York Times Review of Books*, December 12, 2020, https://www.nybooks.com/articles/1990/12/20/the-cabinet-of-dr-seuss/.

4. Ramachandra Guha, *Environmentalism: A Global History* (New York: Longman, 2000).

5. International Global Change Program, "The Anthropocene Era," *IGBP Science* 4 (2001): 11–15.

6. R. F. Hunter, "Hill Sheep and Their Pasture: A Study of Sheep-Grazing in South-East Scotland." *Journal of Ecology* 50, no. 3 (1962): 651–80, https://doi.org/10.2307/2257476; and Louise E. Sweet, "Camel Raiding of North Arabian Bedouin: A Mechanism of Ecological Adaptation," *American Anthropologist* 67, no. 5 (1965): 1132–50, http://www.jstor.org/stable/668360.

7. Lucie M. Bland, David A. Keith, Rebecca M. Miller, Nicholas J. Murray, and Jon Paul Rodriguez, eds., *Guidelines for the Application of IUCN Red List of Ecosystems Categories and Criteria*, International Union for the Conservation of Nature (IUCN), 2016, https://portals.iucn.org/library/node/45794.

8. Marion Pratt, "Useful Disasters: The Complexity of Response to Stress in a Tropical Lake Ecosystem," *Anthropologica* 38, no. 2 (1996): 125–48.

9. Naomi R. Lamoreaux, "The Mystery of Property Rights: A U.S. Perspective," *Journal of Economic History* 71, no. 2 (2011): 275–306, http://www.jstor.org/stable/23018300.

10. Albert S. Thayer, "Adverse Possession," *Journal of the Society of Comparative Legislation* 13, no. 3 (1913): 582–602, http://www.jstor.org/stable/752305.

11. Brian Angelo Lee, "Just Undercompensation: The Idiosyncratic Premium in Eminent Domain," *Columbia Law Review* 113, no. 3 (2013): 593–655, http://www.jstor.org/stable/23479386.

12. Suzan Shown Harjo, ed., *Nation to Nation: Treaties between the United States and American Indian Nations* (Washington, DC: Smithsonian Institution, 2014).

13. Charles F. Wilkinson and John M. Volkman, "Judicial Review of Indian Treaty Abrogation: 'As Long as Water Flows, or Grass Grows upon the Earth'; How Long a Time Is That?," *California Law Review* 63, no. 3 (1975): 601–61, https://doi.org/10.2307/3479850.

14. Montgomery McFate, *Military Anthropology: Soldiers, Scholars and Subjects at the Margins of Empire* (New York: Oxford University Press, 2018).

15. "The Pied Piper of Hamelin." *The Aldine* 4, no. 6 (1871): 90–91, http://www.jstor.org/stable/20636049.

16. Garrett Hardin, "The Tragedy of the Commons," *Science* 162, no. 3859 (December 13, 1968): 1243–48.

17. Hardin, "Tragedy of the Commons."

18. Nathaniel J. Dominy, Sandra Winters, Donald E. Pease, and James P. Higham, "Dr. Seuss and the real Lorax," *Nature Ecology & Evolution* 2 (2018): 1196–98.

19. William G. Robbins, "The Social Context of Forestry: The Pacific Northwest in the Twentieth Century," *Western Historical Quarterly* 16, no. 4 (1985): 413–27. https://doi.org/10.2307/968606.

20. Hardin, "Tragedy of the Commons."

21. Glenn H. Snyder, *Alliance Politics* (Ithaca, NY: Cornell University Press, 1997).

22. Hans Morgenthau, *Politics Among Nations* (New York: McGraw-Hill, 1948).

23. Joseph Lepgold, "NATO's Post-Cold War Collective Action Problem," *International Security* 23, no. 1 (1998): 78–106, https://doi.org/10.2307/2539264.

24. Peter Katzenstein, ed., *Culture of National Security: Norms and Identity in World Politics* (New York: Columbia University Press, 1996).

25. Barbara W. Tuchman, *March of Folly: From Troy to Vietnam* (New York: Knopf, 1984); and Shimon Tzabar, *White Flag Principle: How to Lose a War and Why* (London: Penguin Press, 1972).

26. Wolfgang Schlauch, "American Policy towards Germany, 1945," *Journal of Contemporary History* 5, no. 4 (1970): 113–28, http://www.jstor.org/stable/259868.

27. Thomas Hobbes, *Leviathan*, ed. Edwin Curley (Indianapolis: Hackett, 1994), 76.

28. Hobbes, *Leviathan*.

29. Chris Naticchia, "Hobbesian Realism in International Relations: A Reappraisal," in *Hobbes Today: Insights for the 21st Century* (Cambridge: Cambridge University Press, 2013), 241–63.

30. Simon Harrison, "Skull Trophies of the Pacific War: Transgressive Objects of Remembrance," *Journal of the Royal Anthropological Institute* 12, no. 4 (2006): 817–36, http://www.jstor.org/stable/4092567.

31. Rene Girard, *Violence and the Sacred* (Baltimore, MD: Johns Hopkins University Press, 1972).

32. Jeffrey L. Durbin, "Expressions of Mass Grief and Mourning: The Material Culture of Makeshift Memorials," *Material Culture* 35, no. 2 (2003): 22–47, http://www.jstor.org/stable/29764188.

33. Barbara Ehrenreich, *Blood Rites: Origins and History of the Passions of War* (New York: Metropolitan Books, 1997).

34. Ehrenreich, *Blood Rites*.

35. Constance Casey, "Moose: The Final Frontier," *Landscape Architecture* 104, no. 11 (2014): 66–70, http://www.jstor.org/stable/44796105.

36. Elisabeth S. Clemens, "Of Asteroids and Dinosaurs: The Role of the Press in the Shaping of Scientific Debate," *Social Studies of Science* 16, no. 3 (1986): 421–56, http://www.jstor.org/stable/285026.

37. Gerardo Ceballos, Paul R. Ehrlich, and Peter H. Raven, "Vertebrates on the Brink as Indicators of Biological Annihilation and the Sixth Mass Extinction," *PNAS* 117, no. 24 (2020): 13596–602, https://doi.org/10.1073/pnas.1922686117.

38. David Quammen, "The Weeds Shall Inherit the Earth," *Independent*, November 22, 1998, 30–39.

39. Paul Thompson, "Tempo and Mode in Evolution: Punctuated Equilibria and the Modern Synthetic Theory," *Philosophy of Science* 50, no. 3 (1983): 432–52, http://www.jstor.org/stable/187858.

40. Michael J. Benton and Richard J. Twitchett, "How To Kill (Almost) All Life: The End-Permian Extinction Event," *Trends in Ecology and Evolution* 18, no. 7 (July 2003): 358–65.

41. Jennifer Thomson, "Biocentrism and the Health of the Wild," in *The Wild and the Toxic: American Environmentalism and the Politics of Health* (Durham, NC: University of North Carolina Press, 2019), 71–97, http://www.jstor.org/stable/10.5149/9781469651668_thomson.7.

42. Jens Ivo Engels, "Modern Environmentalism," in *The Turning Points of Environmental History*, ed. Frank Uekoetter (Pittsburgh: University of Pittsburgh Press, 2010), 119–31, https://doi.org/10.2307/j.ctt5hjsg1.11.

43. Arne Naess, "Self-Realization: An Ecological Approach to Being in the World," in *Deep Ecology for the Twenty-First Century: Readings on the Philosophy and Practice of the New Environmentalism*, ed. George Session (Berkeley, CA: Shambhala Press, 1995).

44. Naess, "Self-Realization."

45. Alan Drengson and Yuichi Inoue, *Deep Ecology Movement* (Berkeley, CA: North Atlantic Books, 1995), xix.

46. James Lovelock, *Gaia: A New Look at Life on Earth* (Oxford: Oxford University Press, 1979).

47. Hugh Dyer, "Environmental Security And International Relations: The Case for Enclosure," *Review of International Studies* 27, no. 3 (2001): 441–50.

48. Maria Julia Trombetta, "Environmental Security and Climate Change: Analysing the Discourse," *Cambridge Review of International Affairs* 21, no. 4 (2008): 585–602.

6

The Cat in the Hat and Cyber Warfare

Jon R. Lindsay and Michael Poznansky

The Cat in the Hat was written in 1957, the same year the Soviet Union launched the first artificial satellite—Sputnik 1—into space. It would be another decade or so before the Advanced Research Projects Agency Network (ARPANET) connected geographically separate computers through an effective packet-switching protocol, a feat that paved the way for the creation of the internet. Nevertheless, *The Cat in the Hat* contains numerous lessons and themes that help make sense of the cyber environment and many of the current debates surrounding it.

In this chapter we use the various characters and events from the story to highlight some contentious issues regarding cybersecurity. In particular, we argue that *The Cat in the Hat* addresses the target's cooperation—witting or unwitting—as a permissive condition for cyber operations, as represented by the children who invite the Cat into the house. Moreover, exactly like the Cat's trick, cyber intrusions often result in unintended consequences for the perpetrators carrying them out. In *The Cat in the Hat*, the Cat's initial entry, followed by the unbundling of Thing One and Thing Two, is illustrative of how staged operations are conducted in cyberspace. As we see in *The Cat in the Hat*, cyber perpetrators cover their tracks, giving rise to the challenge of establishing attribution. An ongoing debate in cyber scholarship concerns the relative balance of offense-defense between perpetrators and would-be victims, which here involves relationships between the Cat, Mother, and the architecture of the house. Finally, in addition to the operational and strategic implications of *The Cat in the Hat*, the book sheds light on the broader cultural dimensions of hacking, which concern both deviance and authority within the larger framework of US cultural norms.

UNWITTING COOPERATION

The Cat in the Hat begins on a dreary, rainy day when "it was too wet to play." Two children, Sally and an unnamed male narrator, are stuck inside the house, bored, and are looking for activities to keep themselves occupied. Unsurprisingly, they do not like these conditions: "Not one little bit." Suddenly they hear a thud. In through the front door bursts a bipedal cat, holding an umbrella. He's wearing a red bowtie around his neck and a tall Abraham Lincoln–style top hat with red and white stripes. The Cat tells the children that although the day is dreary, they can still have some fun.

> "I know some good games we could play,"
> Said the Cat.
> "I know some new tricks,"
> Said the Cat in the Hat.
> "A lot of good tricks.
> "I will show them to you.
> "Your mother
> "Will not mind at all if I do."

(How the Cat knows their mother won't mind is anyone's guess.)

The children allow the Cat to enter the house and do not ask him to leave. Thus, the first major theme in *The Cat in the Hat* pertaining to cybersecurity is unwitting cooperation as a permissive condition for cyber operations. Unlike traditional warfighting, which batters down obstacles with brute force, cyber operations rely on secrecy, deception, and manipulation. This makes cybersecurity a lot more like an "intelligence contest" than military combat.[1] It entails the willing but unwitting assistance of computer users and designers.

In this regard, the Cat can be seen as a remote intruder intending to compromise a computer system, symbolized by the house. The first operational task is gaining access to the target. The intruder—er, the Cat—must break into the system (the house) by manipulating the unwitting children (the users of this system) to willingly provide entry. In real life, users routinely ignore the advice of network administrators during mandatory cybersecurity training, just as the children ignore the warnings of the fish:

> But our fish said, "No! No!
> "Make that cat go away!
> "Tell that Cat in the Hat
> "You do NOT want to play.
> "He should not be here.
> "He should not be about.
> "He should not be here
> When your mother is out!"

Users can be lured into unwittingly allowing intruders to penetrate secure systems through confidence games or "social engineering." Consider phishing, in which individuals receive an email from someone they think they know, or a company they may have heard of, asking them to click on a link or attachment. Unbeknownst to the users, these links or attachments install malware. According to one estimate, there were over 255 million phishing attempts in 2022 alone.[2]

Another tactic for getting unwitting users to provide access to a system is tricking them into plugging an infected thumb drive into a computer. In 2008 the US government experienced one of the biggest breaches of its military networks through malware, known as agent.btz, linked to Russia. The compromise of classified and unclassified networks purportedly occurred after a user plugged in a flash drive containing agent.btz at a military base.[3] The initial compromise of the SolarWinds development environment in early 2019 appears to have been through an employee's VPN account, although it is unclear whether this occurred through social engineering or something more nefarious.[4]

A third tactic for gaining access occurs via insider threats. While the two examples already provided (phishing and infected thumb drives) represent insiders who *unwittingly* compromise a system, a more pernicious form of insider threat is when an individual with knowledge of an organization and privileged access causes intentional harm.[5] If we think of the children in *The Cat in the Hat* not as naïve residents who accidentally let a malicious actor in, but rather as insider threats who voluntarily let the Cat in while their mother is away, the arc of the story takes on a different color entirely. Here the children do not simply ignore the fish out of negligence; they are complicit in putting the fish in a teapot where it cannot do anything to help. As Dr. Seuss describes the incident:

> He came down with a bump
> From up there on the ball.
> And Sally and I,
> We saw ALL the things fall!
> And our fish came down, too.
> He fell into a pot!
> He said, "Do I like this?
> "Oh, no! I do not.
> "This is not a good game,"
> Said our fish as he lit.
> "No, I do not like it,
> "Not one little bit!"

In many cybersecurity incidents, retrospective investigation frequently discovers at least one actor who is complicit in the attack. According to news reports, for example, the Stuxnet attack on Iran's nuclear enrichment facility at Natanz was allegedly enabled by an Iranian scientist working for Dutch intelligence, who was able to avoid the suspicions of Iranian security services and plug an infected thumb drive into the "air gapped" computer network at Natanz, thus enabling remote attackers to upload new versions of Stuxnet and to exfiltrate status reports from the intrusion.[6]

In Dr. Seuss's story, the children maintain a conspiracy of silence when Mother inquires about their day. Victims of cybercrime are often reticent to report breaches because they are embarrassed or afraid of business repercussions.[7] If the children are treacherous co-conspirators, they have even more incentive to avoid reporting the incident to their mother. How else can we explain their repeated interactions with the Cat and their ongoing deception of Mother in the *The Cat in the Hat* sequels?

Another cautionary tale in *The Cat in the Hat* related to unwitting cooperation is the system architecture itself. For it is not just the negligence or treachery of the children, or incompetence of the fish, that lets the Cat in the front door. The house is presumably built with back doors as well, which Mother would find convenient for getting into the garage or basement for legitimate maintenance. These doors, however, provide alternative ways into the system. The problem of the unintended access point is illustrated in Dr. Seuss's sequel, *The Cat in the Hat Comes Back*. When the Cat returns to the house, the children are shoveling snow outside. The Cat dashes in through the door they have left open for themselves. Closing one vulnerability just encourages the attacker to find others.

The Cat in the Hat also shines light on the danger of existing system features as a source of compromise. While the Cat does bring in his own tools like Things One and Two, the Cat also appropriates what he finds in the environment to "live off the land," enabling his intrusion and amplifying its effects. As Dr. Seuss explains:

> It is fun to have fun
> But you have to know how.
> I can hold up the cup
> And the milk and the cake!
> I can hold up these books!
> And the fish on a rake!
> I can hold the toy ship
> And a little toy man!
> And look! with my tail
> I can hold a red fan!
> I can fan with the fan
> As I hop on the ball!

Just as the Cat employed toy ships, cake, and fishbowls, the SolarWinds hackers used previously installed software and administrative utilities in the host network to reduce and possibly eliminate detectable signatures. Even worse, these native tools can be used to forge security credentials. The SolarWinds hackers smuggled in a corrupted library file that SolarWinds then compiled on its own built network and signed with its digital certificates before distributing this poison pill to eighteen thousand customers.[8] In this way "zero trust" measures, discussed further later, can provide a false sense of security, if the validated users are unwittingly co-opted or willingly treacherous.

One of the thorniest questions in current cybersecurity debates is who should bear ultimate blame, or responsibility, for compromises. Should the children who willingly allowed the Cat in through the front door, whether out of boredom or complacency, be held responsible for the unwitting breach? The *National Cybersecurity Strategy* released in March 2023, the first one authored by the Office of the National Cyber Director, argues that "end users bear too great a burden for mitigating cyber risks. Individuals, small businesses, state and local governments, and infrastructure operators have limited resources and competing priorities, yet these actors' choices can have a significant impact on our national cyber security. A single person's momentary lapse in judgment, use of an outdated password, or errant click on a suspicious link should not have national security consequences."[9]

Certainly letting the Cat into the house constitutes a lapse of judgment, but should children (or other naïve users) be held responsible? The *National Cybersecurity Strategy* calls for a new approach, shifting the onus from end users to "the most capable and best-positioned actors." As such, "protecting data and assuring the reliability of critical systems must be the responsibility of the owners and operators of the system that hold our data and make our society function, as well as of the technology providers that build and service these systems."[10] In *The Cat in the Hat* it was likely foolish to put the burden for protecting the house (computer systems) on the children (the users) and the fish (a hapless administrator). Instead, Mother (the owner and protector of the system) has that responsibility. Whether it be a babysitter, a smart alarm system, or raising more skeptical children, resilience comes from a holistic and broader approach.

Another way policymakers are addressing the question of responsibility for system security is outlined in the Biden administration's 2021 Executive Order on Improving the Nation's Cybersecurity.[11] One of its key tenets, "zero trust" architecture, begins from the premise that "no actor, system, network, or service operating outside or within the security perimeter is trusted. Instead, we must verify anything and everything attempting to establish access."[12] In terms of Dr. Seuss, anything that goes "BUMP!" at the front door should not be trusted. Credentials must be established through rigorous vetting to avoid the calamity that befalls the children.

UNINTENDED CONSEQUENCES

The Cat's first trick (suitably named "UP-UP-UP with a fish!") involves balancing the poor fish on the handle of his umbrella. The terror in the fish's eyes is evident. The Cat, a real show-off, next balances a cup on his hat, twirls a book in his left hand, and stands on a ball with one foot.

> "Look at me!
> "Look at me now!" said the cat.
> "With a cup and a cake
> "On the top of my hat!
> "I can hold up TWO books!
> "I can hold up the fish!
> "And a little toy ship!
> "And some milk on a dish!
> "And look!
> "I can hop up and down on the ball!
> "But that is not all!
> "Oh, no.
> "That is not all . . .
> "Look at me!
> "Look at me!
> "Look at me NOW!
> "It is fun to have fun
> "But you have to know how."

The Cat's tricks can be interpreted as a variant of malware intended to temporarily hold the children's attention to prevent them from noticing the potential danger. Similar to a distributed denial of service (DDoS) attack that overwhelms servers with requests, the Cat's antics make the entire house (e.g., system) unusable for a period of time. The problem for the Cat (as well as other sorts of cyber operations) is that the situation spins out of control. When the Cat gets greedy and tries to balance more and more objects (cakes, books, fans, toy boats, rakes, etc.), his complex assemblage comes crashing down. Even though the disaster was accidental, the fish nevertheless scolds the Cat:

> "Now look what you did!"
> Said the fish to the cat.
> "Now look at this house!
> "Look at this! Look at that!
> "You sank our toy ship,
> "Sank it deep in the cake.
> "You shook up our house
> "And you bent our new rake."

Exactly like the Cat's trick, cyber intrusions often yield unintended consequences for the perpetrators who carry them out. Several high-profile cyberattacks follow this pattern. In 2017 a Russian military hacking group known as Sandworm targeted the "Linkos Group, a small, family-run Ukrainian software business."[13] This hacker group targeted the company's update servers, which opened a back door into vast numbers of computers that relied on accounting software known as MeDoc.[14] According to a detailed report from *Wired*, "The code that the hackers pushed out was honed to spread automatically, rapidly, and indiscriminately."[15] Although it appeared like a case of ransomware, in reality it destroyed all data.

This event, known as NotPetya, wreaked havoc beyond Ukraine's borders, affecting "multinational companies including Maersk, pharmaceutical giant Merck, FedEx's European subsidiary TNT Express" and more.[16] The estimated damage was $10 billion. A 2018 White House statement identified NotPetya as "part of the Kremlin's ongoing effort to destabilize Ukraine," noting that this "was also a reckless and indiscriminate cyber-attack that will be met with international consequences."[17] Like the Cat, NotPetya represents a "weapon of war released in a medium . . . where collateral damage travels via a cruel and unexpected logic."[18]

In *The Cat in the Hat Comes Back*, the Cat shifts tactics in an effort to avoid past mistakes. Instead of balancing a collection of household objects on his head, he is more measured, while still leaving an utter mess (eating cake in the tub, cleaning up the pink soap ring with Mother's white dress, wiping the dress on a wall, etc.). Similarly, after NotPetya, Russia adjusted its behavior to limit collateral damage in cyberspace. According to a report from Microsoft, "Russian cyber tactics in the war have differed from those deployed in the NotPetya attack against Ukraine in 2017. . . . Russia has been careful in 2022 to confine destructive 'wiper software' to specific network domains inside Ukraine itself."[19] Like the Cat, Russia appears more cognizant of its limits and only performs those tricks it has totally mastered.

STAGED OPERATIONS

After the Cat's initial trick fails spectacularly, he returns to the house with a box and unloads Thing One and Thing Two. The Cat pats them fondly, exclaiming:

> They are tame. Oh, so tame!
> They have come here to play.
> They will give you some fun
> On this wet, wet, wet day.

Thing One and Thing Two begin to perform their own destructive tricks, including flying kites in the house. As they run down the hall, Thing One and Thing Two "Bump their kites on the wall! Bump! Thump! Thump! Bump! Down the wall in the hall." For cybersecurity purposes, these Things are suggestive of a staged operation. The first stage (represented by the Cat's initial entry) is to access and navigate or "move laterally" through a network in search of valuable data or vulnerable targets. The second stage (represented by the Things' subsequent appearance) is to deploy a payload that can steal data or sabotage systems.

The first phase of Stuxnet relied on Windows-based exploits to propagate through Iranian networks in search of computers specifically configured to control centrifuge operations; the second phase then activated custom payloads that manipulated the Siemens-based industrial control system.[20] Developing and testing this payload in the second phase required an entirely different set of tools and techniques (i.e., Things One and Two). Stuxnet's attack payload masked alarms from malfunctioning centrifuges so that Iranian operators would not detect the problems, a clever tool to preserve stealth.[21]

Once discovered, however, due to an accidental detection resulting from configurational friction, the very sophistication of the payload led many to point the finger at nation-state actors with the ability and will to carry out the operation.[22] According to Michael Warner, the command historian at US Cyber Command, "The story broke in 2012 when David Sanger of the *New York Times* claimed that the United States and Israel had created Stuxnet to attack Iran's nuclear weapons with a cyber weapon."[23] Some observers concluded that the sophistication of the attack, including propagation controls suggestive of legal constraints on a covert action, indicated that this must be the work of the United States.[24] The irony here is that the same measures that improve obscurity can also become tools for attribution. With respect to *The Cat in the Hat*, it took a sophisticated actor, the Cat, to perform initial reconnaissance and then deploy additional payloads, the Things, in the house.

LEAVING NO TRACE

Near the end of *The Cat in the Hat* the fish exclaims, "And this mess is so big / And so deep and so tall, / We can not pick it up. / There is no way at all!" Before panic fully sets in among the residents, the Cat returns, riding on a wild machine equipped with advanced tools to clean up the mess.

> He picked up the cake,
> And the rake, and the gown,
> And the milk, and the strings,
> And the books, and the dish,
> And the fan, and the cup,
> And the ship, and the fish.
> And he put them away.
> Then he said, "That is that."
> And then he was gone
> With a tip of his hat.

By the time the Cat leaves, the house is spotless. When Mother returns, she asks, "Did you have any fun? Tell me. What did you do?" As mentioned previously, the children tell her nothing about the shenanigans that have just transpired in her home. Like the Cat, perpetrators in cyberspace often obfuscate their activity. A malicious actor causing mayhem and departing undetected (or at least casting doubt about their complicity) remains one of the most cited challenges associated with detecting cyberattacks and defending against, or even preventing, them. This phenomenon is known as the attribution problem.[25]

In describing the life cycle of a cyberattack, the UK National Cyber Security Centre notes that in many cases, once the desired effect has been achieved, "the more capable actor will exit, carefully removing any evidence of their presence."[26] For example, in May 2023 Microsoft discovered that a Chinese hacking group had conducted a large espionage campaign, including attacks on the military base at Guam. The group, which Microsoft called Volt Typhoon, was designed to avoid attribution in part by "leaving no trace behind," according to a senior official at the National Security Agency.[27] In another infamous case, known as Olympic Destroyer, the Russian GRU (aka Sandworm) masqueraded as a Chinese group pretending to be a North Korean group, potentially as a cautionary message to commercial intelligence firms that had become overconfident about their attribution skills.[28]

The Cat in the Hat portrays another attribution challenge that pertains directly to cyberspace: the use of proxies to conduct cyberattacks.[29] The

Cat opens "a big red wood box . . . shut with a hook," out of which pop Thing One and Thing Two. Earlier we described how they illustrate staged operations. Here they represent something else. Their autonomy from the Cat becomes clear when the Things introduce themselves to the children as new acquaintances:

> And they ran to us fast.
> They said, "How do you do?
> "Would you like to shake hands
> "With Thing One and Thing Two?"

Imagine that Things One and Two do not work directly for the Cat but rather function as contractors or "patriotic hackers." Should the Cat wish to obfuscate his involvement, Thing One and Thing Two would provide a "cut out" or plausible deniability for their actions. Reliance on proxies is a double-edged sword of course, as proxies might have interests misaligned with the Cat's, causing damage that no machine can clean up![30]

Like the situation in *The Cat in the Hat*, many cyber incidents involve proxies. "The prevalent practice of some states [using] proxies or other nonstate agents to undertake cyber attacks on their behalf" presents a major obstacle when attempting to determine attribution.[31] In 2007, for example, the Estonian government announced that it was moving a Soviet-era statue out of the capital. A cyberattack in the form of a DDoS operation followed, lasting nearly three weeks. The operation was "in part attributable to the Nashi youth activist group, but it is unclear whether the Russian Federation had a hand in the group's operations."[32] One of the most important developments to emerge from this incident, however, was a broader discussion of how existing alliances like the North Atlantic Treaty Organization should handle cyberattacks against members, especially when the culpability of a state adversary remains somewhat murky.[33] That said, while the physical location or language spoken by the proxy may indicate which government they serve, establishing a firm connection—particularly one strong enough to serve as a basis for retaliation—can be more difficult, harkening back to the attribution problem.[34]

Another attribution issue that appears in *The Cat in the Hat* concerns forensic clues that may clarify the identity of the attacker. In the story, the operational environment of the domicile proves more complicated than the Cat expects. The Cat's preoperational reconnaissance may have been incomplete or may have failed to anticipate the durability of the furniture and fixtures:

> Thing Two and Thing One!
> They ran up! They ran down!
> On the string of one kite
> We saw Mother's new gown!
> Her gown with the dots
> That are pink, white and red.
> Then we saw one kite bump
> On the head of her bed!

The Cat's operations (especially the antics of Thing One and Thing Two) leave a trail of clues that are sure to point back to him. Similarly, because of the complexity of cyberspace, attackers and defenders alike encounter all sorts of frictions and unexpected circumstances that influence attribution.[35] Friction and operational mistakes on the part of the attacker can provide clues that aid incident response and attribution on the part of the defender. The SolarWinds hackers were very good, for example, but a routine security alert about an employee's smart phone logging in from two locales at the same time triggered an investigation. Subsequently, a disk image on the SolarWinds built network that had not been erased was discovered, which in turn revealed the corrupted library file. The accumulating forensic and contextual evidence painted a convincing picture that this was the work of the Russian SVR, a descendent of the KGB.[36] Technical characteristics of tooling and infrastructure and recurrent patterns of operational behavior create signatures and profiles that can be used to understand the extent of and infer responsibility for intrusions. Attribution assessments are typically the result not of any given smoking gun but rather of the fusion of a mass of individually circumstantial but cumulatively convincing bits of evidence.[37]

As Mother approaches the house, it seems impossible that she will not discover the mess.

> Then our fish said, "Look! Look!"
> And our fish shook with fear.
> "Your mother is on her way home!
> "Do you hear?
> "Oh, what will she do to us?
> "What will she say?
> "Oh, she will not like it
> "To find us this way!"
> "So DO something! fast!" said the fish.
> "Do you hear!
> "I saw her. Your mother!
> "Your mother is near!"

This moment points to an important temporal asymmetry in cybersecurity: the attacker who relies on secrecy has the advantage during the planning and intrusion phase of the operation, but the advantage switches to the defender as detection triggers an investigation and the accumulation of incriminating clues.[38] Yet in this case, miraculously, the Cat performs a meticulous cleanup in the nick of time.

OFFENSE-DEFENSE BALANCE

In *The Cat in the Hat*, who defends the system? The poor fish has been left alone to administer a network whose users behave like children—indeed, they *are* children! He has no power to force them to conform with best practices. He can only plead with them to exercise better "cyber hygiene." But moral suasion is usually *not* an effective way to get people to do the right thing. The fish not only watches helplessly as the children ignore his advice; he becomes pulled into the mayhem himself, trapped in a teapot where he can do no good. The peewee Pisces is powerless before the prowler's prowess. One might even ask why Mother has left the children alone in the first place? Is she embroiled in a custody battle with Father about how to divide responsibility for cybersecurity between society and the state? Is she a single parent under pressure to make ends meet, compromising on time with the children to bring home a paycheck? Is she a negligent vendor, an ineffective regulator, or a distracted sovereign? Actors with the power to improve cybersecurity often lack the incentives to do so. Perverse incentives undermine the effectiveness of cyber defenses.

At first blush, *The Cat in the Hat* seems to support the view that offense has the advantage in cyberspace. In *The Cat in the Hat*, the intruder simply walks into the house, exploits the unwitting cooperation of the children, and trashes Mother's furniture. Network defenses are powerless to stop him. Powerful hacking tools seem to be cheaply available online, and hackers can use them to attack from inaccessible jurisdictions. Networked organizations and societies, full of vulnerable machines and gullible people, create a huge "attack surface" that hackers can exploit.

From the Cat's perspective, however, the offense-defense balance is not so simple. The remote network, the entire target organization, and its ecosystem of counterintelligence defense may be very hard to understand. Detailed intelligence and careful planning are required, as well as a backup in case things go wrong. These tasks are difficult, which means that hacking is not cheap and easy, at least against the targets that matter. The Cat can cope with these contingencies because he is an advanced persistent threat (APT). He uses sophisticated technical tricks and confidence games to gain and maintain access to other people's houses. The feline hacker is confident and prepared, with a collection of exploits used to gain access to the system, escalate privileges to run commands, move laterally through the network, and deploy tailored payloads like Things One and Two. This is obviously not the Cat's first rodeo. He has clearly entertained (or horrified) other children before. In *The Cat in the Hat Comes Back*, we also learn that this APT will revisit the same target with a new arsenal of exploits, trashing not only Mother's things but Father's as well.

An important lesson related to discussions about the relative ease of offense is that the hacker does not work alone. For his part, the Cat relies on a crew of skilled operators like Things One and Two, as well as a supporting cast of administrators and logisticians. Just imagine the expense involved in building the complex cleaning machine! This machine is an example of a world-class exploit infrastructure on a par with leading hacking organizations like Equation Group.[39] The cleanup process is nothing short of miraculous, conducted right as Mother's incident response team is arriving outside the window. No amateurs could have pulled this off. Clearly the Cat had access to sophisticated resources and rehearsal ranges to develop and deploy the exquisite machine. In *The Cat in the Hat Comes Back*, moreover, it is revealed that the Cat has another Little Cat in his hat, which has yet another Little Cat in its hat, and so on for twenty-six levels of recursion, from A to Z. This suggests a mastery of the recursive algorithms and data structures used throughout computer science. The littlest cat ("the Z you can't see") is especially skilled at operational security. Little Cat Z also has exquisite skills and tooling called Voom, which "is so hard to get, / You never saw anything / Like it, I bet." Voom can clean up embarrassing spots (actually, "Voom cleans up anything"), which is especially useful when operating in the presence of multiple intelligence threats on the same host network.[40]

Deterrence is thought to be difficult if not impossible in cyberspace because operations are secret, temporary, or irreversible. As a result, very little escalation occurs as a result of cyberattacks.[41] But *The Cat in the Hat* asks us to question this premise. After all, it is the threat of Mother's impending arrival that encourages the Cat to go to great lengths and expense with the cleanup. If the Cat was not afraid of the adverse consequences of exposure (at least for the children if not himself), then why bother with all the stealth and subterfuge? The threat of cyberwar may not improve deterrence, but deterrence does shape cyber operations. Hackers pull their punches to avoid compromising valuable tools and to preserve the opportunity to hack another day.[42]

DEVIANCE AND AUTHORITY

Thus far in this chapter we have focused mainly on the operational and strategic implications of *The Cat in the Hat*. But Dr. Seuss also sheds light on the broader cultural dimensions of hacking, which concern both deviance and authority within the larger framework of cultural norms. Ever since Matthew Broderick played a teenager in a hoodie hacking into US nuclear command and control systems in *WarGames* (1983), the archetype of the deviant hacker genius has been inextricably linked to cybersecurity. Theodor Geisel can be seen as a teenage hacker of the mind. Having adopted the pen name "Dr. Seuss" as a Dartmouth undergraduate, Geisel has become universally recognized by his hacker name, through which he projects a mythos of playful mastery and idiosyncratic genius, inspiring delight in children, nostalgia in parents, and the analysis of academics. As we have seen, of course, high-end hacking is often a more bureaucratized affair. So too with Dr. Seuss, as Geisel relied on publishers and administrators to get his books out, and he eventually licensed his brand for stories and screenplays by other authors. Dr. Seuss is less of a single person and more of a hacking collective like Anonymous or Guccifer.

The archetype of the deviant hacker has some truth behind it. The term *hacking* just refers to creative, unconventional problem-solving. Hacker conferences like Defcon and Black Hat celebrate this creative spirit, and the cybersecurity industry rewards it handsomely. Indeed, civilian actors play a vital role in cybersecurity, as technology vendors, network operators, security professionals, and of course victims.[43] These factors create cultural mismatch for military organizations like US Cyber Command, which aspire to recruit talented hackers. Not only is it difficult for the government to pay hackers as much as they can earn in industry, but commercial threat intelligence is actually the front line of many significant cyber incidents, drawing on unique commercial data and access to targets. In addition, the people who excel at hacking may not conform to or endorse military virtues of obedience, physical fitness, and avoidance of controlled substances. This is known as the military's "blue haired hacker problem."[44] Isn't it interesting, then, that Thing One and Thing Two, the expert hackers that the Cat calls in to deliver his payload, have blue hair? Indeed, the cat in *The Cat in the Hat* is something of a deviant character himself.

Is the Cat a malicious hacker bent on breaking systems to burnish his reputation? Is he a nation-state APT subverting the house to gain subtle political advantages? Is somebody paying him? Or is playing with children its own reward? Or, more charitably, is he trying to teach the children and the fish a lesson to improve their cybersecurity posture? We also must consider the possibility that Mother has hired the Cat as a red team to test her own counterintelligence defenses. One cannot say anything conclusive about the Cat's motivations, which is par for the course in the ambiguous world of cybersecurity.[45] It is difficult to tell whether the Cat is a white hat or black hat hacker. Indeed, he is something different yet: a white and red hat. How about that?[46]

The Cat extends a hand to the children, male and female alike, to ask them to have fun in the deconstruction of the system. The Cat even smuggles in the deviants who have been excluded from the system, embodied in the androgenous blue-haired figures of Thing One and Thing Two. In a constructed community built right under the nose of societal authority, these liberated Things offer the children a choice between living authentically and perpetuating domination. The hacking that looks like destruction from the outside is deeply generative from the inside. *The Cat in the Hat* is thus a highly subversive story, clothing this archetypally feminist revelation in a male character to liberate the children from the oppressive advice of the male fish, the slippery servant of absent authority.

And this brings us to Mother, who is the ultimate authority figure in the story, structured in opposition to the deviance of the Cat. It is Mother's house. Father is absent, at least until *The Cat in the Hat Comes Back*. And it is Mother who strikes fear in the hearts of everyone in the story: the Cat, the Things, the fish, and the children alike. What are we to make of this? The pillar of patriarchy in the national security house is . . . Mother. It is as if the feminist message of *The Cat in the Hat* is afraid of itself. This may be the most profound lesson of the story: cybersecurity is a paradox. The very social institutions that nurture liberal order are the ultimate enablers of their subversion. In traditional international relations, war is from Mars and peace is from Venus. But in the world of secret statecraft, peace encourages conflict in the form of espionage, subversion, covert action, and disinformation.[47]

IN CONCLUSION: WARNINGS TO PONDER

The Cat in the Hat was followed by at least nine sequels, a 1971 TV special, a 2003 live action movie starring Mike Myers and Alec Baldwin, an Xbox video game that same year, a 2010 musical series featuring Martin Short, another movie in the works, licensed spin-offs by other authors like Tish Rabe, and of course, lucrative merchandise. The Cat became Dr. Seuss's most beloved character and a big moneymaker for his estate.

Like cyber conflict, *The Cat in the Hat* is with us for the long run. However, the capabilities, users, "hacks," associated technology, and political implications of cyber change constantly and rapidly. Hacking began with innocent fun and games, breaking into secure systems for the "lulz." Yet as it enters its adolescence, hacking has assumed an ominous aspect resulting from the economic stakes of information technology innovations and dangerous new forms of conflict. *The Cat in the Hat* embodies this transformational moment, as innocence gives way to experience.[48]

This chapter has argued that the various characters and events in the story illustrate many contemporary issues regarding cybersecurity. We began by demonstrating that *The Cat in the Hat* addresses unwitting cooperation as a permissive condition for cyber operations, followed by a discussion of unintended consequences of cyber intrusions for perpetrators. We made the case that *The Cat in the Hat* provides an example of staged operations, exemplifies the challenge of establishing attribution, and illuminates the relative balance of offense-defense between perpetrators and would-be victims. In the final section we discussed how *The Cat in the Hat* sheds light on the broader cultural dimension of hacking, specifically deviance and authority.

The Cat in the Hat illuminates many of the contemporary issues of cybersecurity, but the book also offers important warnings about the future. The first warning concerns accelerated change. *The Cat in the Hat* franchise must keep coming up with new hijinks to keep children entertained. In the world of cyber, the pattern of invention and counterinvention of offense-defense capabilities means that cyber technology grows ever more sophisticated at an accelerated rate. We no longer worry about internet worms like Code Red, Slammer, and the I Love You virus because technology has already transcended those simple types of attacks.

The second warning concerns liminality. The waning of a state system based on patriarchal combat between sovereign kings, so to speak, encourages the waxing of a more interdependent, transnational, intersectional, and indeed intimate constellation of conflicts. *The Cat in the Hat* (because it occurs within the safety of "home") embodies how these more understated forms of conflict work within collective institutions, and cyber conflict likewise depends on systems of trust embodied in shared information infrastructure. Cybersecurity is not like a war between autonomous military organizations (the canonical image of traditional international relations) but rather the paradoxical subversion and amplification of domestic institutions from the inside.

The third warning concerns power. We admire *The Cat in the Hat* just as we enjoy heist films like *Ocean's Eleven* (2001), because the cons get away with something impossible. Even when there are unexpected events, they are skilled and lucky enough to get away with the loot in the end. These tales are entertaining precisely because they are so improbable. In the real world, the actors that carry out impressive feats like Stuxnet and SolarWinds are typically nation-state intelligence agencies. They have ample resources, decades of experience in clandestine collection and covert action, and the protection of national security organizations in case things go wrong. This calls into question assumptions about the offense-defense balance in cyberspace or the ease with which offensive cyber capabilities can proliferate.[49]

The final warning concerns tragedy. In many cases the strong prey upon the weak in cybersecurity, inverting popular assumptions that cyberspace provides an asymmetrical advantage to weaker actors. For example, the cybersecurity firm NSOGroup, founded by veterans of Israeli signals intelligence, produces spyware known as Pegasus, which employs some of the most sophisticated exploits seen, including "zero-click" access to targets' iPhones. NSOGroup licenses Pegasus to authoritarian regimes to target dissidents and journalists, with lethal consequences in several cases, including the dismemberment of *Washington Post* journalist Jamal Khashoggi.[50] While *The Cat in the Hat* is a lot of fun, it is important to bear in mind that hacking has produced several tragic outcomes, especially when digital surveillance is combined with coercive regime security forces.

Is this really a book for children? Perhaps not.

NOTES

1. Robert Chesney and Max Smeets, eds., *Deter, Disrupt, or Deceive: Assessing Cyber Conflict as an Intelligence Contest* (Washington, DC: Georgetown University Press, 2023).

2. Bob Violino, "Phishing Attacks Are Increasing and Getting More Sophisticated. Here's How to Avoid Them," *CNBC*, January 10, 2023, https://www.cnbc.com/2023/01/07/phishing-attacks-are-increasing-and-getting-more-sophisticated.html.

3. Council on Foreign Relations, *Agent.btz* (report), November 2008, https://www.cfr.org/cyber-operations/agentbtz#:~:text=of%20USB%20drives.-,Agent.,operations%20for%20the%20U.S.%20military.

4. Kim Zetter, "The Untold Story of the Boldest Supply-Chain Hack Ever," *Wired*, May 2, 2023, https://www.wired.com/story/the-untold-story-of-solarwinds-the-boldest-supply-chain-hack-ever/.

5. Cybersecurity and Infrastructure Security Agency, "Defining Insider Threats." Accessed September 20, 2023, https://www.cisa.gov/topics/physical-security/insider-threat-mitigation/defining-insider-threats.

6. Kim Zetter and Huib Modderkolk, "Revealed: How a Secret Dutch Mole Aided the U.S.-Israeli Stuxnet Cyberattack on Iran," *Yahoo News*, September 2, 2019, https://news.yahoo.com/revealed-how-a-secret-dutch-mole-aided-the-us-israeli-stuxnet-cyber-attack-on-iran-160026018.html.

7. Eamon Javers, "Cyberattacks: Why Companies Keep Quiet," *CNBC*, February 25, 2013, https://www.cnbc.com/id/100491610.

8. Zetter, "Untold Story."

9. White House, *National Cybersecurity Strategy* (March 1, 2023), 4, https://www.whitehouse.gov/wp-content/uploads/2023/03/National-Cybersecurity-Strategy-2023.pdf.

10. White House, *National Cybersecurity Strategy*, 4–5.

11. White House, Executive Order [14028] on Improving the Nation's Cybersecurity, May 12, 2021, https://www.federalregister.gov/documents/2021/05/17/2021-10460/improving-the-nations-cybersecurity.

12. Shalanda D. Young, "Memorandum for the Heads of Executive Departments and Agencies," Office of Management and Budget, January 26, 2022, 2, https://www.whitehouse.gov/wp-content/uploads/2022/01/M-22-09.pdf.

13. Andy Greenberg, "The Untold Story of NotPetya, the Most Devastating Cyberattack in History," *Wired*, August 22, 2018, https://www.wired.com/story/notpetya-cyberattack-ukraine-russia-code-crashed-the-world/.

14. British Broadcasting Corporation, "Ukraine Cyber-Attack: Software Firm MeDoc's Servers Seized," *BBC News*, July 4, 2017, https://www.bbc.com/news/technology-40497026.

15. Greenberg, "Untold Story."

16. Greenberg, "Untold Story."

17. White House, "Statement by the Press Secretary on the 'NotPetya' Cyber-Attack," February 2018, https://www.presidency.ucsb.edu/documents/statement-the-press-secretary-the-notpetya-cyber-attack.

18. Greenberg, "Untold Story."

19. Microsoft, "Defending Ukraine: Early Lessons from the Cyber War," June 22, 2022, 2, https://query.prod.cms.rt.microsoft.com/cms/api/am/binary/RE50KOK.

20. Nicolas Falliere, Liam O. Murchu, and Eric Chien, "W32.Stuxnet Dossier, Version 1.4," *Wired*, February 4, 2011, https://www.wired.com/images_blogs/threatlevel/2011/02/Symantec-Stuxnet-Update-Feb-2011.pdf.

21. Ben Buchanan, *The Hacker and the State: Cyber Attacks and the New Normal of Geopolitics* (Cambridge, MA: Harvard University Press, 2020), 135–37.

22. Jon R. Lindsay, "Stuxnet and the Limits of Cyber Warfare," *Security Studies* 22, no. 3 (2013): 396n101; and Kim Zetter, *Countdown to Zero Day: Stuxnet and the Launch of the World's First Digital Weapon* (New York: Broadway Books, 2014), 378.

23. Michael Warner, "A Brief History of Cyber Conflict," in *Ten Years In: Implementing Strategic Approaches to Cyberspace*, ed. Jacquelyn Schneider, Emily O. Goldman, and Michael Warner (Newport, RI: Naval War College Press, 2020), 20.

24. David E. Sanger, *Confront and Conceal: Obama's Secret Wars and Surprising Use of American Power* (New York: Broadway Paperbacks, 2012), 205.

25. Thomas Rid and Ben Buchanan, "Attributing Cyber Attacks," *Journal of Strategic Studies* 38, nos. 1–2 (2015), https://doi.org/10.1080/01402390.2014.977382.

26. National Cyber Security Centre, "How Cyber Attacks Work," July 23, 2023, https://www.ncsc.gov.uk/information/how-cyber-attacks-work.

27. Zeba Siddiqui and Christopher Bing, "Chinese Hackers Spying on US Critical Infrastructure, Western Intelligence Says," *Reuters*, May 25, 2023, https://www.reuters.com/technology/microsoft-says-china-backed-hacker-targeted-critical-us-infrastructure-2023-05-24/.

28. Greenberg, "Untold Story."

29. On proxies in cyberspace, see Tim Maurer, *Cyber Mercenaries: The State, Hackers, and Power* (Cambridge: Cambridge University Press, 2018).

30. Florian J. Egloff, *Semi-State Actors in Cybersecurity* (New York: Oxford University Press, 2021).

31. Ariel E. Levite and June Lee, "Attribution and Characterization of Cyber Attacks," in *Managing U.S.-China Tensions Over Public Cyber Attribution*, ed. Ariel E. Levite, Lu Chuanying, George Perkovich, and Fan Yang (Washington, DC: Carnegie Endowment for International Peace, 2022), 37.

32. Michael N. Schmitt and Liis Vihul, "Proxy Wars in Cyberspace: The Evolving International Law of Attribution," *Fletcher Security Review* 1, no. 2 (2014): 56.

33. Associated Press, "A Look at Estonia's Cyber Attack in 2007," *NBC News*, July 8, 2009, https://www.nbcnews.com/id/wbna31801246.

34. Justin Sherman, "Untangling the Russian Web: Spies, Proxies, and Spectrums of Russian Cyber Behavior," *Atlantic Council Issue Brief*, September 19, 2022, https://www.atlanticcouncil.org/in-depth-research-reports/issue-brief/untangling-the-russian-web/.

35. Matthew Monte, *Network Attacks and Exploitation: A Framework* (Indianapolis, IN: Wiley, 2015),73–92.

36. Zetter, "Untold Story."

37. Rid and Buchanan, "Attributing Cyber Attacks."

38. Juan Andrés Guerrero-Saade, "Draw Me Like One of Your French APTs—Expanding Our Descriptive Palette for Cyber Threat Actors" (paper presented at *Virus Bulletin* conference, Montreal, October 3–5, 2018), https://www.virusbulletin.com/uploads/pdf/magazine/2018/VB2018-Guerrero-Saade.pdf.

39. Kaspersky Labs, "Equation: The Death Star of Malware Galaxy," Securelist, February 16, 2015, https://securelist.com/equation-the-death-star-of-malware-galaxy/68750.

40. Juan Andres Guerrero-Saade, "King of the Hill: Nation-State Counterintelligence for Victim Deconfliction" (paper presented at *Virus Bulletin* conference, London, October 2–4, 2019), https://www.virusbulletin.com/virusbulletin/2020/01/vb2019-paper-king-hill-nation-state-counterintelligence-victim-deconfliction/.

41. Erica D. Lonergan and Shawn W. Lonergan, *Escalation Dynamics in Cyberspace* (New York: Oxford University Press, 2023).

42. Jon R. Lindsay, "Restrained by Design: The Political Economy of Cybersecurity," *Digital Policy, Regulation and Governance* 19, no. 6 (2017): 493–514.

43. Nina A. Kollars, "Taking Nonstate Actors Seriously (No, Seriously)," in *Deter, Disrupt, or Deceive: Assessing Cyber Conflict as an Intelligence Contest*, ed. Robert Chesney and Max Smeets (Washington, DC: Georgetown University Press, 2023), 261–72.

44. Jacquelyn Schneider, "Blue Hair in the Gray Zone," *War on the Rocks*, January 2018, https://warontherocks.com/2018/01/blue-hair-gray-zone/; and Rebecca Slayton, "What Is a Cyber Warrior? The Emergence of U.S. Military Cyber Expertise, 1967–2018," *Texas National Security Review* 4, no. 1 (January 11, 2021), http://

tnsr.org/2021/01/what-is-a-cyber-warrior-the-emergence-of-u-s-military-cyber-expertise-1967-2018/.

45. Leonie Maria Tanczer, "50 Shades of Hacking: How IT and Cybersecurity Industry Actors Perceive Good, Bad, and Former Hackers," *Contemporary Security Policy* 41, no. 1 (January 2, 2020): 108–28, https://doi.org/10.1080/13523260.2019.1669336.

46. The Cat's color scheme is white, red, and black. These colors have a deep cultural resonance in myth and fairy tales across the world. They are often associated with life transitions, with white traditionally representing the innocent purity of youth, red the coming of age through menses and childbirth (for women) or battle and sacrifice (for men), and black the wisdom of age in the female crone or male sage. This is why Snow White, a parable of losing our way in young adulthood, has white skin, red lips, and black hair. As such, *The Cat in the Hat* is a parable of the cultural transformation of hacking, from innocence to experience.

47. Michael Poznansky, *In the Shadow of International Law: Secrecy and Regime Change in the Postwar World* (New York: Oxford University Press, 2020).

48. Yet the Cat's body is drawn in black, suggesting a more mature understanding of cybersecurity on the horizon. A wiser understanding of cybersecurity, perhaps, might accept that the Cat will come back again and again, as sure as the seasons, and our calling is to surf this chronic chaos without overreacting. And maybe, to have a little fun with it.

49. Rebecca Slayton, "What Is the Cyber Offense-Defense Balance? Conceptions, Causes, and Assessment," *International Security* 41, no. 3 (January 1, 2017): 72–109; and Max Smeets, "Cyber Arms Transfer: Meaning, Limits and Implications," *Security Studies* 31, no. 1 (2022): 65–91.

50. Ronen Bergman and Mark Mazzetti, "The Battle for the World's Most Powerful Cyberweapon," *New York Times*, January 28, 2022, https://www.nytimes.com/2022/01/28/magazine/nso-group-israel-spyware.html.

7

Private Snafu and Political Propaganda

Kevin P. Eubanks

During World War II, years before he would become known the world over as Dr. Seuss, Theodor Geisel and a group of well-known creative artists made a series of twenty-six animated training and orientation films for the US Army called *Private Snafu*.[1] These short films were shown to new recruits on a biweekly basis from 1943 to 1945 and were intended to drum up troop morale, encourage responsible decision-making on and off the battlefield, and generate support for the US war effort at home and abroad.[2] Geisel and his team had enlisted to serve their country by bringing their unique talents to bear on the war effort,[3] and by all accounts *Private Snafu* was a very popular and successful propaganda campaign.[4]

In *The Goldbrick* (1943), for example, Private Snafu's sleep is interrupted by the piercing blare of the bugle at morning reveille. A weary Snafu rubs his eyes and complains about the mundane demands of a soldier's life: "Another day, nuts!" Snafu exclaims. "If only I could get outta drill." As if on demand, Goldie the Goldbrick (a fairy godfather figure) appears, and through a series of simple rhymes shares with Snafu the closely held secrets to avoiding one's duties and responsibilities: "When there's cold and there's rain /and you don't want to train /just pretend that you're sick / and that your poor back is sore." Snafu embraces the ill-advised fantasy and begins to shirk his obligations, including falling asleep in an unfinished trench he's been tasked with digging. Only when Snafu finds himself in the midst of combat in the South Pacific do the serious consequences of his poor choices become apparent. Snafu becomes cut off from his unit, is isolated behind enemy lines, and ultimately dies when his foxhole is destroyed by Japanese planes. In the final scene, Goldie the Goldbrick is unmasked as an enemy Japanese agent charged with encouraging such demoralizing behavior among American troops.

As this example illustrates, the *Snafu* films follow a predictable formula in which careless or selfish acts result in a cascade of second- and third-order effects with dire consequences for national security. Undoubtedly these films were crafted deliberately to persuade the intended audience of young, male US Army recruits to obey military orders and regulations. Contrary to the views of some critics, however, the *Private Snafu* films are more than just tools of military training and indoctrination. This chapter suggests that Theodor Geisel's *Private Snafu* reinforces the same set of core liberal democratic principles advocated for in Dr. Seuss's better known postwar children's literature. Whether standing his ground against authoritarianism, extolling the virtues of individual freedom and responsibility, or emphasizing the "civil" in civil society, Theodor Geisel dutifully served his country during World War II in much the same way as Dr. Seuss so effectively served the nation's children after it. *Private Snafu* pulls this off not by seeking to turn young citizens into soldiers, as some scholars insist, but rather by seeking to convert young soldiers into democratic citizens who carry with them onto the battlefield those same democratic convictions the United States entered World War II to defend.

HOLLYWOOD AT WAR

The origins of the *Private Snafu* film series can be traced directly to the establishment in 1942 of the Office of War Information (OWI) by President Franklin Delano Roosevelt.[5] To win over the population at home and gain an advantage over the adversary, the OWI was dedicated to the creation and dissemination of US military propaganda, or, to phrase it more diplomatically, "the development of an informed and intelligent understanding, at home and abroad, of the status and progress of the war effort and of the war policies, activities, and aims of the government."[6] These US government efforts marked the formal emergence of the information environment as a domain of total war and established the military's task of leveraging what is known today as the "informational arm" of US national power.

Of course it made good sense, given US government goals at the time, to look to Hollywood for help in shaping and reinforcing its prowar, prodemocratic narrative. Hollywood, after all, was in the narrative business. Leading the charge to utilize the power of film in World War II was Army Chief of Staff George C. Marshall,[7] who famously said near the end of the war that it had produced two new formidable weapons: the airplane and the motion picture.[8] Marshall wisely looked to the talent of established civilian filmmakers, artists, animators, and storytellers whose work had already been proven capable of influencing the public and shaping individual and collective norms, values, and behavior.[9]

The OWI, under Marshall's orders, quickly set about producing professional, carefully curated propaganda films like Frank Capra's *Why We Fight* (1942) series and John Huston's *Winning Your Wings* (1942).[10] The OWI reviewed most scripts ahead of production and issued strict guidelines to ensure the alignment of the artistic content with US national objectives. Among other things, films produced during the war were expected to unite the American audience against the adversary, to praise the family of democratic nations in their anti-fascist struggle, to inspire the American public to contribute to the war effort, and to inspire both the public and the troops themselves to collectively identify with the just cause of the war. By 1943 these films had already succeeded in shifting public sympathy toward US strategy and policy goals.[11]

Now that Hollywood was in the business of stamping out fascism and defending democracy at home and abroad, recruiting cartoonists made clear sense to the OWI. As influential as live-action propaganda films like *Why We Fight* were in advancing US national interests, animation played an equally critical role from the earliest days of the war. The day after the Pearl Harbor invasion, the US Navy approached Walt Disney about producing twenty training films on the subject of aircraft and warship identification, and by 1943 95 percent of Disney's creative output was devoted to wartime propaganda films.[12] Not to be left out, Warner Brothers Studios created one of the very first propaganda films of the war, *Any Bonds Today?* (1942), an animated short scored by Irving Berlin and featuring Bugs Bunny that encouraged the American people to purchase war bonds to help fund the country's mobilization for war.[13]

Geisel's day job since 1928 had been as an illustrator and cartoonist for popular magazines, including *Life* and *Vanity Fair*. He also worked on national advertising campaigns for corporate clients large and small, including Standard Oil, Esso, and Narragansett Brewing, effectively persuading US citizens to buy everything from marine oil to bug repellent and beer.[14] In addition to advertising, Geisel had also begun to make a name for himself as a children's author. Before he joined "Fort Fox," the name given to Hollywood's Fox Studios during the war, Geisel published several children's books under the soon-to-be famous pseudonym Dr. Seuss, including *The Pocket Book for Boners* (1931), *And to Think That I Saw It on Mulberry Street* (1937), *The 500 Hats of Bartholomew Cubbins* (1938), and *Horton Hatches an Egg* (1940).[15]

With the outbreak of war, Geisel's desire to use his artistic talents to engage with American political life became more and more urgent. From 1939 to 1941 Geisel drew political cartoons for New York's *PM Magazine*, an influential publication that advocated for the progressive views of the US Popular Front, encouraging readers to adopt anti-isolationist, anti-racist, and anti-authoritarian stances.[16] As the war raged on in Europe and the Pacific, Geisel's passionate anti-fascist and anti-isolationist positions led him to join the US Army as a commissioned officer in 1943.[17] At the time, Capt. Theodor Geisel was just shy of forty years old.

Led by Capra, the Oscar-winning director of *It Happened One Night* (1934) and *Mr. Smith Goes to Washington* (1939), the Education and Information Division of the US Army Signal Corps brought on a motivated Geisel "to make educational films that would run in the *Army-Navy Screen Magazine*, a biweekly newsreel shown to the troops."[18] Capra made Geisel head of the animation branch and provided him with an elite supporting cast consisting of fellow children's authors Philip D. Eastman and Munro Leaf as well as the Warner Brothers artists behind *Looney Tunes* (and beloved characters such as Bugs Bunny, Daffy Duck, Elmer Fudd, and Porky Pig), including animators and directors Chuck Jones and Friz Freleng, vocal impressionist Mel Blanc, and composer Carl Stalling.[19]

PROPAGANDA AND ITS DISCONTENTS

Following Philip M. Taylor's broadest possible definition of propaganda in *Munitions of the Mind: A History of Propaganda* as any "communication of ideas designed to persuade people to think and behave in a desired way,"[20] Geisel was and had always been a propagandist. As propaganda, *Private Snafu* uses humor to cultivate positive morale among the young troops; offers a shared experience around which its audience can unite, whether the shared drudgery of KP duty or the shared goal of vanquishing the German and Japanese enemies; and relies on its comic protagonist to shape the audience's collective understanding of its role and purpose in war.

Given these rather innocuous goals, why does *Private Snafu* remain controversial? In the United States, the primary objection to domestic propaganda concerns the moral imperative and political value of truth in a democracy.[21] Does a democratic government have a right to deceive or manipulate the opinions of its citizens, even if doing so serves its national interests? In contrast to foreign propaganda, domestic propaganda generally seeks to persuade citizens of the justness of a political cause and to rally the population around the nation's political and military objectives. Foreign propaganda, on the other hand, targets foreign military organizations, political leadership, and civilians.

The US military has used domestic propaganda since the American Revolution, when, for example, George Washington had the colonial press exaggerate patriot victories and enemy casualties to prevent the colonists from becoming discouraged by setbacks on the battlefield.[22] Until World War II, print media was the primary means of disseminating propaganda to a mass audience.[23] The World War I "I Want You" recruitment poster, in which a stern Uncle Sam points directly at the viewer, was popular and effective enough to be reprised during World War II and remains one of the most memorable images in American print culture.[24] Though print media remains in use today, photography and film emerged as the principal propaganda tools during World War II.

In addition to being antithetical to democracy, another common objection to the use of propaganda concerns its goal, namely, to influence the mind. By employing propaganda, war becomes more than a mere physical contest, but takes place on a mental or psychological front as well (which recently has become known as war's "cognitive dimension").[25] As Taylor notes, "Psychological warfare has always rested as an uneasy activity in democracies, even in wartime. It is partly to do with the suspicion that using the mind to influence the mind is somehow unacceptable. But is it more unacceptable to shoot someone's brains out rather than to persuade that brain to drop down their weapon and live?"[26]

Reflecting the general public's aversion to propaganda since the end of the Vietnam War, the US Joint Chiefs of Staff and all branches of the armed services have avoided using the terms "propaganda," "psychological warfare," and "psychological operations" to describe the use of information as a weapon of war. Although the North Atlantic Treaty Organization continues to employ the term psychological operations to describe the use of information to influence adversaries and host nations, the United States remains a stubborn terminological outlier.

Joint Publication 3.0, *Joint Campaigns and Operations* (2022) not only establishes how US military forces can use information as a weapon of war, but also shapes their understanding of and behavior toward its use as such: to *influence* adversaries and host nation populations ("to change or maintain perceptions, attitudes, and other elements that drive desired relevant actor behaviors"), to *create a positive perception* of the United States ("to influence foreign audiences and affect the legitimacy, credibility, and influence of the USG, joint force, allies, and partners), and to *counter adversary propaganda* ("to counter, discredit, and render irrelevant the disinformation, misinformation, and propaganda of other actors").[27] Instead of referring to these activities as propaganda, the official terminology in the United States is the much less problematic (and less transparent) "information operations." Nested under JP 3.0 is the pertinent doctrine JP 3.13, *Information Operations* (2012), which defines information operations as "the integrated employment, during military operations, of IRCs [information-related capabilities] in concert with other lines of operation to influence, disrupt, corrupt, or usurp the decision making of adversaries and potential adversaries while protecting our own."[28] In the US military framework there are many types of information operations, including information assurance, public affairs, strategic communications, military deception, and so on.

The specific category of "information operation" that most closely tracks with the generally accepted understanding of propaganda is what the Joint Chiefs of Staff call "military information support operations" (MISO). Joint Publication 3-13.2, *Military Information Support Operations* (2014), defines MISO as "planned operations to convey selected information and indicators to foreign audiences to influence their emotions, motives, and objective reasoning and ultimately induce or reinforce foreign attitudes and behavior favorable to the originator's objectives."[29]

Unlike MISO, however, propaganda is referred to consistently in joint doctrine as an enemy activity, not as something in which the United States might itself engage. While it does not use the term propaganda, JP 3-13.2 actually describes the same functions. In this case, MISO is used to "develop and convey messages and devise actions to influence select foreign groups and promote themes to change those groups' attitudes and behaviors."[30] Even though the offensive words propaganda, psychological operations, and psychological warfare are not employed, the unfortunate uniformed military personnel tasked with the dull job of writing doctrine apparently made little effort to conceal the obvious, though unspoken, truth: that MISO and propaganda are the same thing with different names. Thus on the first page of *Military Information Support Operations*, readers encounter US Army general Dwight D. Eisenhower's use of the *verboten* terminology: "In this war, which was total in every sense of the word, we have seen many great changes in military science. It seems to me that not the least of these was the development of *psychological warfare* as a specific and effective weapon."[31] Although the exact nature and scope of information operations within US military doctrine is in disarray,[32] what remains clear is that the functions of propaganda, psychological warfare, and information operations remain more or less the same: to develop and convey messages that influence or change a target audiences' attitudes and behaviors.

PROPAGANDA AND DEMOCRACY

Undeniably, *Private Snafu* is military propaganda or, in contemporary terminology, an information operation designed to expedite the young recruit's transformation into a responsible and capable soldier.[33] Precisely what such expedience involves, however, and whether it is radically different in nature and kind from the approach taken by Geisel in the postwar children's literature are questions that remain unresolved in the debate over *Private Snafu*'s place within the Dr. Seuss canon. One may argue that *Private Snafu* is not an anomaly, but rather a continuity wholly consistent with the progressive politics of the postwar Dr. Seuss. The source of this continuity lies in Geisel's "democratic imagination"[34]—his devotion to the ideological premises of liberal democracy and social contract theory—which not only informs but was substantially influenced by his work on *Private Snafu* during World War II.[35]

While it bears some similarity to Dr. Seuss's later work, *Private Snafu* can certainly be distinguished on many grounds. First, these films were a collaborative effort, which reflected the contributors' belief that their collective service outweighed the contributions of any single individual in this special military operation.[36] Moreover, *Private Snafu* was intended for young adults and therefore appeals to adult prejudices, values, and experiences.[37] As a product of their time, the films reflect the predominant cultural ethos of the 1940s. Women tempt soldiers away from their duties and are represented as enemy agents or deadly weapons, for example, in *Spies* (1943) and *Booby Traps* (1944).[38] Racist depictions of the Japanese enemy in *Operation Snafu* (1945) and other films in the series exploit an emergent and distinctly American nationalism.[39] Geisel not only grew to regret such characterizations, but also repudiated the adult audience for which they were intended when after the war he famously declared adults to be "obsolete" and promised from that point forward to only write books for children.[40] Finally, *Private Snafu* is the only Dr. Seuss work under consideration in this volume that was produced under the immediate supervision of the US government and in direct support of US national security efforts.

This array of tensions lies behind critics' struggles to align Geisel's work during World War II with his postwar books for children. One perspective views Fort Fox as a wartime laboratory in which Geisel experimented with the medium (animation) and the message (storytelling) alongside other masters of the craft.[41] These scholars focus on the formal and rhetorical connections linking elements of *Private Snafu* and Geisel's postwar literary aesthetic and note evolutions in Seuss's verse, characterization, and narrative style during his time at Fort Fox. Critics from this school of thought nevertheless draw a firm line between the state-sponsored propaganda of *Private Snafu* and the more subversive "emancipatory practice" for which the pre- and postwar Dr. Seuss is most well-known.[42]

Another perspective views Geisel's work for the US Army as a profound, life-changing experience that shaped his more absolute postwar commitment to peace, tolerance, education, and, most importantly, children. These scholars argue that Geisel ultimately rejected the questionable motives and blatant prejudices of *Private Snafu* in favor of adopting a broader ideological mission as he finally came into his own as Dr. Seuss after the war.[43] From this point of view, the later books for children function as a form of atonement for Geisel's wartime propaganda.[44]

Finally, a much less forgiving interpretation describes Geisel as being thoroughly complicit in the US Army's efforts to indoctrinate young military recruits and abandoning, if not just temporarily suppressing, the progressive values for which Geisel fought before the war and that notoriously inspired so many of his works in the postwar era. According to this reading, Geisel's propaganda efforts during the war served the authoritative, deceptive, mechanical ends of the US government and its military arm.[45] *Private Snafu* is offered up as an example of what happens "when children's literature goes to war" and reluctantly chooses security over freedom.[46] As Christopher Dow argues, "Private Snafu cartoons wield a double-edged sword that not only *incised* indoctrinational and informational material but attempted to *excise* any attachments the soldier might have that could impair his single-minded attention to fighting and winning the war."[47] For these scholars, the sharpest distinctions should be drawn between the Geisel who produced *Private Snafu* and the author of *Horton Hears a Who* (1954) and *The Cat in the Hat* (1957).

To be sure, *all* of Geisel's work exposes a more than passing interest in the dynamics of social messaging and mass persuasion, and there's no dearth of scholarship citing Dr. Seuss's active engagement with the emerging social sciences of education, parenting, and child psychology.[48] There is in fact widespread agreement that the bulk of Geisel's artistic and literary output is propagandistic in nature.[49] Those who would deny the fact point to the fragile distinction between propaganda and education. In 1937, political theorist Carl Joachim Friedrich (1901–84) offered this more or less contemporary definition of the difference: "Propaganda always aims at getting people either to do or not to do some very particular thing. Education, on the other hand, is fundamentally concerned with moulding [sic] and developing a human being in terms of an ideal, as far as his nature allows it."[50] After all, the Beginner Books, including *The Cat in the Hat, One Fish, Two Fish, Red Fish, Blue Fish* (1960), and *Green Eggs and Ham* (1960), were intended to serve as elementary reading primers, and much of Dr. Seuss's work for children notably endeavored to extend the scope of nature's "allowance" by equipping young readers with the moral-ethical insight and courage essential to achieving Friedrich's (and, one can certainly argue, Geisel's) sociopolitical "ideal."

On the surface, Friedrich's simple contrast lends support to the claim that Dr. Seuss's children's literature escapes the propaganda label because it transcends propaganda's limited function and coercive means by placing education and childhood development at the center of its purpose and aesthetic. In addition to supporting literacy (the educational bedrock of liberal democracy and the antidote to falling victim to certain kinds of propaganda), Dr. Seuss's children's literature routinely encourages positive social values, such as courage (*Horton Hears a Who*, 1954), compassion (*Thidwick the Big-Hearted Moose*, 1948), and tolerance (*The Sneetches*, 1961), while discouraging negative social behaviors like greed and materialism (*How the Grinch Stole Christmas*, 1957), envy (*Gertrude McFuzz*, 1958), pride (*The Big Brag*, 1958), and abuse of power (*Yertle the Turtle*, 1958). Moreover, the kind of moral courage imparted in many of Dr. Seuss's children's books and certainly all the books about the limitless possibilities of a child's imagination foster resistance to various forms of social and political authority, which are the traditional sources of propaganda. Following Friedrich's distinction, then, it is easy enough to see how children's literature "is fundamentally concerned with moulding [sic] and developing a human being in terms of an ideal," while *Private Snafu*, on the other hand, merely instructs young soldiers "to do or not to do some very particular thing" in the context of their everyday life on and off the battlefield, from reading their field manual (*Snafuperman*, 1944) to keeping US troop movements to themselves (*Spies*) and wearing mosquito repellant (*Private Snafu vs. Malaria Mike*, 1944). While the pre- and postwar children's literature's educational contribution to the individual and society may be said to constitute the "emancipatory practice" alluded to by Philip Nel, *Private Snafu*, it would seem, remains military propaganda and thus something utterly different.

However, the distinction between education and propaganda barely holds up to scrutiny. Even Friedrich acknowledges its limits: "The purposes of the educator and of the propagandist . . . are yet linked by unbreakable ties to the common creed. . . . Neither the educator nor the propagandist can do without the other."[51] And J. Fred MacDonald goes much further in collapsing the distinction between education and propaganda with his insistence that propaganda is "an omnipresent, intrusive aspect of modern life . . . an integral part of all modern civilization."[52] Because propaganda functions as a kind of "social adhesive," which through education, advertising, media, government, religion, and culture enables national unity and identification, it is both ubiquitous and critical to the preservation of order in both democratic and authoritarian societies.[53]

Consequently, one way of resolving the interpretive tension between propaganda and education vis-à-vis *Private Snafu* is to recognize that the broad category of propaganda has significant distinctions within it. In his seminal study, *Propaganda: The Formation of Men's Attitudes* (1965), French philosopher Jacques Ellul (1912–94) carefully distinguishes between many *kinds* of propaganda. At one end of the spectrum, Ellul identifies a vertical, political *propaganda of agitation* built around an opposition to individuality and through which a mechanical, unquestioning adherence to an organization's mission and values is produced.[54] Ellul associates a propaganda of agitation with authoritarian, antidemocratic methods of social control and behavioral engineering, such as those infamously practiced by Nazi Germany to manufacture total loyalty to Hitler as well as a new and potent form of German nationalism.[55]

Scholars who view *Private Snafu* as an obvious example of Ellul's propaganda of agitation cite the cartoons' official function as government-sanctioned military communication and the apparent goals of containing "those desires, drives, and fears that form the foundation of human thought and emotion,"[56] and of presenting the young would-be soldier, via the character of Snafu, as a "potential threat" to US national security.[57] From this perspective, in order to mitigate the threat, and rather unlike children's literature, *Private Snafu* seeks conspicuously to transform its youthful and naïve audience into a lethal military force comprised of mindless, undesiring automatons that exist solely to carry out the orders of the US government. The more extreme implication, of course, is that Geisel and his creative team employed the same coercive, undemocratic methods to influence young US military recruits during World War II as the fascist and imperialist adversary against whom they were fighting on behalf of US and global democracy. As a result, Geisel's *Private Snafu* period may indeed expose a moralist kowtowing to the status quo rather than the progressive advocate for a more, not less, democratic world that everyone came to know him as after the war.[58]

The other pole of Ellul's taxonomy, however, is a horizontal, social *propaganda of integration*, which not only reinforces the equivalence between propaganda and education but also aligns with Ellul's special account of democratic propaganda.[59] As such, the concept of a propaganda of integration proves useful for more accurately characterizing the kind of propaganda one encounters across Geisel's oeuvre, including in his work on *Private Snafu.* For Ellul, a horizontal and socially integrative propaganda, such as the kind advanced by a public education system, "aims at stabilizing the social body, at unifying and reinforcing it."[60] Like MacDonald's "social adhesive," Ellul's propaganda of integration, which is at home in both totalitarian and democratic contexts, weaves the individuals of the social fabric together into a harmonious whole.

Unlike authoritarian propaganda, however, a socially integrative and democratic propaganda also offers "solution[s] that [are] 'found' by the individual rather than imposed from above,"[61] and thus it promotes a uniquely democratic (and especially American) social unity predicated on the preservation of individual agency and freedom.[62] Moreover, while an authoritarian propaganda can be said to produce a fantasy in which the powers that be neither believe nor participate, Ellul points out that a democratic propaganda promotes a fantasy in which everyone is invited to participate, including both the creators of the propaganda and those targeted by it.[63] This "communal action" suggests that an identification and alignment between the objectives of the government and its people is what makes it possible for Geisel and his team to successfully navigate the formidable demands of national security, on the one hand, and those of a free, democratic civil society, on the other.[64]

The next few sections venture to demonstrate what Ellul might call the socially integrative, democratic nature of *Private Snafu*. Despite having been composed under the direct supervision of the US Army and US Department of Defense, the animated films that make up *Private Snafu* unify and reinforce the uniquely American social body through their categorical anti-authoritarian stance and emphasis on individual and social freedom and responsibility.

ANTI-AUTHORITARIANISM

The Cat in the Hat is arguably the book most often associated with Dr. Seuss and the most representative of the author's trademark anti-authoritarianism. In the book a mother has left her two children at home on a rainy day. Feeling bored, which is arguably *the* cardinal sin of the Seussian moral cosmos, the children are delighted when the Cat suddenly appears and promises them a day filled with adventure and "lots of fun that is funny." The pet fish (who monitors the children's behavior while the mother is away and fears her response to the alleged misbehavior and the unknown consequences of the Cat's visit) orders the Cat to "go away," but the Cat and his minions, Things One and Two, ignore the warnings and keep their promise to entertain the children. In the end *The Cat in the Hat* confers the power of authority normally reserved for the mother and pet fish upon the rebellious Cat. Instead of the parental authority suppressing the children's creative mischief, Dr. Seuss endorses the anarchy unleashed by the Cat's visit.

On the surface, *The Cat in the Hat* bears no resemblance to the films in the *Private Snafu* series. Nel writes, "While the SNAFU cartoons share *The Cat in the Hat*'s irreverent tone, SNAFU ultimately endorses the idea that its audience should submit to authority. *The Cat in the Hat* does not."[65] In Nel's view, Snafu and the soldiers he represents must submit to military authority in order to become the effective fighting force the US government wants them to be; therefore, with each installment Snafu's quirky, inadvertent revolution *must* be put down and order must be restored among the rank and file.[66] Snafu's success as propaganda is thought to depend upon the effectiveness of this indoctrination effort, insofar as an obedient and compliant soldier serves national military interests in a way that a freewheeling, mistake-prone Private Snafu does not.

Like *The Cat in the Hat*, *Private Snafu* cleverly subverts the authoritarian impulse of the military's hierarchical, chain-of-command culture, especially as it is reflected in the genre of military training and orientation films. This subversion begins with the words in the title of the films: "Situation Normal . . . All . . . All *Fouled* Up." Many scholars view the title as a mechanical-rhetorical device that only acknowledges the authentic lived experience of the soldier and registers their disorientation and resentment toward the conditions of their wartime employment in order to break down their defenses.[67] Like the films' treatment of women, drinking, and carousing, the title acts as a kind of "Foucauldian pressure valve designed to vent the average soldier's discontent, releasing just enough anarchic energy to keep the system functioning,"[68] but ultimately posing no real threat to the system itself. According to this reading, and in keeping with the films' function as military propaganda, the title seeks to deliver a humorous and emotional appeal that will most effectively incline its audience toward the intended message.[69]

When the title acknowledges the soldiers' discontent, however, it also registers the lack of control over the circumstances that called the films into existence in the first place. Specifically, the acronym SNAFU calls to mind the global mobilization required to fight the war and the uncertainty of the world into which Snafu (and, by extension, every other soldier) has been thrown. Consequently, the "moral lesson" at the end of each installment is conveyed with the implicit understanding that in reality things *are* actually quite fouled up (or "out of whack," as Dr. Seuss might put it) and so not easily tamed by strategy and policy or official training and orientation efforts. Thus Snafu's name—along with the names of his brothers, the inept dog trainer, FUBAR (Fouled Up Beyond All Recognition), and the incompetent pigeon keeper, TARFU (Totally and Royally All Fouled Up)—functions as a consistent reminder of the chaos and disorder lurking just outside the door of the theater.[70] Unlike US government and military authority, which would like to impose permanence and discipline upon a transitory and undisciplined reality, the humorous use of a vulgar synonym both as a title and the principal character's name underscores the military authority's lack of control over the tragic, unmanageable reality of war in which every member of *Private Snafu*'s audience was situated.

Luck and chance also figure prominently in *Private Snafu* and, along with the title, further destabilize the agency and decision-making authority associated with military training and readiness. For example, Snafu does things right only by chance in *Outpost* (1944) and *No Buddy Atoll* (1945). In the former, Snafu's unremarkable discovery of a can of pickled fish eyes leads to US victory at sea against the Japanese Navy. Snafu reports the location of the enemy accidentally and therefore never actually learns the twofold lesson of the installment: that no piece of information is insignificant, and that what may appear unimportant to one may be critical to another. That Snafu succeeds despite himself challenges both the assumptions and the clear-cut, rational objectives of military planning and training.

Indeed, the character meant to represent military authority in the film,[71] the Technical Fairy First Class, also functionally undermines that authority. The Technical Fairy is an unshaven, battle-hardened, cigar-smoking stereotype of the career noncommissioned officer who guides Snafu in many of the short films. As such, he is the figure in the films that symbolizes the ideal product of the military system.[72] However, not only does the Technical Fairy skip back and forth between moral-ethical poles (e.g., when he encourages Snafu to betray his duties in *The Goldbrick*), but he also never forces the lesson through coercive or deceptive measures. Instead, the Technical Fairy acquiesces to Snafu's every desire and grants him his every wish, a decidedly unmilitary mode of instruction that reveals the genuine source of authority is neither the Technical Fairy nor the military training apparatus, but rather Snafu himself. Snafu *learns* through his mistakes; he is not coerced through the familiar tricks of propaganda, but rather driven to alter his behavior on his own, voluntarily, in order to avoid the negative consequences of his actions. Traditionally, one expects propaganda to preserve the *appearance* of volition on the part of the target while the sponsors of the information operation remain the chief influence on decision-making. In the case of *Private Snafu*, however, the opposite is not only true but also obvious in the simple and repetitive plot structure.

Finally, *Private Snafu* chips away at the foundational faith in authority through the medium of animation itself. Unlike the live-action films, which were to a large extent restricted from breaking the rules of the real world they were designed to imitate, animation, in which there are no rules and nothing is what it seems, allowed for the expansion of the government's propaganda toolset. As several scholars note, it is this fantastical element of animation (alongside its ability to maintain an identity with the real world) that made it so effective as propaganda.[73] The animated fantasy and the infinite pliability of the form effectively conceal the ulterior motives of the propagandist, whether the US government or the creative artists behind *Private Snafu*. While the ideological messaging of the live-action film was always relatively overt and easy for censors and audiences to spot, "animation was able to develop as an unhindered ideological force within Hollywood" during this time by routinely escaping the attention of government censors.[74]

To be sure, the genre of animation always presents the viewer with at least two contradictory impulses, both of which undercut any absolute demands of authority the films are thought to advance. On the one hand, animation reminds us that *everything* is possible. The films have Snafu at various moments impossibly managing multiple tasks at once, transforming into tanks and planes, and, of course, dying over and over again only to reappear alive and well in the next episode. These possibilities suggest a radical and boundless freedom that vastly exceeds the limited interests and influence of military propaganda. On the other hand, animation also reminds viewers that what it makes possible is so *only* within the animated world of the cartoon, a dynamic that always (though always only "provisionally") mocks the real world, in which such things simply cannot happen.[75] In *Hot Spot* (1945), to take one of many examples, as two trucks pass each other on a narrow mountain road, a truck avoids Snafu's vehicle by driving off the cliffside and around Snafu. Moments later, Snafu encounters a similar obstacle on the same road and attempts to mimic the truck by driving off the cliffside, too, only to learn that *he* cannot do that.[76] Such a sharp distinction between animated and real possibility likewise serves to consistently challenge the power and potency of the military authority that oversaw the films' production.

As the preceding discussion indicates, *Private Snafu* differed radically in form and content from the military training materials used in the past.[77] Ironically, while critics acknowledge that *Private Snafu*'s irreverence was precisely what made it so effective as military propaganda, they tend to overlook how this irreverence authentically challenges the propaganda function. By acknowledging the reality of the soldiers' lived experience amid the chaos of war, representing Snafu as the author of his own destiny, and exposing the tensions between what is and is not under the military's control, *Private Snafu* directly challenges the government authority that solicited its creation as well as the film series' alleged goal of converting man into machine since, as the films rightly suggest, such an impossible conversion can *only* happen within the animated world of the cartoon.

INDIVIDUAL FREEDOM

In *Horton Hears a Who*, Horton the elephant discovers an entire civilization called Who-ville living on a speck of dust. Horton defends the Whos against obliteration by the Jungle of Nool, whose inhabitants do not believe in the Whos' existence because they can neither see nor hear them. Despite the ridicule he endures from his fellow citizens for believing in something that apparently does not exist, Horton remains steadfast in his commitment to rescue the Whos from annihilation. Horton responds to the ambivalence on the part of his fellow jungle animals by uttering the story's most famous refrain, that "a person's a person, no matter how small"; urging the Whos to speak up more loudly for their right to exist; and imploring the Jungle of Nool to recognize Who-ville as inherently worthy of being saved. In the end, the voice of Jo-Jo, not coincidentally the smallest Who of all, is the catalyst that carries the sound of the Whos all the way to the ears of the jungle animals and alerts them to the Whos' existence and vulnerability.[78] The story of Horton the elephant insists upon the value of every individual, whether Horton or Jo-Jo, as well as upon the impact one individual can have on the lives of others and the world around them. Moreover, in the figure of Horton, and not unlike the Cat in the Hat and other Seussian protagonists, the story highlights the heroism of not submitting to social authority, which, of course, would have done away with the speck of dust.

Superficially, Snafu also bears no resemblance to Horton. Whereas Horton represents the moral triumph of the noncompliant individual, Snafu (in the eyes of many critics) represents the utilitarian benefit of obedience to authority. From the conventional scholarly perspective, the character of Private Snafu is a mere device, a negative example that only has value in its failure. Insofar as this is the case, critics interpret Private Snafu's reckless behavior as proof of the rule and as the key to the films' function as propaganda. However, it may be argued that *Private Snafu*'s goal is not to limit but to evolve the individual. Snafu, like Horton, Jo-Jo, and each and every Who, is ultimately *someone*, no matter how small, and makes his effective appeal to the audience as such. In this sense, the eponymous hero of the film is not merely a rhetorical device but also a compelling reference to the thing that is being shaped, namely, the individual. As everyman, Snafu *is* his audience, and thus he is fundamentally good.

Didactically, *Private Snafu* relies on the audience's identification with the protagonist not simply to persuade the troops to follow military guidelines, but also to recognize their inherent value as free and capable individuals. Snafu has been described variously as "a careless, ugly bumpkin," "a deadbeat soldier," and the "company jerk."[79] If one invokes the rarer, more sympathetic reading of Snafu as a "boyish, hapless soldier,"[80] however, then Snafu may be seen as something more or, at least, *other*, than "the worst soldier in the Army."[81] Each film presents Snafu with an opportunity to come into his own potential and to improve his skills as a soldier through education and experience. By extension, each soldier watching *Private Snafu* may then imagine the choice they would make in Snafu's place between different courses of action. The correct choice generally recapitulates liberal democratic norms, such as contributing to the social good through individual sacrifice and hard work. In *Gripes* (1943), for example, Snafu decides to shirk the menial, mundane task he has been assigned because he is jealous of other soldiers who are out "a jabbin' the Japs and huntin' the Hun." As he sweeps the kitchen floor, peels potatoes for chow, and washes up pots and pans, Snafu reasons, "If I ran this army / boy, I'm tellin' you / I'd make a few changes / that's just what I'd do." After the Technical Fairy inevitably gives Snafu the opportunity to "run the Army," his lack of preparation and leadership skill puts his troops in grave danger when the Germans launch a surprise attack.

In most cases, Snafu makes responsible attempts to do the right thing. However, he is often thwarted by his ignorance and lack of understanding of his value and role as soldier *and* citizen. In order to practice the kind of self-discipline that would prevent repetition of the mistake, Snafu must first develop the self-awareness to comprehend his own responsibilities. For example, several films in the *Private Snafu* series reinforce one of the US military's more obvious priorities at the time: the need to keep military intelligence secret and out of the hands of the adversary. In each scenario, Snafu's careless sharing of information about US military strategy, operations, logistics, and so on results in devastating losses when the enemy intercepts leaked information. The moral of these films concerns personal responsibility, not just the military requirement for secrecy, a parallel made evident in *Censored* (1944) when the lesson is revealed to be "Every man his own censor!" and in *Spies* when Snafu ironically promises to hold accountable the person responsible for the leak, namely, himself. Indeed, the vast majority of the lessons conveyed in the films do not require military discipline or obedience to the orders of an external authority, but rather depend upon the progressive development of Snafu's (and the soldiers') self-awareness and self-discipline; as such, the lessons imparted by *Private Snafu* are not so different from those found in the postwar children's books.

Other installments in the *Private Snafu* series impart commonsense lessons and themes that invoke the everyday knowledge and discipline required of a productive and responsible member of American society. In *Pay Day* (1944), for instance, Snafu ignores the Technical Fairy's advice to invest his wartime earnings in a postwar family, home, and car. Instead, he allows himself to be lured by foreign merchants at a Middle Eastern bazaar into spending his hard-earned wages on alcohol and souvenirs instead of saving for the bright, middle-class future imagined for him (or with which he is also lured) by the US Army. Though the emphasis on foreign markets speaks directly to the US soldier's experience abroad, the lesson of responsible spending and saving reflects a standard aspiration of the American working class at the time and the image of social and economic mobility for young men living in mid-twentieth-century America. In *Gas* (1944), in which Bugs Bunny makes a well-known cameo, Snafu learns that the advice to keep one's gas mask ready at hand is not empty guidance, and in *Fighting Tools* (1943) the protagonist discovers the importance of keeping his equipment in working order. Following these recommendations, one would be hard pressed to say what the object of the military indoctrination might be.

Viewing Snafu as an active subject rather than as a mere literary device clarifies a number of important claims about the nature of the propaganda coming out of Fort Fox that are also reinforced across Geisel's entire body of work. First, Snafu's value and the source of his authority lie much less in his mistakes than in his freedom to make those mistakes and to subsequently learn from them. This perspective invites an interpretation of *Private Snafu*'s function as educational and emancipatory rather than as transactional and authoritative. Second, Snafu never submits to anyone's authority other than his own, and this particularly American kind of freedom remains as intact throughout the *Snafu* series of films as it does in the pre- and postwar children's literature. Third, Snafu can freely choose whether to remain "a careless, ugly bumpkin" or to become the "deep thinker" called for in the opening sequence of *Coming [Soon]!!!* (1943), the first film in the *Private Snafu* series. Such a dichotomy likewise points to notions of individual freedom and a thinking, imaginative, and ultimately empowered citizen-subject, and certainly echoes Dr. Seuss's preoccupation with the imagination and freedom of thought on either side of the war in works such as *And to Think That I Saw it on Mulberry Street* and *If I Ran the Zoo* (1950).

This undeniably more positive reading does not comport with the interpretation of *Private Snafu* as a venture into the "dark depths that social and behavioral conditioning can plumb."[82] Instead, *Private Snafu* reinforces that Geisel's anti-authoritarianism is fundamentally rooted in his faith in the freedom of the individual to realize, unobstructed, its potential within the democratic milieu. Each film in the series is constructed to support this ideological premise even as its representation is grounded in the concrete reality of the soldiers' everyday lives. In *Private Snafu*, as throughout Geisel's oeuvre, individual freedom (and respect for that freedom) always precedes national security. The implication is that the latter should be set into motion always and *only* on the former's behalf and not the other way around; otherwise, one is obliged to accept the paradoxical irony that freedom must be suppressed in order to defend it. However, this rule of exception was likely an inadmissible logic for the much more pragmatic, anti-fascist Theodor Geisel, as he was too swept up in the much less abstract exception already underway in history (i.e., the world war itself) to seriously question the theoretical basis of his own motives.

CIVIL SOCIETY

Many Dr. Seuss books, such as *Thidwick the Big-Hearted Moose*, *Yertle the Turtle*, and *The Sneetches*, encourage readers to be kind and tolerant and to treat others with dignity and respect. In *Yertle the Turtle*, the king of the turtles demands his fellow turtles allow him to climb atop their backs so that he can survey his territory and subjects. As Yertle climbs higher and higher up the mountain of turtles, he assesses that everything he sees must belong to him. As the mountain grows higher and Yertle's kingdom expands, the suffering of the other turtles worsens. In the end, the mountain of turtles comes crashing down when those on the bottom can no longer endure the painful burden. Yertle is deposed, and the turtles are finally able to swim free, "as turtles, and maybe all creatures, should be."

As the previous section demonstrates, *Private Snafu* reinforces the value of personal freedom through its representation of the origin and consequences of Snafu's choices. Like *Yertle the Turtle*, however, the *Private Snafu* films also emphasize the impact those consequences have on *others*, thus linking the development of Snafu's (and the soldiers') self-consciousness to the development of his (and their) social consciousness and role in securing the collective freedom of the whole. Put another way, in addition to its list of things a soldier should or should not do, the films convey important lessons about responsible democratic citizenship. In fact, scholars have noted that Geisel's experiences during the war dramatically shaped this trajectory, and we can find evidence of this teleology in the *Private Snafu* series of films.[83]

Theoretically, the social consequences of Snafu's actions—collective victory or defeat in World War II, for instance—are already implicit in each lesson outlined in the previous section about Snafu's individual freedom, but several of the short films explicitly address the social nature of these consequences. Much of what Snafu learns are basic lessons in the kind of social behavior that supports a democratic civil society and that would be perfectly at home in Dr. Seuss's later works. In *The Infantry Blues* (1943), for instance, Snafu learns the grass is not always greener on the other side, not to covet the circumstances of others, and that all his fellow soldiers are fighting the same fight. Lamenting the drudgery of being an infantryman while on a long and arduous service hike, Snafu dreams of being an airman or a sailor: "The Air Force gets the glory / and the Navy gets the cheers /but all a dogface ever gets / is mud behind the ears." While the immediate recommendation applies to Snafu's, or any individual soldier's, morale, the analogy to liberal democratic culture and norms is just as obvious: *everyone* has a role to play in the standing up for and preservation of these norms and *ought* to fulfill that role. Likewise, in *The Homefront* (1943), Snafu is disabused of the notion that those who remained at home during the war had it easy and took for granted the sacrifices of young soldiers abroad like him. In sharp contrast to all of Snafu's selfish assumptions, he discovers while he is fighting overseas that his father is sweating away at a tank factory back home, his mother is tilling the family's land to supply food to the troops, his grandfather is employed in the local shipyard, and his girlfriend has joined the Women's Army Corps.

Similarly, in *The Chow Hound* (1944), a patriotic bull leads itself to slaughter to feed Snafu and his fellow soldiers. Snafu, on the other hand, does not sacrifice any of his time and labor (and certainly not his *whole* self) in the service of others. In fact, he consumes more than he needs and wastes much of what is provided to him. While the bull is most often and rightly linked to the prewar *The Story of Ferdinand* (1936) by Munro Leaf, who worked alongside Geisel on *Private Snafu*, the situation in *The Chow Hound* also anticipates Dr. Seuss's popular postwar tale *Thidwick the Big-Hearted Moose*. Just as Snafu selfishly ignores his soldierly duties and responsibilities to his squadron, branch, military, and nation, Thidwick's uninvited "guests" exploit his unconditional kindness for their own gain until Thidwick is finally exiled from his herd. The guests and the herd abandon their social responsibility to defend and care for Thidwick in the same way Snafu ignores his individual obligation to his fellow soldiers, the war effort, and the democratic society in defense of which the war is being fought.

The bull's readiness to sacrifice himself is no doubt essential to maintaining an effective fighting force, but it is also a concrete manifestation of the moral courage required to cultivate the collective good and of society's responsibility to nourish the moral courage of the individual. According to social contract theory, the fairness of this exchange between the individual and the general (social) will and the means (the contract) through which to achieve it is the foundation of any democratic society. As Jean-Jacques Rousseau wrote, "In short, whoso gives himself to all gives himself to none."[84] Moreover, social contract theory requires that in "sacrificing" one's freedom, one *gains* something in return. Specifically, those who adhere to the social contract gain political liberty and a democratic civil society in which to realize the full potential of their individual and collective freedom. The social contract not only allows for the preservation and maintenance of that freedom over time, but through this exchange and preservation also suppresses the more dangerous ideological expressions of individual *and* social abuses of power. According to Katrin Froese's reading of this unique, transformative surplus, "Freedom is not simply the ability to determine the course of one's life without the interference of others [as Snafu clearly thinks it is], but it is also an act of creation through which the boundaries of the self are continuously transformed as a result of social interaction."[85] Following Froese's optimism, one may argue that what viewers witness more than anything else in *Private Snafu* and across Geisel's oeuvre *is* this act of creation, or rather, a series of creative and cathartic acts through which the spectator (self) is transformed into something much more than itself as a result of the fictional sacrifice and the social and civic belonging earned through it.

Many of the *Private Snafu* films ask the audience of young recruits to adopt a more selfless perspective. By practicing self-discipline not only for one's own sake or for the sake of military victory but also for the sake of others, the spectator is encouraged to become a better democratic citizen and, perhaps only coincidentally, a better soldier. As a result, those who view the films today may find themselves thinking less about information operations and national security as elements of war and more about social identity and civic duty as elements of national security. As such, *Private Snafu* doesn't seek to "excise any attachments the soldier might have that could impair his single-minded attention to fighting and winning the war,"[86] as Dow, for example, claims; instead, the films demonstrate that Snafu's failure very often lies precisely in his *not* acting on behalf of those attachments.

In conclusion, *Private Snafu* should not be read as a propaganda of agitation, but rather as an example of Ellul's democratic propaganda of integration. Geisel "despised the 'indoctrination' practices associated with the German and Japanese educational systems during the war,"[87] precisely because these authoritarian systems embraced the coercive strategies of agitation as described by Ellul. Geisel, on the other hand, advocated for a democratic, progressive, and "permissive" pedagogy as well as an open and transformative educational system based on self-discovery, mutual tolerance, and a free and secure civic life supported by whimsical stories that aligned with these values.[88] As *Private Snafu* shows, a genuinely democratic propaganda must subvert itself, or at least allow for its own subversion, in order *to be* what it is. Thus, after going to such great pains to distinguish between democratic and antidemocratic types of propaganda, Ellul concludes that democratic propaganda is an impossible contradiction: "There is, therefore, no 'democratic' propaganda."[89] Nevertheless, few today would argue that *Private Snafu* is not military propaganda, but neither does Geisel abandon in the series the democratic values for which he became so well-known after the war. Likewise, with its many winks and nods, *Private Snafu* does not shed much light on the darker side of the genre, as some scholars suggest it does.

While *Private Snafu* and Dr. Seuss's children's books were created under drastically different conditions, they nevertheless employ similar means toward similar ends. *Private Snafu*'s objective is to develop the young recruit into a mature and responsible participant in American democracy. Consequently, instead of suspending democratic values to gain a military advantage over the adversary, *Private Snafu* reinforces these values for the same reason. From this perspective, what emerges is another reading in which the *Private Snafu* series of films is neither paradox nor hypocrisy, in which there is no remarkable tension between propaganda and education or between Geisel's work during World War II and Seuss's postwar children's literature. Rather, what is revealed is a logically consistent evolution that suggests a different interpretation of *Private Snafu*, not as departure but as synthesis.

NOTES

1. Two additional films in the series were never shown during the war—the war ended before *Coming Home* (1946) could make it to the screen, and the fictional scenario in *Going Home* (1945) was pulled from release because it too closely resembled contemporary top-secret developments in US atomic power and hinted at war termination plans in the Pacific. See Michael Shull and David Wilt, "Private Snafu Cartoons," in *Doing Their Bit: Wartime American Animated Short Films, 1939–1945* (Jefferson, NC: McFarland, 2004); and Christopher Dow, "Private Snafu's Hidden War: Historical Survey and Analytical Perspective," *Bright Lights Film Journal* no. 42 (November 2003), https://brightlightsfilm.com/private-snafus-hidden-war-historical-survey-analytical-perspective/.

2. Philip Nel, "Children's Literature Goes to War: Dr. Seuss, P. D. Eastman, Munro Leaf, and the Private SNAFU Films (1943–46)," *Journal of Popular Culture* 40, no. 3 (June 2007): 468; and Mark David Kaufmann, "Ignorant Armies: Private Snafu Goes to War," *Public Domain Review*, March 25, 2015, https://publicdomainreview.org/essay/ignorant-armies-private-snafu-goes-to-war.

3. Nel, "Children's Literature Goes to War," 468–69; Brian Jay Jones, *Becoming Dr. Seuss: Theodor Geisel and the Making of an American Imagination* (New York: Dutton, 2019); and Michael Birdwell, "Technical Fairy First Class? Is This Any Way to Run an Army? Private Snafu and WWII," *Historical Journal of Film, Radio, and Television* 25, no. 2 (June 2005): 203.

4. Dow, "Private Snafu's Hidden War."

5. Dow, "Private Snafu's Hidden War."

6. Steve M. Barkin, "Fighting the Cartoon War: Information Strategies in World War II," *Journal of American Culture* 7, nos. 1–2 (Spring/Summer 1984): 114.

7. Dow, "Private Snafu's Hidden War."

8. Arthur Mayer, "Fact into Film," *Public Opinion Quarterly* 8, no. 2 (Summer 1944): 206.

9. Dow, "Private Snafu's Hidden War."

10. Geisel also worked with Capra on the *Why We Fight* films. See Jones, *Becoming Dr. Seuss*.

11. Kaufmann, "Ignorant Armies."

12. Tracy Louise Mollet, *Cartoons in Hard Times: The Animated Shorts of Disney and Warner Brothers in Depression and War 1932–1945* (New York: Bloomsbury, 2019).

13. Mollet, *Cartoons in Hard Times*.
14. Jones, *Becoming Dr. Seuss*.
15. Jones, *Becoming Dr. Seuss*.
16. Nel, "Children's Literature Goes to War," 468–69.
17. Nel, "Children's Literature Goes to War," 468–69.
18. Nel, "Children's Literature Goes to War," 469.
19. Nel, "Children's Literature Goes to War," 469.
20. Phillip M. Taylor, *Munitions of the Mind: A History of Propaganda* (Manchester: Manchester University Press, 1995).
21. Dennis M. Murphy and James F. White, "Propaganda: Can a Word Decide a War?," *Parameters* 37, no. 3 (August 2007): 15; and Chad W. Seagren and David R. Henderson, "Why We Fight: A Study of U.S. Government War-making Propaganda," *Independent Review* 23, no. 1 (Summer 2018): 69–70.
22. Robin K. Crumm, *Information Warfare: An Air Force Policy for the Role of Public Affairs* (Maxwell AFB, AL: Air University Press, 1996).
23. Crumm, *Information Warfare*.
24. George L. Vogt, "When Posters Went to War: How America's Best Commercial Artists Helped Win World War I," *Wisconsin Magazine of History* 84, no. 2 (2000–2001): 43.
25. Herbert Lin, "Doctrinal Confusion and Cultural Dysfunction in the DoD," *Cyber Defense Review* 4, no. 2 (Summer 2020): 92.
26. Taylor, *Munitions of the Mind*.
27. Joint Chiefs of Staff, *Joint Campaigns and Operations*, Joint Publication 3.0 (Washington, DC: Government Printing Office, 2022), xv, iii–16.
28. Joint Chiefs of Staff, *Information Operations*, Joint Publication 3.13 (Washington, DC: Government Printing Office, 2012), ix.
29. Joint Chiefs of Staff, *Military Information Support Operations*, Joint Publication 3-13.2 (Washington, DC: Government Printing Office, 2014), iii-20.
30. *Military Information Support Operations*, vii.
31. *Military Information Support Operations*, vii (emphasis added).
32. Lin, "Doctrinal Confusion," 90.
33. Dow, "Private Snafu's Hidden War."
34. Henry Jenkins, "'No Matter How Small': The Democratic Imagination of Dr. Seuss," in *Hop on Pop: The Politics and Pleasures of Popular Culture*, ed. H. Jenkins, T. McPherson, and J. Shattuc (Durham, NC: Duke University Press, 2002), 204.
35. Nel, "Children's Literature Goes to War," 479.
36. Jones, *Becoming Dr. Seuss*.
37. Dow, "Private Snafu's Hidden War."
38. Nel, "Children's Literature Goes to War," 481–82.
39. Nel, "Children's Literature Goes to War," 478–79.
40. Jenkins, "No Matter How Small," 194; and Fiona Macdonald, "The Surprisingly Radical Politics of Dr. Seuss," *BBC*, March 2, 2019, https://www.bbc.com/culture/article/20190301-the-surprisingly-radical-politics-of-dr-seuss.
41. Nel, "Children's Literature Goes to War," 472–73.
42. Nel, "Children's Literature Goes to War," 485.

43. Nel, "Children's Literature Goes to War," 479–80.
44. Macdonald, "Surprisingly Radical Politics."
45. Nel, "Children's Literature Goes to War," 474.
46. Nel, "Children's Literature Goes to War," 485.
47. Dow, "Private Snafu's Hidden War" (emphasis added).
48. See, for example, Nel, "Children's Literature Goes to War," 469–73; and Jenkins, "'No Matter How Small,'" 187–208.
49. See, for example, Dow, "Private Snafu's Hidden War" and Jenkins, "'No Matter How Small,'" 187–208.
50. Carl Joachim Friedrich, "Education and Propaganda," *Atlantic Monthly*, June 1937, 694–95.
51. Friedrich, "Education and Propaganda," 696.
52. J. Fred MacDonald, "Propaganda and Order in Modern Society," in *Propaganda: A Pluralistic Perspective*, ed. Ted J. Smith III (New York: Praeger, 1989), 23.
53. Crumm, *Information Warfare*.
54. Jacques Ellul, *Propaganda: The Formation of Men's Attitudes* (New York: Alfred Knopf, 1971).
55. Ellul, *Propaganda*. Ellul notes that democratic nations may also turn to agitative strategies to shape mass attitudes during wartime.
56. Dow, "Private Snafu's Hidden War."
57. Birdwell, "Technical Fairy First Class," 203.
58. Philip Nel, "Dada Knows Best: Growing Up 'Surreal' with Dr. Seuss," *Children's Literature* 27, no. 1 (1999): 151. See also Nel, "Children's Literature Goes to War," 469–85.
59. Ellul, *Propaganda*.
60. Ellul, *Propaganda*.
61. Ellul, *Propaganda*.
62. Unlike Ellul, who questions whether a democratic propaganda doesn't produce a subject that is only "apparently" (*Propaganda*, 82) free, Geisel would appear to take the ideological premise of liberal democracy and the social contract at face value. In Geisel's universe, there is nothing apparent about the special place occupied by democracy on the political spectrum or about its ideological superiority, and Geisel never gives voice to Ellul's skepticism that even a democratic propaganda may be fundamentally disingenuous.
63. Ellul, *Propaganda*.
64. Ellul, *Propaganda*.
65. Nel, "Children's Literature Goes to War," 474.
66. Nel, "Children's Literature Goes to War," 478.
67. Kaufmann, "Ignorant Armies."
68. Kaufmann, "Ignorant Armies."
69. It is ironic that the very means through which *Private Snafu* is said to indoctrinate its viewers, namely, by integrating the lived experience of the soldier, simultaneously cancels the original authority and undermines the allegedly coercive goals of the films by declaring that lived experience *SNAFU*, in what amounts to a permanent, subtextual protest.

70. These characters appear in *Three Brothers* (1944).

71. Meredith Fischer, "Capturing the Animated Soldier: Private Snafu and the Docile Body Assemblage," *Studies in Popular Culture* 41, no. 1 (2018): 116.

72. The Technical Fairy anticipates the hard-nosed Non-commissioned Officer (NCO) that appears in many cartoons and films of the period, a character perhaps best captured in John Wayne's role as Sergeant John Stryker in *Sands of Iwo Jima* (1949).

73. Mollet, *Cartoons in Hard Times*; and Larry Margasek, "Hollywood Went to War in 1941—And It Wasn't Easy," National Museum of American History, Behring Center, May 3, 2016, https://americanhistory.si.edu/blog/ hollywood-went-war-1941.

74. Mollet, *Cartoons in Hard Times*.

75. Fischer, "Capturing the Animated Soldier," 108–9.

76. Fischer also cites this example in "Capturing the Animated Soldier," 109.

77. Dow, "Private Snafu's Hidden War." See also David Culbert, "'Why We Fight': Social Engineering for a Democratic Society at War," in *Readings in Propaganda and Persuasion: New and Classic Essays*, ed. Garth S. Jowett and Victoria O'Donnell (Thousand Oaks, CA: Sage Publications, 2006), 169–75.

78. Jenkins, "'No Matter How Small,'" 188–89.

79. Fischer, "Capturing the Animated Soldier," 96. See also Dow, "Private Snafu's Hidden War"; Norman M. Klein, *Seven Minutes: The Life and Death of the American Animated Cartoon* (New York: Verso, 1998); and Shull and Wilt, *Doing Their Bit*.

80. Fischer, "Capturing the Animated Soldier," 96. See also Bishnupriya Ghosh, "Animating Uncommon Life: Malaria Films (1942–1945) and the Pacific Theater," in *Animating Film Theory*, ed. Karen Beckman (Durham, NC: Duke University Press, 2014), 275.

81. Fischer, "Capturing the Animated Soldier," 96, and Dow, "Private Snafu's Hidden War." See also Birdwell, "Technical Fairy First Class," 206; and Jerry Beck and Will Friedwald, *Looney Tunes and Merrie Melodies: A Complete Guide to the Warner Bros. Cartoons* (New York: Holt, 1989).

82. Dow, "Private Snafu's Hidden War."

83. Indeed, Nel ("Children's Literature Goes to War," 479) cites the striking differences between *Horton Hatches an Egg* (1940) and *Horton Hears a Who!* (1954) in order to track this same evolution in Dr. Seuss's writings before and after World War II: "Seuss's message books are a distinctly postwar phenomenon. In the prewar *Horton Hatches the Egg* [. . .], Horton has to take care of one egg; in the postwar *Horton Hears a Who!* [. . .], he has to save an entire planet."

84. Jean-Jacques Rousseau, *The Origin of Civil Society* (1762), in *A World of Ideas: Essential Readings for College Writers*, ed. Lee A. Jacobus (Boston: Bedford/ St. Martin's, 2010), 68.

85. Katrine Froese, "Beyond Liberalism: The Moral Community of Rousseau's Social Contract," *Canadian Journal of Political Science* 34, no. 3 (September 2001): 579.

86. Dow, "Private Snafu's Hidden War."

87. Jenkins, "'No Matter How Small,'" 188.

88. Jenkins, "'No Matter How Small,'" 188–89.

89. Ellul, *Propaganda*, 241.

8

Yertle the Turtle
and Authoritarianism and Resistance

Katherine Blue Carroll

At the opening of his 1958 story *Yertle the Turtle*, Theodor Geisel (better known as Dr. Seuss) shows us a community of smiling turtles cavorting in a lime-green pond watched over by a craggier, larger turtle perched on a small rock. This is Yertle, the king of the pond. He is "the king of them all," we are told, but there is little menace in this.[1] Children's literature is full of benevolent kings.[2] But fourteen pages later the pond's turtle-citizens are piled one on top of the other in a precarious tower, their eyes wide with alarm. Only Yertle is smiling now, balanced on top of the stack of living turtles, his arm raised in a showman's gesture of display. "What a throne! What a wonderful chair!" he crows.[3]

Yertle has decided that the higher his throne is the more he can see and that, being "the ruler of all that I see," the more he rules. "I rule from the clouds! Over land! Over sea! / There's nothing, no, NOTHING, that's higher than me," says King Yertle at the literal and figurative height of his powers.[4] Like King Derwin in his high castle in *The 500 Hats of Bartholomew Cubbins* (1938), Yertle embodies a consistent theme in Geisel's work of physical height both symbolizing and giving power, an equation surely comprehensible to children.[5]

Below Yertle the unhappy turtles of the throne are silent, as are those who come meekly when summoned to fulfill his delusions of grandeur. Only Mack, the bottommost turtle, dares to complain. He first asks how long the situation will last and then gets straight to the point. "I know on the top you are seeing great sights / But down at the bottom we, too, should have rights." When Yertle belittles and silences Mack, telling him that he has "no right to talk to the world's highest turtle," Mack gets mad.[6] His small act of defiance is simply a burp, but it brings down Yertle's turtle throne and with it, his despotic rule.

Yertle was modeled on Hitler, the most frequent target of the hundreds of political cartoons Geisel published in *PM Magazine* between 1941 and 1943.[7] Like Hitler in Geisel's cartoons, Yertle has a tendency to overreach (and he nearly had an identical mustache).[8] But *Yertle the Turtle* can be read as a cautionary tale about dictatorship anywhere. This chapter examines three important and interrelated questions about which Yertle sets us wondering: (1) How do dictators stay in power? (2) When do citizens rise up together to resist dictators? (3) Why do some individuals take great personal risks to resist dictatorship? Any fruitful discussion of theory benefits from examples to make it comprehensible. Thus, this chapter uses the Middle East to explore the theories of authoritarianism raised in *Yertle*, but the concepts are applicable anywhere.

My discussion of what scholars of the Middle East have had to say about these questions is not comprehensive, but I hope it highlights just how much *Yertle the Turtle* has to teach us about the nature of dictatorship and the importance of individual acts of resistance. I conclude that whereas much attention has been paid to the Yertles and turtle communities of the Middle East, perhaps we should focus more on the Macks of the region, its everyday underdog (underturtle) heroes who act alone without the benefit of an existing social movement or organization. Such lone dissidents should interest us regardless of the outcome of their actions, but in the current era of social media they have proven more important to broader protest than ever anticipated.

The Middle East offers a rich and appropriate arena for the reconsideration of Geisel's tale. This is not just because *Yertle the Turtle* is set on the Island of Sala-ma-Sond, a distinctly Middle Eastern sounding place (though one that looks in Geisel's illustrations as if it is located off the coast of Maine). Despite this, the story is not plagued by the Orientalism of *If I Ran the Zoo* (1950), *Scrambled Eggs Super!* (1953), and *On Beyond Zebra!* (1955), which exoticize Middle Eastern "others."[9] *Yertle* includes no such wrenching distractions, being focused entirely on almost identical turtles, as one of a set of books full of positive messages that represent Geisel's work after the end of World War II.[10] Rather, it is fruitful to consider *Yertle* against the backdrop of the Middle East because the region has been and remains a bastion of authoritarianism, and especially of authoritarian monarchy. As of 2019, all but three of the world's seven remaining absolute monarchies were in the region.[11] The recent weakening of democracy worldwide has made the Middle East's seeming resistance to what Samuel Huntington called the "third wave"—the worldwide spread of democracy after the fall of the Soviet Union—less singular.[12] Yet for decades the region has offered ample fodder for the study of authoritarianism and, since the advent of the Arab Spring in 2010, mass resistance.

STAYING IN POWER

Scholars of the Middle East have spent much more time investigating the persistence of authoritarianism in the region than the causes of popular protest or the motivations of lone dissidents. Indeed, in the aftermath of the Arab Spring, a wave of criticism and self-criticism among scholars emerged arguing that the excessive attention paid to what dictators did to stay in power had distracted attention from the changes in popular strength and opinion that fueled the uprisings.[13]

Much of the literature on persistent Middle Eastern authoritarianism focuses on factors such as oil wealth and foreign support as key to maintaining dictators.[14] These resources and others, including personal ruthlessness and creativity, allow authoritarians to buy off dissent, co-opt elites, undermine the development of potentially threatening institutions and social groups (including from within their own families), manipulate elections, construct and disseminate legitimating ideologies, and (a perennial favorite of despots everywhere) physically threaten those who challenge their dominance.

A powerful web of coercive institutions helps carry out these strategies.[15] The authoritarians of the Middle East have some of the largest per capita standing armies in the world, and all of these are used to suppress internal dissent. Although the region's regimes guard figures on military spending, we do know that between 2000 and 2019 45 percent of American arms sales were to the Middle East.[16] By some estimates, certain Gulf States spend over 10 percent of their GDP on their armed forces alone, siphoning essential funds from economic development and reform efforts.[17]

During the Arab Spring these militaries held the key to the fate of their regimes.[18] When faced with mass protest, some armies remained loyal (the countries of the Gulf); some defected to support protesters, bringing down the regime (Tunisia and Egypt, and this was also true in Iran in 1979); and some split, often along tribal lines (Libya and Yemen).[19] What explains the differences? Scholars were unprepared for this question, as Middle Eastern militaries had been all but ignored since the coups of the 1950s.[20] Regime-military relations take place behind closed doors and are difficult to study.[21] In any case most scholars assumed there was little "daylight" between regimes and their armed forces or that deep networks of surveillance and repression would keep soldiers in line.[22]

When this assumption proved inaccurate, Arab militaries became the subject of intense investigation. Studies now point to myriad factors that influence military loyalty to authoritarian regimes. These include whether the military was empowered politically under the regime,[23] whether militaries were exposed to western ideals of civil-military relations,[24] whether militaries were institutionalized or operated by strong rules and norms and not merely the whim of the ruler,[25] and whether the regime channeled resources to other security organizations,[26] to name a few. In Tunisia, the most influential Arab Spring case of disloyalty, the military was professional, unified, and losing funding to security services. Having "no special stake in the regime's survival," it abandoned Tunisian strongman Zine al Abedine Ben Ali relatively quickly, sparking hope for similar outcomes across the region.[27] In Egypt, a cohesive and powerful military long supported by the authoritarian Hosni Mubarak had become frustrated with his rule, and in particular with the idea that his son might succeed him.[28] After backing him for nearly thirty years, Egypt's senior officers pushed Mubarak out the door.

Many Middle Eastern dictators not only cultivated and monitored their militaries, they also guarded against them by vesting some coercive power in other organizations, an element of a "coup-proofing" strategy.[29] Saddam Hussein's Republican Guard, Saudi Arabia's National Guard, Libya's Revolutionary Guard Corps, and Iran's Islamic Revolutionary Guard Corps are examples. These organizations generally have strong familial, ideological, tribal, or economic ties to the regime, encouraging loyalty.[30] Often they alone are allowed to get physically close to the regime. Some paramilitary bodies, such as Iran's Basij (Organization for the Mobilization of the Oppressed), resemble social movements, organizing study groups and ceremonies and distributing charity, yet when called upon they too unleash violence on behalf of authoritarian rulers.[31]

Of course despots also house the power that maintains them in police forces, secret police forces, and party organizations. In Tunisia before the Arab Spring there were over 100,000 policemen in uniform and only 10.4 million citizens.[32] (To compare, in 2010 New York City's approximately eight million residents were policed by just under thirty-five thousand police officers.)[33] Ben Ali's political party the Constitutional Democratic Rally monitored, obstructed, and harassed citizens, as did certain Ba'ath party organizations in both Syria and Iraq.[34]

Together these organizations repressed the region's citizens in ways both subtle and heavy-handed. One need only read the reports of Freedom House, Amnesty International, or Human Rights Watch to understand the ubiquity of politically motivated detention, torture, and death in the region.[35] Over time, citizens internalized fear and habits of caution that in themselves limited opportunities for resistance.

But wait. Yertle has no army, no secret police, and indeed no mechanisms of coercion at all that we can see beyond "bellowing" and "braying."[36] In Geisel's world and for Geisel's audience, however, we may read this behavior as coercive. For the innocent, joyful turtles of the pond and for the children who are Geisel's readers, bellowing and braying are both real threats (who knows what Yertle might do?) and a punishment in and of themselves. All this is to say, it matters little whether citizens are more or less interested in democracy or have more or fewer resources with which to pursue it. If the state maintains an overwhelming *apparent* capacity for repression, authoritarianism will endure.

Geisel begins *Yertle the Turtle* by informing his readers that the kingdom's turtles have a clean pond, ample food, and generally "everything turtles might need."[37] This suggests that Yertle may be running a rentier state. In rentier states a high percentage of state funding comes as a windfall from external sources straight into government coffers, free to be distributed by the authoritarian regime to maintain itself. This may mean spending more money than taxpayers would normally allow on internal security, or it may mean purchasing political quiescence with handouts in a "no taxation, no representation" bargain.[38] Middle Eastern rulers with enough wealth to use this strategy often have access to oil or natural gas revenues, but other external funding sources (foreign aid or transit fees) may be sufficient to maintain a rentier state.[39]

In Saudi Arabia, the classical model of the rentier state, well over half of the citizenry is employed by the state, many unproductively.[40] Fuel, water, and electricity are subsidized, and princes hand out patronage. In immediate response to the Arab Spring, which saw relatively small but still unprecedented protests in the kingdom, Saudi Arabia repurchased some of its citizens' loyalty with $37 billion in the form of benefits such as pay raises and loan forgiveness.[41]

Rentier theory has its critics.[42] What if citizens don't like *how* the state distributes funds? And what if, while you may be comfortable, you can't help noticing how much *more* comfortable the rulers are? The scheme also depends on maintaining a stream of revenue that grows with the population and its expectations. And even non-rentier states like Tunisia may, by controlling scarce economic resources such as bank loans or state contracts, purchase political quiescence. (Tunisians called their "deal" *Khubzism* ["bread-ism"]. They let the leader have politics, and he let them have bread.)[43] Yet in all, scholars return again and again to rentier theory to explain authoritarianism, most finding that rents do play some role.

Middle Eastern autocrats also make arguments about why they are the best people for the job, why authoritarianism is preferable for their citizens, or why their states are in fact *not* dictatorships, despite massive evidence to the contrary. In arguing for its legitimacy, the Saudi royal family historically pointed to its commitment to governing in accordance with Wahhabism, which it presented as the only legitimate interpretation of Islam. Yet over time as the Saudi royal family's decisions have seemingly deviated from common Wahhabist interpretations, the family has begun to highlight its historical role as founders and cautious stewards of the kingdom. The regime also argues that citizens' ability to bring individual grievances to the monarch and informal consultation with social groups in fact makes the country democratic.[44] During the 1980s Saddam Hussein fashioned himself heir to the great historical rulers of Iraq's territory, a modern-day Nebuchadnezzar, the ancient Babylonian king. But he also traced his lineage back to the Prophet Mohammad, just in case.[45] After standing up to the West in the First Gulf War, his emphasis on himself as a great Muslim leader intensified. He added "God is Most Great" to the Iraqi flag.[46] Libya's longtime strongman Muammar Qaddafi bludgeoned his population with a shifting, peculiar ideology that rejected the notion of the state itself.[47] As Lisa Anderson put it in 2001, "Several generations of Libyans have come to maturity seeing themselves reflected in a sort of intellectual funhouse mirror."[48] Sometimes, as Lisa Wedeen has elegantly shown, these ideologies are promoted with statements so "patently absurd" that they cannot possibly produce real legitimacy. In such cases, the regime's ability to force citizens to repeat pro-regime statements they do not believe or engage in physical acts of support for the regime functions as a form of domination, meant to deter resistance.[49]

As these examples illustrate, legitimating authoritarian ideologies evolve, reformulated with a scoop more religion, a smidgen more "great man" theory, or, if residents are fortunate, as they have been in some Gulf States and monarchies, with reference to truly successful policies. Almost all such authoritarian formulas in the Middle East did, however, maintain a there-but-for-the-grace-of-God-go-we subtext. In the past this was often targeted at flawed but relatively democratic Lebanon or at portrayals of potentially extremist Islamic governance. Today there are more frightening examples available of what might happen if states become (Iraq) or try to become (Syria) more democratic.

What is King Yertle's legitimating ideology? He has none. Geisel will not allow Yertle to muddy the waters of the pond with justifications. In this way, he drives home the illegitimacy of all authoritarian rulers, whom Geisel consistently portrayed as "irrational tyrants," their lust for power stemming not from some larger goal or ideology but only from the fact that they are bullies.[50] The Sala-ma-Sond monarchy, as Ron Novy points out in "Rebellion in Sala-ma-Sond," is "a government by, for, and about the whims of one person." As such, Mack is within his rights to overthrow it.[51] Apart from King Birtram's care for the trees that keep the ocean out of his kingdom, the subjects of Geisel's kingdoms do not benefit at all from their leaders; they merely suffer from the inevitable failure of their "grandiose" designs.[52] In *Yertle the Turtle* (1958), *The 500 Hats of Bartholomew Cubbins* (1938), *The*

King's Stilts (1939), and *Bartholomew and the Oobleck* (1949), Geisel pokes fun at these kings' "pretentions, foolishness, and arbitrary power."[53] Yertle, for example, has constructed a regime so fragile that it can be brought down with a burp.[54] Having toppled from his throne into the pond, he ends as "King of the Mud."[55] We may read this as a taunt, but taken with Yertle's earlier eruptions ("I'm king of a cow! And I'm king of a mule! . . . Oh, marvelous me!"), it is more convincing as Geisel's suggestion that authoritarian kingship is so irrelevant to true power that one may as well be king of anything.[56]

EXPLAINING MASS PROTEST

Yertle has built his empire literally on the backs of his citizens.[57] The turtles of the throne are "feeling great pain," in fear their shells will crack, and starving. Yet all but Mack are silent. There are many of them and only one Yertle, so why do they allow themselves to be treated this way? Likewise, why did the people of the Middle East for so many years not revolt against the region's dictators? And then why did they suddenly do so in 2011? Obviously, the reasons people bore despotic rule are linked to regime strategies for survival, but repression may stop or spark protest. And what a regime may do to stay in power and what a population may do to challenge it are different objects of analysis, potentially interrelated, but also capable of changing independently. In this section I introduce some of the "bottom up" explanations scholars have given for the presence or absence of popular protest in the Middle East.

Mass popular resistance in the Middle East received relatively little attention prior to the Arab Spring because, apart from the famous case of the Iranian Revolution and lesser-known instances such as the Cedar Revolution in Lebanon in 2005, in which peaceful protests forced Syria to withdraw from the country, there was not much of it to study.[58] What resistance did take place was small scale, informal (such as signing petitions to rulers, joining loose networks, disregarding laws), or quickly repressed.[59] Consider the Damascus Spring movement of 2000–2001 following the death of longtime leader Hafez al-Asad, which you have probably not heard of. Syrian intellectuals signed statements and met privately to discuss reforms to the authoritarian state. Ultimately the regime stepped in and repressed the movement, arresting many of the organizers.[60]

For some, culture explained the region's years of passivity in the face of authoritarianism. People did not protest against dictatorship because their sense of identity made them ambivalent about democracy or about revolt.[61] Some interpretations of Sunni Islam insisted that the sovereignty of God could only be operationalized through a ruler answerable solely to religious scholars. L. Carl Brown locates quiescence in the region's "political pessimism," rooted in the deep experience of empire and in Islam. Islamic law, he argues, gives people the tools to solve their own problems, and historically this is what they have done. Government is a necessary evil. It is pointless to make demands of it.[62] In his analysis of authoritarianism in Tunisia, published nine years before the Arab Spring, Larbi Sadiki cites a culture of political deference (which he allowed was changing), a preference for consensus over confrontation, and a belief in gradualism as factors that kept the population from challenging Ben Ali's rule directly.[63]

Such cultural explanations for passivity were losing ground even before the Arab Spring. Today they are nearly obsolete. For two decades polling data have shown Arab Muslim support for democracy. Moreover, Islamist and other organizations have made demands for democracy, and citizens have shown every evidence of having expectations of their states.[64] Slow-moving culture is in any case a difficult variable to use to explain sudden change.[65]

A classic approach to understanding mass uprisings focuses on the role of organizations in planning and promoting protest, an approach grounded in part in an appreciation of black churches' centrality to the American civil rights movement.[66] Churches, sports leagues, unions, and the like are part of civil society, bodies outside of family, state, and market that can advance shared interests.[67] Such organizations also promote trust within society, considered a key ingredient of social action.[68] Where civil society is strong, protest may be more likely, with groups contributing leadership, organization, and trust.

Is civil society strong in the Middle East? For some it is not, a consequence in part of rulers suppressing social organizations and sowing distrust. Saudi Arabian Islamist Mohammad al-Duwish, lecturing fellow Saudis on "Social Institutions and Youth" in 2005, summarizes this view: "Because the Islamic world suffers from an excess of individualism, institutions do not work well. At the state level, for instance, power is personalized. There is no continuity in government. . . . Even the institutions of preaching are limited, and when they succeed, this success is due to individuals. Our society is governed by individualism. Even in families, it is an individual, the father, who governs all the other individuals."[69]

In Arab culture, Lawrence Rosen has argued, individuals create and recreate flexible webs of personal and reciprocal obligation and interdependency throughout their lives. Whatever role they may be acting in (politician, police officer, judge), they are never free of these ties.[70] But does personal power necessarily limit the potential for mass mobilization? If civil society groups can organize and direct political action, then certainly so can personal networks. If civil society groups promote trust, then so do these personal networks, which may include people from all walks of life. Together these webs, which are not civil society but may function like it, imbue society with both independence from the state and durability in the face of it.

Yet civil society did emerge in the 1990s in the Middle East, with an explosion of new civic organizations, and scholars began to focus on their potential to challenge authoritarianism, perhaps through facilitating mass protest.[71] These scholars have been disappointed. Civic organizations have been controlled by authoritarian regimes or repressed, or their membership has been ambivalent about democracy.[72] Tellingly, these organizations can take little credit for the Arab Spring. While Tunisia's large trade union and even soccer fan clubs and Egypt's charity organizations played some role in organizing and maintaining protest on the ground, overall the Arab Spring protests were both spontaneous and leaderless.[73]

On the first page of *Yertle the Turtle* we see turtles swimming in a row, but in general turtles are solitary animals.[74] The nature of Yertle's demands reinforces this isolation, as they must climb the throne one by one, facing Yertle's rule alone. The turtles reflect a theme common in Geisel's books: individuals may cooperate, but larger groups do not. Seussian civil society is nonexistent, argues Doni Gewirtzman at the conclusion of a symposium examining it through Geisel's works: "No one is trying to keep the peace. There is no law. There is no God. There is no philosopher-king who is trying to step in and manage the situation. In particular, there are very few mediating institutions to help manage the many group conflicts that arise in Seuss's world. There is no bowling league for star-bellied and plain-bellied Sneetches to get to know one another and build trust."[75]

Instead, Geisel celebrates the individual, true to his conscience, equal to all others, and potentially heroic.[76] This classic liberal individualism is a deeply American theme, but in Geisel's books this seems to come with an undercurrent of disdain for "the people" who accept domination (like the turtles) or hate one another for small differences (in bread-buttering techniques or belly designs). "Groups of people or indeed the people in general," Steven A. Cook notes of Geisel's work, "are seen as specifically gullible and easily cowed." Cook is concerned that Geisel's low expectations for civil society may be an antidemocratic theme in his works.[77] But we might also read this as reflecting the world of the young child in the nuclear or even extended family. There are no mediating subgroups between the child and parental authority.

Alternatively, we may choose not to interpret Geisel's affirmation of individual vision, conscience, and even eccentricity as coming at the expense of society. Individualism is not the problem in his worlds but the fundamental remedy. The individual is the world's "moral center," with a responsibility to act for the common good.[78] While he may be exalted, the individual (who is always a male) may not pursue his own ends in ways that hurt the community and still remain a hero.[79] The community, then, is the ultimate object of salvation, and what Geisel loathed was not society, but conformity. Gewirtzman also concludes that as parents recite and children memorize Geisel's stories, they are essentially expressing consent for our common humanity and equality, as well as commitment to limited government and civil society. Through his works these values are in fact revitalized.[80]

It is hard not to ask: Can the resources bestowed by an active civil society really overcome the stark calculus facing most potential protesters? Many scholars viewed the Middle East's years of passivity as the result of rational actors making informed cost-benefit analyses along the lines of "If I protest, I will be killed and nothing will change." If the purpose of protest was to achieve change, why bother? For protest to occur, these scholars argued, either authoritarian regimes have to weaken or citizens have to amass significant resources. In both cases calculating citizens also have to *perceive* the shift in the balance of power, although they can make mistakes in their calculations. Theda Skocpol's classic analysis of the Iranian Revolution of 1979 is an example of this approach. She argues that revolution occurs when there is a crisis of the economy or national security that the regime cannot address, which weakens the regime and opens an opportunity to challenge it.[81] Alternatively, protests may arise from factors that empowered society vis-à-vis the regime. Some studies of the Iranian Revolution mention the cassette tape as a new technology that, by allowing Iranians to spread unifying and directing messages, specifically the sermons of Ayatollah Khomeini, made mass protest more likely.[82] Likewise the role of social media and cell phones, both of which made mass, leaderless organization easier, has been a focus of some in explaining the Arab Spring.[83]

A nuanced understanding of individual rationality and protest is offered by "cascade models," in which an unfolding protest movement changes people's calculations about the risks of participation. A swelling crowd may make a hesitant protester think, "What are the chances the army's bullets would hit me?" At a certain point the calculating individual may even fear later reprisals as a result of *not* participating.[84] In his study of the Iranian Revolution, Charles Kurzman recounts Iranians deciding to participate based on the growing, persistent, and diverse crowds in the streets. Once the revolution appeared to be viable, Iranians not only assessed the risk of participating to be lowered, they also did not want to miss out on the experience.[85]

Critics of the rational protester explanation point out that in 2011 the dictators of the Middle East were at the height of their coercive powers. Also, while later protesters may have joined after movements appeared viable, this does not explain the initial wave. In Syria and elsewhere, as Eva Bellin points out, the military repeatedly shot at crowds, "yet the protestors continued to reassemble, facing down the bullets."[86] In the winter of 2021, protesters against a military coup in Myanmar were planning growing demonstrations even though more than seven hundred were in detention and at least one had been killed by the police, who were using escalating force. "I heard they have permission to shoot," one protester told a reporter, not indicating any intention of going home.[87]

Either protesters are very bad calculators, or they choose to take the risk of revolt for other reasons. For some, is protesting simply the right thing to do, regardless of the perceived cost? In this approach to understanding protest, strongly held values such as the desire for dignity, unity, and truth motivate people to take risks. Individual citizens act not to achieve political change but to preserve these internal values from new threats by the state or to be true to themselves.[88]

We do not know what Mack's life was like before Yertle decided his kingdom was too small, but the people of the Middle East had endured indignities (sham elections, official lies, bureaucratic aggression or inaction, arbitrary arrest . . . the list is long) for years. They had also endured unemployment, bouts of violence, failing states, and other types of misery. What changed in 2011? The values approach to explaining protest asks us to believe either that there were new, motivating affronts to protesters' deeply held principles or that their principles changed before the Arab Spring in a way that made business as usual unbearable. Neither of these explanations seems plausible on such a large scale.

What if, instead of deeply held values or rational calculations, emotions explain mass protest? These feelings of anger, joy, fear, or hope may be fleeting or enduring, but they differ from values in that they are not central to someone's sense of right and wrong. The Tunisian people were motivated by anger at the story of the indignities suffered by Mohammad Bouazizi, whose self-immolation in December 2010 sparked protests first in his city, then across Tunisia, and then throughout the region. But once crowds poured into the streets, they felt joy at their own unity and courage. When Ben Ali fell, this hope and joy spread across the region, sparking protests.[89] A study of the 2009 Green Movement in Iran, which was characterized by mass protest after a rigged election, found guilt to be the key emotion for second-wave protesters. They went into the streets, knowing the danger they faced, out of a sense of responsibility to friends and family members who had been hurt in earlier protests.[90]

Wendy Pearlman argues convincingly that emotional states, as narrated by individuals, correlated with the presence or absence of protest in the region during the Arab Spring: "Emotions of fear, sadness, and shame encourage individuals to avoid risk, prioritize security, and pessimistically submit to circumstances that they find threatening. Emotions of anger, joy, and pride increase risk acceptance, prioritization of dignity, and an optimistic readiness to engage in resistance."[91]

Emotions changed people's priorities, gave new meaning to events and stories, and altered people's calculations about the future.[92] Unlike shifts in resources, shifts in emotions are unpredictable. Mass revolt is, then, unpredictable, contingent, and complex.[93]

Fear, as one might imagine and as dictators understand, is an especially important emotion in reducing dissent. Over and over across the Middle East during the Arab Spring, citizens explained their protest with the phrase "the barrier of fear has been broken."[94] Citizens may fear their own regime, or they may fear outside actors or forces, making them embrace the direction, strength, and order that dictators claim to offer.[95]

Geisel offers us no clue to the turtles' calculations about risk or to their values, but he tells us a lot about their fear. When Yertle bellows, they cower in the water, watching him anxiously with big, round eyes. "The turtles 'way down in the pond were afraid," Geisel tells us. "They trembled. They shook. But they came. They obeyed."[96] Where fear is present, as in Sala-ma-Sond, mass protest is absent.

LONE DISSIDENTS (EXPLAINING MACK)

The wall of fear that Yertle has constructed with his bellowing and braying remains firm for the turtles. Only Mack, the bottommost turtle in Yertle's throne, stands up to the dictator, denouncing his cruelty, asserting his own rights and those of the community, and then taking action to end Yertle's rule. There is no doubt that Mack's burp is an intentional act of political rebellion.[97] Geisel prefaces it by telling us that Mack "decided he'd taken enough. And he had."[98]

Mack's burp makes us wonder about Middle Eastern citizens who stood up to dictatorship entirely alone. Their actions may have inspired a social movement or protest, or they may have eventually drawn around them a group of like-minded supporters. However, at least initially, they acted independently of an organization or movement, Too many of these individuals' names and stories are unavailable to us, but some we do know. For example, Loujain al-Hathloul was a Saudi Arabian activist jailed from 2018 to 2021 for opposing the regime's restrictions on women. She began her campaign alone in 2012 by uploading a video in which she both identified herself and criticized the regime's driving ban for women.[99] Another example is Galal al-Behairy, an

Egyptian poet and lyricist jailed since 2018. Al-Behairy was arrested for his role in writing a song critical of the regime that involved other Egyptians in its writing, production, and release, but he remains in prison because of a prior book of poetry written under only his name.[100] Yet another example is Tawakkol Karmān, the Yemeni journalist, activist and Nobel Prize winner, whose activism began when she was threatened by the Yemeni government after publishing pieces it perceived as critical. These initial restrictions on her led her, with eight colleagues, to found the organization Women Journalists Without Chains in 2005.[101]

These are just a few of the surely thousands of Middle Eastern "Macks."[102] What calculations, values, or emotions motivated these individuals? Are they in some way similar? How can we understand them?

The literature explaining mass protest is uninterested in these lone dissidents. Apart from acts by specific political elites and lone wolf terrorists (distinguished from lone dissidents by their violence), most social scientists ignore individual actions entirely.[103] Likewise, while scholars of the Middle East have been moved by these individuals' fates, they have not made their actions a specific object of analysis. This is understandable. Apart from political psychology, which focuses mostly on leaders and mass behavior, in political science the study of individual acts risks producing a set of stories rather than a generalizable theory of political behavior. Better a structural model in which, ideally, people are interchangeable.

But this omission is unfortunate. Lone dissenters provide their fellow citizens with information (at the very least about the cost of dissent) and ideas, both of which may contribute to change.[104] Their actions may also lead to broader protest by taking on a symbolic meaning that changes an emotional climate or by inspiring the creation of organizations or a movement. If they survive, lone dissidents may turn into movement leaders. If Mohammad Bouazizi taught us anything, it is that in the current era of social media the actions of everyday individuals may inspire others. Within two months of Bouazizi's death the Tunisian dictator Ben Ali and Egypt's dictator Hosni Mubarak had fallen, and across the Middle East people were taking to the streets, shouting demands. Democracy, altered state-society relations, and long-standing wars resulted from his act. Of course as Marc Lynch has noted, the efficacy of Bouazizi's act "will always stand as a great historical mystery," but that does not mean we will see no more like him.[105] Indeed, if social media plays a role in future mass protest there is every reason to think we will. What is more likely to inspire a mass action *by* everyday individuals *acting as* everyday individuals than another everyday individual free of the baggage of the well-known? As sociologist James Jasper points out: "Sometimes individual actions precede recruitment into more organized protest movements, but not always. By defining such actions as outside their purview and then ignoring them, mobilization and process theorists cannot even ask when these acts do and do not lead individuals into social movements."[106]

If people are recruited through personal networks into protest, or if they join because existing crowds reassure or inspire them, we cannot truly understand protest until we understand the first person to act.[107] But even in the absence of any of these contributions to understanding broader protest, lone dissidents' actions—brave or foolhardy—should interest us simply as puzzling phenomena.

Geisel would certainly encourage the study of lone dissidents. In his worlds these autonomous, isolated heroes are always politically effective. While their adversaries, who are generally political leaders, struggle to achieve their goals, Geisel's lone dissidents succeed easily, often through some prosaic act such as burping or wearing a hat.[108] They are, in Cook's words, "a potent political force against which the most seemingly fearsome political authority is no match."[109]

Geisel would also want us to value their acts as objects of analysis because these dissidents are ordinary. He tells us no fewer than six times that Mack is a "plain little turtle," just as at the start of *500 Hats* he tells us that both the hero, Bartholomew, and his hat are entirely common. The message is that ordinary people should understand their own power and stand up to the strong to rid their communities of unjust rule, taking leadership of movements for change.[110] Here we see Geisel's commitment to egalitarianism: everyone is equal in his worlds, and thus anyone can lead. Indeed, Geisel hesitates to "posit any political or social order whatsoever, even a meritocratic one." When his heroes are extraordinary it is in that they are uniquely small, weak, or low in a hierarchy. For example, King Yertle has created a literal hierarchy with Mack at the bottom.[111] In *500 Hats* the hero Bartholomew also lives at the very bottom of a hill and the king at the very top. He sees that king's great view "backward," which makes him feel "mighty small."[112] The heroes' physical and hierarchical positions reinforce Geisel's message of the fundamental equality of all citizens but also show his championing of the underdog.

What clues does Geisel give us to explain Mack's acts and thus to guide us as we embark on the study of other lone dissidents? Mack's position in the turtle hierarchy might be read as a relevant factor. "I've pains in my back and my shoulders and knees," Mack tells Yertle.[113] He is bearing more weight than any other turtle, and he has borne it for longer. He even groans. What role does suffering play in motivating lone dissidents? Mack is also the farthest turtle in the throne from Yertle and from whatever punishments he may mete out, the equivalent perhaps of living in a distant province. Is close contact with the repressive state more or less likely to be present in cases of lone dissent? Mack is also closer than other turtles to the beloved but denied waters of the pond. What role does frustration play in motivating those who rebel alone?

Mack only acts when Yertle begins to talk of ruling the moon. We may see this as Mack trying to protect himself before "'bout five thousand, six hundred and seven" more turtles are stacked on him, or we may see it as a moral or philosophical response to a ruler who has placed himself in the role of God. ("I'll stack 'em to heaven!" Yertle insists.)[114] King Didd in *Bartholomew and the Oobleck* (1949) also seeks control of nature, demanding a new kind of weather, which results in a sticky green goo falling from the sky. It is these "grandiose designs" that bring down Geisel's villains. His political cartoons targeting Hitler portrayed a similar, fatal overreach.[115]

In his work on whistleblowers, Jasper argues that normal people may experience a shock to their moral principles that spurs them into individual action. These principles may be religious beliefs, professional ethics, ideological beliefs, or community solidarity. Jasper cites research that finds individuals are more likely to act the more explicitly they hold the violated principle, the more committed an individual is to the rules they feel have been broken, and the more "prickly" their personality is.[116] Whereas the "values" explanation for mass protest is unsatisfying, at least in the case of the Middle East, we can easily imagine how specific events might challenge an individual's moral principles, spurring them to action.

In an authoritarian environment, individuals who confront the ruler often, but not always, make demands for policy changes like greater freedom and justice that would benefit the community as a whole. Mack begins his revolt by complaining to Yertle about his own pain, but all his other remarks refer to the suffering of his fellow turtles. "'We turtles can't stand it. Our shells will all crack! Besides, we need food. We are starving!' groaned Mack."[117] There is a substantial literature on altruism ranging from economics to neuroscience, and it may provide clues to the actions of lone dissidents. In *Heart of Altruism: Perceptions of a Common Humanity*, Kristen Monroe finds that what she calls "perspective" or "a feeling of being strongly linked to others through a shared humanity" is all that the various altruists she interviewed (who took great risks to help others) have in common. The altruists describe their actions as not calculated but as "obvious" or normal.[118] But if Mack has perspective, where does it come from? Monroe does not say, but other studies have pointed to upbringing, religion, genetics, and education as possible sources of altruistic behavior.[119]

Early in the story, Mack speaks deferentially to Yertle, calling him "Your Majesty" and prefacing his remarks by saying, "I don't like to complain." But over time Mack becomes more confrontational. Just before Mack's revolutionary burp, Geisel tells us that he "got a little bit mad."[120] Has anger driven away Mack's fear, as Pearlman suggests may happen in mass protest? Undoubtedly Geisel's remark that Mack was only *a little bit mad* is facetious. But choosing to take this phrase literally opens the door for another potential approach to understanding how lone dissidents might lose their fear. In *Graveyard of Clerics: Everyday Activism in Saudi Arabia* Pascal Menoret notes that, at least for the dissidents of the Islamic Awakening (who sometimes act together and sometimes take risks alone, such as speaking out against the regime), the emotion that banishes fear is, paradoxically, indifference: "In the face of state repression, they had to cultivate indifference and apathy to reach another level of political consciousness. This was not calculated deception; it was the cultivation of an ethical (and political) attitude of detachment from the state and the political sphere. Political indifference (*la mubala*) was the paradoxical condition for autonomous politics."

"Where did I get this audacity from?" one Saudi who has taken political risks explains to Menoret. "Was it from my self-assurance? No: it was from my apathy." Many Saudis are taking more risks, he continues, because "they don't give a fuck anymore."[121] How, Menoret asks, could Saudis "forget about the reality of arrests and prison, torture and pain?" In an empty place of complete apathy, boredom, and despair, courage may take root.[122] In the face of intense indifference, rational calculations are meaningless, because nothing matters. One acts only so that something will happen, whatever that may be. Such emotions are personal, unrelated to the actions of crowds, to political events, or even to acute suffering.

CONCLUSION: PLEA FOR A SEUSSIAN RESEARCH AGENDA

Many American children still receive their early political socialization through the books of Theodor Geisel, whose sixty works of children's literature may be considered a "civic institution."[123] Geisel himself certainly viewed childhood as a fertile field in which the seeds of liberal political values could be sown. *Yertle the Turtle and Other Stories* is one of six of his books that have been banned or censored (including recently *Yertle*, by a British Columbia School District, for being "too political" for children).[124] Four of Geisel's other children's books also have overtly political messages, most about kings.[125] For this if no other reason we should focus keenly on what he can teach us about politics.

In *Yertle the Turtle*, Mark West finds a parable, "a simple story about standing up to despots."[126] But I suspect that for children, as for many adults, *Yertle* is more. In particular, it holds the deep puzzle of why Mack alone stands up to the despot. (Is he hurting more? Did he go to a different school? Did his parents have different rules? Is he originally from another pond?) Mack is a rebel, and rebels are intriguing to both children and adults. Also, Mack's rebellious act is particularly appealing to Geisel's audience. What child doesn't appreciate the inappropriate burp? Yet as a social scientist considering the puzzle of Mack, I find very little scholarship to help me understand his actions.

Incredible work has been done, especially since the Arab Spring, on social movements in the Middle East. My intention here is not to preach about the value of structure or agency, but only to point the way to what I feel is the missing link in the study of authoritarianism and resistance: explaining lone acts of dissent. Geisel himself was wary of being overly preachy. In the concluding couplet of *Yertle the Turtle* we learn that after Yertle's fall, "of course . . . all the turtles are free / As turtles and, maybe, all creatures should be." When asked "why 'maybe' and not 'surely'?" Geisel responded, "I wanted other persons to say 'surely' in their minds instead of my having to say it."[127]

NOTES

1. Dr. Seuss, *Yertle the Turtle and Other Stories* (London: HarperCollins, 1986).

2. Images of political authority figures in children's literature are overwhelmingly positive, but not in Geisel's works. Of the five books that feature kings, only one monarch is not of bad character. This is the ineffective King Birtram of *The King's Stilts* (New York: Random House, 1939), who falls into melancholy and fails to rule when his stilts are stolen. Timothy Cook, "Another Perspective on Political Authority in Children's Literature: The Fallible Leader in L. Frank Baum and Dr. Seuss," *Western Political Quarterly* 36, no. 2 (June 1983): 326–36.

3. Seuss, *Yertle the Turtle*.

4. Seuss, *Yertle the Turtle*.

5. Dr. Seuss, *The 500 Hats of Bartholomew Cubbins* (New York: Vanguard Press, 1965).

6. Seuss, *Yertle the Turtle*.

7. Richard H. Minear, *Dr. Seuss Goes to War: The World War II Editorial Cartoons of Theodor Seuss Geisel* (New York: New Press, 2001).

8. In a 1987 interview Geisel said he felt adding the mustache would be "gilding the lily a bit." Jonathan Cott, *Pipers at the Gates of Dawn: The Wisdom of Children's Literature* (New York: Random House, 1983), 29. On Yertle's physical stance see Martha Brennan, "Subversive as Hell: Political Satire in the Works of Dr. Seuss," *Waterloo Historical Review* 9 (April 2017), https://doi.org/10.15353/whr.v9.153.

9. Li-chung Yang, "Globetrotters and Exotic Creatures: The Imaginary Others in Dr. Seuss," *Wenshan Review of Literature and Culture* 12, no. 2 (June 2019): 165–86. Orientalist and even racist characterizations in Geisel's books are dis-

cussed in Katie Ishizuka and Ramón Stephens, "The Cat Is Out of the Bag: Orientalism, Anti-Blackness, and White Supremacy in Dr. Seuss's Children's Books," *Research on Diversity in Youth Literature* 1, no. 2 (2019), https://sophia.stkate.edu/rdyl/vol1/iss2/4; and Philip Nel, *Was the Cat in the Hat Black? The Hidden Racism of Children's Literature and the Need for Diverse Books* (New York: Oxford University Press, 2017).

10. In particular *Horton Hears a Who* (New York: Random House, 1954) is dedicated to a Japanese friend and has been interpreted as atonement for racist wartime cartoons. Geisel later referred to these as full of "snap judgments." Richard Minear and Sopan Deb, "The Dr. Seuss Museum and His Wartime Cartoons about Japan and Japanese Americans," *Asia-Pacific Journal* 15, no. 3 (August 15, 2017), https://apjjf.org/2017/16/Minear.html. See also Sophie Gilbert, "The Complicated Relevance of Dr. Seuss's Political Cartoons," *Atlantic*, January 31, 2017, https://www.theatlantic.com/entertainment/archive/2017/01/dr-seuss-protest-icon/515031/.

11. *Oxford Constitutional Law Online Encyclopedia*, s.v. "Absolutism," by Manuel Brunner, April 2019, https://oxcon.ouplaw.com/view/10.1093/law-mpeccol/law-mpeccol-e18. Including authoritarian monarchies such as Jordan, Morocco, and Bahrain would bring this to six out of ten.

12. Freedom House, "Freedom in the World 2019: Democracy in Retreat," accessed September 18, 2020, https://freedomhouse.org/report/freedom-world/2019/democracy-retreat; and Samuel P. Huntington, *The Third Wave: Democratization in the Late Twentieth Century* (University of Oklahoma Press, 1991).

13. F. Gregory Gause III, "Why Middle East Studies Missed the Arab Spring: The Myth of Authoritarian Stability," *Foreign Affairs* 90, no. 4 (July/August 2011): 81–84, 85–90.

14. See, for example, F. Gregory Gause, "The Persistence of Monarchy in the Arabian Peninsula: A Comparative Analysis," in *Middle East Monarchies: The Challenge of Modernity*, ed. Joseph Kostiner (Boulder, CO: Lynne Rienner Publishers, 2000), 167–87.

15. These institutions are described in Eva Bellin, "The Robustness of Authoritarianism in the Middle East: Exceptionalism in Comparative Perspective," *Comparative Politics* 36, no. 2 (January 2004): 139–57; and Eva Bellin, "Reconsidering the Robustness of Authoritarianism in the Middle East: Lessons from the Arab Spring," *Comparative Politics* 44, no. 2 (January 2012): 127–49.

16. Pieter D. Wezeman et al., "Trends in International Arms Transfers, 2019," Stockholm International Peace Research Institute, March 2020, https://www.sipri.org/publications/2020/sipri-fact-sheets/trends-international-arms-transfers-2019.

17. Anthony H. Cordesman and Nicholas Harrington, *The Arab Gulf States and Iran: Military Spending, Modernization, and the Shifting Military Balance* (Washington, DC: Center for Strategic and International Studies, December 12, 2018. https://www.csis.org/analysis/arab-gulf-states-and-iran-military-spending-modernization-and-shifting-military-balance. For more information on military spending in the region see the US State Department's *World Military Expenditures and Arms Transfers* reports, which were published through 2021, at https://www.state.gov/world-military-expenditures-and-arms-transfers/.

18. Bellin, "Reconsidering the Robustness of Authoritarianism," 128; and Zoltan Barany, "Comparing the Arab Revolts: The Role of the Military," *Journal of Democracy* 22, no. 4 (October 2011): 24–35.

19. Robert Springborg, "Arab Militaries," in *The Arab Uprisings Explained: New Contentious Politics in the Middle East*, ed. Marc Lynch (New York: Columbia University Press, 2014).

20. One pre–Arab Spring study that did document military-regime relations is Steven A. Cook's *Ruling But Not Governing: The Military and Political Development in Egypt, Algeria, and Turkey* (Baltimore, MD: Johns Hopkins University Press, 2007).

21. Barany, "Comparing the Arab Revolts."

22. Gause, "Why Middle East Studies," 84.

23. Kevin Koehler, "Political Militaries in Popular Uprisings: A Comparative Perspective on the Arab Spring," *International Political Science Review* 38, no. 3 (June 2017): 363–77.

24. Kenneth M. Pollack, "The Arab Militaries: The Double-Edged Swords," in *The Arab Awakening: America and the Transformation of the Middle East*, ed. Kenneth M. Pollack, Daniel Byman, and Akram Al-Turk (Washington, DC: Brookings Institution, 2011), 58–65.

25. Bellin, "Reconsidering the Robustness of Authoritarianism"; and Gause, "Why Middle East Studies."

26. David. S. Sorenson, "Civil–Military Relations in North Africa," *Middle East Policy* 14, no. 4 (December 2007): 99–114.

27. Barany, "Comparing the Arab Revolts," 31.

28. Barany, "Comparing the Arab Revolts," 27–28.

29. On this and other "coup-proofing" strategies see Holger Albrecht, "Does Coup-Proofing Work? Political–Military Relations in Authoritarian Regimes amid the Arab Uprisings," *Mediterranean Politics* 20, no. 1 (2015): 36–54; and James T. Quinlivan, "Coup-Proofing: Its Practice and Consequences in the Middle East," *International Security* 24, no. 2 (Fall 1999): 131–65.

30. Mehran Kamrava, "Military Professionalization and Civil-Military Relations in the Middle East," *Political Science Quarterly* 115, no. 1 (Spring 2000): 67–92.

31. Ali Alfoneh, "The Basij Resistance Force: A Weak Link in the Iranian Regime?," The Washington Institute for Near East Policy, February 5, 2010, https://www.washingtoninstitute.org/policy-analysis/view/the-basij-resistance-force-a-weak-link-in-the-iranian-regime.

32. Nadia Marzouki, "Tunisia's Wall has Fallen," MERIP, January 20, 2011, https://merip.org/2011/01/tunisias-wall-has-fallen/.

33. Mike Maciag, "Law Enforcement Officers Per Capita for Cities, Local Departments," Governing, August 31, 2012, https://www.governing.com/gov-data/safety-justice/law-enforcement-police-department-employee-totals-for-cities.html.

34. For an in-depth description of repression in Tunisia see Béatrice Hibou, *The Force of Obedience: The Political Economy of Repression in Tunisia*, trans. Andrew Brown (Cambridge, UK: Polity Press, 2011); for Iraq see Kanan Mikaya, *Republic of Fear: The Politics of Modern Iraq* (Berkeley: University of California Press, 1998).

35. For a quantitative view of state-sanctioned killings, torture, disappearances, and political imprisonment worldwide from 1976 see Mark Gibney et al., "The Political Terror Scale," accessed September 21, 2023, http://www.politicalterrorscale.org.

36. Dr. Seuss, *Yertle the Turtle*.

37. Dr. Seuss, *Yertle the Turtle*.

38. Michael Herb, "No Representation without Taxation? Rents, Development, and Democracy," *Comparative Politics* 37, no. 3 (April 2005): 297–316; Hazem Beblawi, "The Rentier State in the Arab World" *Arab Studies Quarterly* 9 no. 4 (Fall 1987): 383–98; and Michael Ross, "Does Oil Hinder Democracy?," *World Politics* 53, no. 3 (April 2001): 325–61.

39. Adeel Malik, "Rethinking the Rentier Curse," in *Combining Economic and Political Development: The Experience of MENA*, ed. Giacamo Luciani (Leiden: Brill, 2017), https://journals.openedition.org/poldev/2266.

40. Harvard Kennedy School of Government, "The Labor Market in Saudi Arabia: Background, Areas of Progress, & Insights for the Future," 2019, https://epod.cid.harvard.edu/sites/default/files/2019-08/EPD_Report_Digital.pdf.

41. "To Stave Off Arab Spring Revolts, Saudi Arabia and Fellow Gulf Countries Spend $150 Billion," Knowledge at Wharton, September 21, 2011, https://knowledge.wharton.upenn.edu/article/to-stave-off-arab-spring-revolts-saudi-arabia-and-fellow-gulf-countries-spend-150-billion/.

42. Mohammed Hachemaoui and Michael O'Mahony, "Does Rent Really Hinder Democracy? A Critical Review of the 'Rentier State' and 'Resource Curse' Theories," *Revue Française De Science Politique* (English ed.) 62, no. 2 (2012): 207–30; and Gwenn Okruhlik, "Rentier Wealth, Unruly Law, and the Rise of Opposition: The Political Economy of Oil States," *Comparative Politics* 31, no. 3 (April 1999): 295–315.

43. Larbi Sadiki, "Ben Ali's Tunisia: Democracy by Non-Democratic Means," *British Journal of Middle Eastern Studies* 29, no. 1 (May 2002): 57–78.

44. Madawi Al Rasheed, *A History of Saudi Arabia*, 2nd ed. (New York and Cambridge: Cambridge University Press, 2010).

45. Phebe Marr, *The Modern History of Iraq*, 3rd ed. (Boulder, CO: Westview Press, 2012).

46. Marr, *Modern History of Iraq*, 235.

47. Dirk Vandewalle, *A History of Modern Libya*, 2nd ed. (New York: Cambridge University Press, 2012).

48. Lisa Anderson, "Muammar al-Qaddafi: The 'King' of Libya," *Journal of International Affairs* 54, no. 2 (Spring 2001): 516.

49. Lisa Wedeen, "Acting 'As If': Symbolic Politics and Social Control in Syria," *Comparative Studies in Society and History* 40, no. 3 (July 1998): 506.

50. Philip Nel, "'Said a Bird in the Midst of a Blitz': How World War II Created Dr. Seuss," *Mosaic: An Interdisciplinary Critical Journal* 34, no. 2 (June 2001): 75.

51. Ron Novy, "Rebellion in Sala-ma-Sond," in *Dr. Seuss and Philosophy: Oh the Thinks You Can Think!*, ed. Jacob M. Held (Lanham, MD: Rowman & Littlefield, 2011), 172, 177.

52. Cook, "Another Perspective on Political Authority," 330.

53. Peter Dreier, "Dr. Seuss's Progressive Politics," *Tikkun* 26, no. 4 (Fall 2011): 28–47.

54. Henry Jenkins, "'No Matter How Small': The Democratic Imagination of Dr. Seuss," in *Hop on Pop: The Politics and Pleasures of Popular Culture*, ed. Henry Jenkins, Tara McPherson, and Jane Shattuc (Durham, NC: Duke University Press, 2002).

55. Seuss, *Yertle the Turtle*.

56. Seuss, *Yertle the Turtle*.

57. Philip Nel, "Children's Literature Goes to War: Dr. Seuss, P. D. Eastman, Munro Leaf, and the Private SNAFU Films (1943–46)," *Journal of Popular Culture* 40, no. 3 (June 2007): 150–84.

58. Marc Morjé Howard and Meir R. Walters, "Explaining the Unexpected: Political Science and the Surprises of 1989 and 2011," *Perspectives on Politics* 12, no. 2 (June 2014): 394–408.

59. Asef Bayat cautions against viewing the poor, in particular, as only either revolutionary or passive. Actions they may take for survival such as squatting and other "quiet encroachments" are meaningful challenges to the authoritarian states of the Middle East. Bayat, "Un-civil Society: The Politics of the Informal People," *Third World Quarterly* 18, no. 1 (March 1997): 61.

60. Carnegie Middle East Center, "Syria in Crisis: The Damascus Spring," *Diwan*, April 1, 2012, https://carnegie-mec.org/diwan/48516?lang=en.

61. Most western scholars skeptical about democracy in the Muslim world acknowledge at least the possibility of more democratic interpretations of Islam. For examples see Martin Kramer, "Islam vs. Democracy," *Commentary* 95 (January 1993): 35–42; Bernard Lewis, *Islam and the West* (New York: Oxford University Press, 1993); and Samuel Huntington, "Religion and the Third Wave," *National Interest*, no. 24 (Summer 1991): 29–42.

62. L. Carl Brown, *Religion and State: The Muslim Approach to Politics* (New York: Columbia University Press, 2001).

63. Sadiki, "Ben Ali's Tunisia."

64. For such cultural arguments see Elie Kedourie, *Democracy and Arab Political Culture* (London: Frank Cass, 1994); P. J. Vatikiotis, *Islam and the State* (New York: Croom Helm, 1987); and Hisham Shirabi, *Neopatriarchy: A Theory of Distorted Change in Arab Society* (New York: Oxford University Press, 1988).

65. John Esposito and John Voll, *Islam and Democracy* (Oxford: Oxford University Press, 1995).

66. Douglas McAdam, *Political Process and the Development of Black Insurgency: 1930–1970* (Chicago: Chicago University Press, 1999).

67. "State of Civil Society, 2011," CIVICUS, April 2012, https://www.civicus.org/downloads/2011StateOfCivilSocietyReport/State_of_civil_society_2011-web.pdf.

68. On the links between civil society and trust see, of course, Robert Putnam, *Bowling Alone: The Collapse and Revival of American Community* (New York: Simon and Schuster, 2000).

69. Pascal Menoret, *Graveyard of Clerics: Everyday Activism in Saudi Arabia* (Stanford, CA: Stanford University Press, 2020), 30.

70. Lawrence Rosen, "Expecting the Unexpected: Cultural Components of Arab Governance," *Annals of the American Academy of Political and Social Science* 603 (January 2006): 163–78.

71. Richard Norton, *Civil Society in the Middle East* (Leiden: E. J. Brill, 1995).

72. Vincent Durac, "A Flawed Nexus? Civil Society and Democratization in the Middle East and North Africa," Middle East Institute, October 15, 2015, https://www.mei.edu/publications/flawed-nexus-civil-society-and-democratization-middle-east-and-north-africa.

73. Katia Pilati et al., "Between Organization and Spontaneity of Protests: The 2010–2011 Tunisian and Egyptian Uprisings," *Social Movement Studies* 18, no. 4 (January 2019): 463–81.

74. Geisel modeled the turtles on an earlier cartoon that satirized slow wartime producers as a teetering "V" of turtles. Fiona Macdonald, "The Surprisingly Radical Politics of Dr. Seuss," *BBC*, March 2, 2019, https://www.bbc.com/culture/article/20190301-the-surprisingly-radical-politics-of-dr-seuss. This earlier idea may have been inspired by the phrase "turtles all the way down," referring to the idea of infinite regression.

75. Doni Gewirtzman, "The Seussian Dead Hand: Concluding Remarks to *Exploring Civil Society through the Writings of Dr. Seuss*," *New York Law School Law Review* 58, no. 701 (2013–14): 702, https://digitalcommons.nyls.edu/nyls_law_review/vol58/iss3/11/.

76. Shira Wolosky, "Democracy in America: By Dr. Seuss," *Southwest Review* 85, no. 2 (2000): 167.

77. Cook, "Another Perspective on Political Authority," 332–33.

78. Wolosky, "Democracy in America," 182.

79. Cook, "Another Perspective on Political Authority," 332.

80. Gewirtzman, "Seussian Dead Hand."

81. Theda Skocpol, *States and Social Revolutions: A Comparative Analysis of France, Russia, and China* (Cambridge: Cambridge University Press; 1979).

82. Stephen Zunes, "The Iranian Revolution: 1977–1979," International Center for Non-Violent Conflict, April 2009, https://www.nonviolent-conflict.org/wp-content/uploads/2016/02/The-Iranian-Revolution-1.pdf.

83. Howard, Philip, and Muzammil M. Hussain, "The Role of Digital Media," *Journal of Democracy* 22, no. 3 (July 2011): 35–48.

84. Wendy Pearlman, "Moral Identity and Protest Cascades in Syria," *British Journal of Political Science* 48, no. 4 (October 2018): 878.

85. Charles Kurzman, *The Unthinkable Revolution in Iran* (Cambridge, MA: Harvard University Press, 2004).

86. Bellin, "Reconsidering the Robustness," 140.

87. Rebecca Ratcliffe, "Woman Reportedly Shot Dead As Myanmar Police Escalate Crackdown," *Guardian*, February 27, 2021, https://www.theguardian.com/world/2021/feb/26/myanmar-envoy-urges-un-to-use-any-means-necessary-to-restore-democracy.

88. Wendy Pearlman summarizes the values approach to protest in "Emotions and the Microfoundations of the Arab Uprisings," *American Political Science Review* 11, no. 2 (June 2013): 387–409. An example of this approach is Ashutosh Varshney, "Nationalism, Ethnic Conflict, and Rationality," *Perspectives on Politics* 1, no. 1 (March 2003): 85–99.

89. Pearlman, "Emotions and Microfoundations."

90. Mahtub Mochanloo, "Grief, Obligation and Connection in the Iranian Green Movement" (BA honors thesis, Vanderbilt University, 2016).

91. Pearlman, "Emotions and Microfoundations," 388.

92. Pearlman, "Emotions and Microfoundations."

93. Nikkie Keddie, "Can Revolutions Be Predicted? Can Their Causes be Understood?," in *Debating Revolutions*, ed. Nikkie Keddie (New York: New York University Press, 1995).

94. Pearlman, "Emotions and Microfoundations," 389.

95. Madeline Albright and William Woodward, *Fascism: A Warning* (New York: Harper, 2018).

96. Seuss, *Yertle the Turtle*.

97. In the debate over Bouazizi's intentionality, Marc Lynch asserts that "Bouazizi's act of desperation was not the inarticulate expression of rage that it has been portrayed to be. It was a calculated political act designed to provoke precisely the kind of popular response it achieved." Lynch, *The Arab Uprising: The Unfinished Revolutions of the New Middle East* (New York: PublicAffairs, 2012), 75.

98. Seuss, *Yertle the Turtle*.

99. Alia Al-Hathloul and Lina Al-Hathloul, "Loujain Al-Hathloul Is Not a Terrorist," *Marie Claire*, January 11, 2021, https://www.marieclaire.com/politics/a35122269/loujain-alhathloul-prison-sentence-essay/.

100. "Artist Profile: Galal Al-Behairy," Artists at Risk Connection, November 17, 2020, https://artistsatriskconnection.org/story/galal-el-behairy.

101. Nobel Price Laureate Tawakkol Karman, "About," accessed August 18, 2023, https://www.tawakkolkarman.net/enabout.

102. Many others may be found on the blogs and websites of dissident groups or in reports of Amnesty International, Human Rights Watch, Freedom House, and other monitoring organizations.

103. A useful work for deriving hypotheses for explaining lone dissidents might be Paul Gill, *Lone-Actor Terrorists: A Behavioural Analysis* (London: Routledge, 2015).

104. Cass R. Sunstein, *Why Societies Need Dissent* (Cambridge, MA: Harvard University Press, 2003).

105. Lynch, *Arab Uprising*, 73.

106. James M. Jasper, *The Art of Moral Protest: Culture, Biography, and Creativity in Social Movements* (Chicago: University of Chicago Press, 1997), 150.

107. James M. Jasper, *Protest: A Cultural Introduction to Social Movements* (New York: Polity Press, 2014).

108. Cook, "Another Perspective on Political Authority."

109. Cook, "Another Perspective on Political Authority," 334.

110. Dreier, "Dr. Seuss's Progressive Politics," 47; and Jenkins, "'No Matter How Small,'" 198.
111. Cook, "Another Perspective on Political Authority," 331–32.
112. Seuss, *500 Hats*.
113. Seuss, *Yertle the Turtle*.
114. Seuss, *Yertle the Turtle*.
115. Nel, "Said a Bird," 77.
116. Jasper, *Art of Moral Protest*, 137.
117. Seuss, *Yertle the Turtle*.
118. Kristen Renwick Monroe, *The Heart of Altruism: Perceptions of a Common Humanity* (Princeton, NJ: Princeton University Press, 1996), 234.
119. For an overview of these theories of the sources of altruism see Svetlana Feigin, Glynn Owens, and Felicity Goodyear-Smith, "Theories of Human Altruism: A Systemic Review," *Annals of Neuroscience and Psychology* 1, no.1 (2014), http://dx.doi.org/10.7243/2055-3447.
120. There has been speculation about Bouazizi's emotional state at the time of his death, with assumptions that he felt some combination of wounded pride, desperation, and anger. Banu Bargu, "Why did Bouazizi Burn Himself? The Politics of Fate and Fatal Politics," *Constellations* 23, no. 1 (March 2016): 27–36.
121. Menoret, *Graveyard of Clerics*, 13.
122. Menoret, *Graveyard of Clerics*, 13–14.
123. Wolosky, "Democracy in America," 167.
124. E. Wayne Ross, "Dr. Seuss and Dangerous Citizenship" (keynote address to the 6th Annual Equity and Social Justice Conference, SUNY New Paltz, March 2, 2013), https://blogs.ubc.ca/ross/files/2013/03/Dr-Seuss-and-Dangerous-Citizenship-Talk.pdf.
125. *500 Hats*, *The King's Stilts* (1939), *Bartholomew and the Oobleck* (1949), and *Horton Hears a Who*.
126. Mark West, "Dr. Seuss's Responses to Nazism: Historical Allegories or Political Parables," *Jabberwocky* 19, no. 1 (July 2016): 4.
127. Interview with Jonathan Cott, quoted in Michael Winship, "Yertle the Commander in Chief: Dr. Seuss Shows Us How Protest Can Sometimes Topple a Tyrant; A Children's Classic Rings True Today and Offers an Answer to Authoritarian Rule," *Salon*, January 16, 2017, https://www.salon.com/2017/01/16/yertle-the-commander-in-chief-dr-seuss-shows-us-how-protest-can-sometimes-topple-a-tyrant_partner/.

9

Hunches in Bunches

Intelligence and National Security Decision-Making

Genevieve Lester, John Nagl, and Montgomery McFate

Published in 1982, *Hunches in Bunches* surprisingly never attained the success of other Dr. Seuss books. This underappreciated story begins with a boy and his dog sitting awkwardly in an olive-green chair. With an anxious expression on his face, the boy twiddles his fingers, suffering from indecision about the day's activities. "My trouble was I had a mind," notes the boy. "But I couldn't make it up." As the story progresses, a variety of colorfully attired Hunches appear, each making suggestions about potential courses of action. Some Hunches encourage the boy to be responsible, while others recommend slacking off. Some Hunches appear to have the boy's best interests in mind, while others appear to be unhelpful, potentially dangerous, or clearly psychotic. The conflicting recommendations create confusion in the boy's mind, causing him to constantly reevaluate his decisions.

While superficially *Hunches in Bunches* appears to tell the story of a boy and his dog, the underlying themes of the book have direct relevance to intelligence and national security decision-making. Most of the classic literature in the field concerns exactly the same themes that appear in *Hunches in Bunches*: conflicting information, indecision, bureaucratic pressure and politicization, untrustworthy sources, mutually exclusive courses of action, unintended consequences, cognitive fallacies, and so on.[1] Although *Hunches in Bunches* has received less attention and acclaim than other Dr. Seuss books, the wisdom it contains regarding intelligence and national security decision-making should establish *Hunches in Bunches* as *the* seminal work in the field.

The primary theme of *Hunches in Bunches* concerns decision-making with limited information and foresight. In terms of national security, the story reflects how policymakers and other national security leaders are constantly faced with ambiguous information and conflicting incentives upon which they must make an "informed decision." Characterizing most of these decisions as "informed" would be a stretch. In fact, serious decisions that involve the fate of thousands or even millions of individuals (such as whether to preemptively destroy North Korea's nuclear capability) must frequently be made based on ambiguous information.[2] Moreover, US national security decisions must often be made within a short time frame; in the highly politicized environment of Washington, D.C.; and with an incomplete assessment of consequences. Decision-making in national security, particularly at the strategic level, requires a clear delineation of the various options—or Hunches—to be successful. Yet as *Hunches in Bunches* demonstrates, every choice has consequences, is constrained by external factors, imposes opportunity costs, and precludes other paths.

In this chapter we examine how each Hunch in the Dr. Seuss book can be linked to a topical area in intelligence and national security decision-making. Using the Wrong Hunch, we provide an overview of how intelligence analysis contributes to national security decision-making through a work process known as the "intelligence cycle." Next, using the example of the Happy Hunch, we examine how incorrect problem framing can create problematic interpretations in the foreign policy arena. We then interpret the Real Tough Hunch from the context of strategic intelligence, which requires deep sociocultural knowledge to properly support high-level decision-making. The Better Hunch raises the subject of how national security decisions may result in second- and third-order unintended consequences. Next, the Sour Hunch illuminates the critical separation of signals from noise in the intelligence domain. The Very Odd Hunch captures how intelligence can be politicized by decision-makers and the intelligence community. The Unhelpful Hunch suggests those situations when the intelligence community has been less than helpful. We interpret the Spookish Hunch as an allegory for the analytical fallacy of *mirror imaging*, which can impair accuracy of analysis and lead to costly, erroneous decisions. The Nowhere Hunch speaks to the dangers of chasing one's own analytic tail, while the Up Hunch represents the "rational actor" fallacy. The Down Hunch suggests the connection between deception and intelligence. Wild Hunches in Big Bunches demonstrates the dynamics of bureaucratic conflict within large organizations. The Super Hunch exemplifies the overreliance on technology within the intelligence community. Finally, the Munch Hunch provides a happy conclusion to the strange journey of *Hunches in Bunches*. The strangely precise correspondence between the Hunches in this book and core topics in the intelligence domain suggests that Ted Geisel had a deep knowledge of national security affairs.

THE WRONG HUNCH

At the beginning of *Hunches in Bunches*, the young protagonist reclines in his green chair while concentric rippling circles of ink spiral out from his head, presumably representing his confusion regarding the bevy of potential options he must consider:

> Do you want to kick a football?
> Or sit there on your behind?
> Do you want to go out skating?
> Fly a kite? Or climb a tree?
> Do you want to eat a pizza?
> Take a bath? Or watch TV?

The young protagonist appears to be overwhelmed and confused by his choices. His predicament mirrors the reality of the national security decision-making environment, in which the enormous volume of information increases exponentially every day and choices must be made. Amid this confusion, according to Tom Fingar, "the mission of intelligence analysis is to evaluate, integrate, and interpret information in order to provide warning, reduce uncertainty, and identify opportunities."[3] In other words, intelligence analysis creates order from the chaotic cacophony facing decision-makers (and our protagonist).

Like the questions that the boy ponders, intelligence analysis begins with a *question*. The question defines the problem for which the decision-maker needs a solution and therefore drives the entire analytic cycle.[4] According to the standard model of the intelligence cycle, the first step occurs when the consumer (such as the president, the National Security Council, or the Joint Chiefs of Staff) issues a requirement for information. For example, one of these decision-makers might ask for the status of nuclear development in Iran. Raw intelligence is collected through a variety of means, including human intelligence (HUMINT), signals intelligence (SIGINT), and open-source intelligence (OSINT), among others. Then this raw intelligence is "processed," meaning refined through translation, decryption, or interpretation into information that can be used by the analyst. In the analysis phase (the "thinking part" of the intelligence process), disparate pieces of information are integrated into a coherent picture.[5] Much like our protagonist's conflicting options, intelligence at this stage can be contradictory and confusing. As the young boy notes:

> Oh, you get so many hunches
> that you don't know ever quite
> if the right hunch is a wrong hunch!
> Then the wrong hunch might be right!

Unraveling complex knots of information requires expertise and patience on the part of the analyst. From the aggregated bits of information, the analyst develops a "story," which helps explain the current situation and forecast what might occur.[6] During the dissemination phase, the story is delivered to the original decision-maker, in the form of hard copy reports, oral summaries, a PowerPoint presentation, or whatever format the decision-maker prefers. The decision-maker should now have enough context and information about the topic or situation identified in the initial question to make a decision regarding the course of action. The never-ending cycle then begins again.[7]

The Wrong Hunch provides assurance that order can be found within the chaos.

THE HAPPY HUNCH

In the next illustration in *Hunches in Bunches*, the protagonist and his dog look out of a large window in his living room, watching a merry creature in a yellow costume (the Happy Hunch) leap across a hill covered with whimsical trees. As he looks out of the window, the boy has "a Happy Hunch / that I shouldn't be *in* . . . but OUT!" The living room seems dim compared to the bright light outside, suggesting that the boy has the wrong perspective: he's looking at the world from the inside out. Instead, he should be looking at the world from the outside in.

The protagonist's perception has been shaped by how the situation is framed. Frames can be defined as naturally occurring cognitive structures in the human mind that organize the interpretation of reality. The concept originated with anthropologist Gregory Bateson (although it is usually attributed to Erving Goffman),[8] who noted that "psychological frames are exclusive, i.e., by including certain messages (or meaningful actions) within a frame, certain other messages are excluded. Psychological frames are [also] inclusive, i.e., by excluding certain messages certain others are included."[9] In other words, frames are the "schemata of interpretation" that "organize experiences and guide action,"[10] thereby creating an overarching organization for a group of ideas.

Much of the research on framing has concerned how actors use frames to mobilize support for their respective positions,[11] producing a competition between different versions of the same event. For example, one perspective on globalization takes the view that the economic forces of capitalism increase equality, while another perspective holds that the result is destruction of indigenous societies. In studying social movements, sociologist David Snow and his colleagues coined the term "frame resonance" to explain why some frames seem to be more effective at mobilizing followers than others. "Does the framing suggest answers and solutions to troublesome situations and dilemmas that resonate with the way in which they are experienced? Does the framing build on and elaborate existing dilemmas and grievances in ways that are believable and compelling? Or is the framing too abstract and even contradictory? In short, is there some degree of what might be conceptualized as frame resonance?"[12] Frames resonate when individuals perceive the proffered frame as credible and salient, meaning that the frame resonates with the targets' cultural narratives and corresponds to (rather than contradicts) preexisting norms, values, and beliefs.[13] In psychological terms, "the frame that a decision-maker adopts is controlled partly by the formulation of the problem and partly by the norms, habits, and personal characteristics of the decision-maker."[14]

In the realm of intelligence analysis, preexisting frames may cause misperception of people and events, resulting in intelligence failure.[15] When Geisel was writing *Hunches in Bunches,* political events taking place in the wider world demonstrated quite dramatically the consequences of misperception and the cost of intelligence failure. In the fall of 1978, massive unrest was occurring in Iran, eventually resulting in the fall of the Shah, armed conflict between revolutionary factions, assumption of power by the Ayatollah Khomeini, Americans being taken hostage, and so on. Such political chaos might seem hard to overlook, but the Central Intelligence Agency (CIA) believed (erroneously) that nothing much was amiss in Iran, reporting to President Jimmy Carter, "Iran is not in a revolutionary or even a pre-revolutionary situation."[16] US analysts came to the wrong conclusion because they were looking in the wrong direction (at Iran's international relations instead of domestic issues) and using the wrong sources (e.g., SAVAK instead of college protestors). Like the young protagonist in *Hunches in Bunches* gazing at a desert landscape (which, in fact, resembles certain deserts in Iran), the US intelligence community framed the problem in the wrong way and therefore looked in the wrong direction.

Thus, the Happy Hunch suggests that reframing the problem may reveal the solution.

REAL TOUGH HOMEWORK HUNCH

In the next illustration, a Real Tough Hunch with green skin, yellow eyes, and an evil grin speeds around the corner on a pair of roller skates. The boy and his dog had been intending to follow the Happy Hunch, but this ill-tempered Real Tough Hunch prevents them from executing their initial plan. "You're not going ANYwhere!" says the Real Tough Hunch. "There is homework to be done, Bub! / Sit your pants down on that chair!" Shocked by the vehement directive of the Real Tough Hunch, the boy and his dog stare with open mouths and drop their books on the floor. The Real Tough Hunch looks mean!

The phrase "doing the homework" refers to the deep contextual understanding necessary to support good quality intelligence analysis. Sherman Kent (regarded as the father of analysis) called this "strategic intelligence," meaning "the kind of knowledge our state must possess regarding other states in order to assure itself that its cause will not suffer nor its undertakings fail because its statesmen and soldiers plan and act in ignorance."[17] Strategic intelligence implies a depth of knowledge about other societies beyond mere technical knowledge of weapons systems or order of battle. Understanding ancient and traditional societies, according to Adda Bozeman in her masterpiece *Strategic Intelligence and Statecraft*, should begin with the historical record, which functions as an "encrypted" repository of the "firmly embedded cultural patterns that prescribe or proscribe thought and action."[18]

Consider the al-Qaeda hijackers who attacked targets in the United States on September 11, 2001 (9/11). After the attacks, forensic investigators found a four-page document written in Arabic in three different locations associated with the event (Muhammad Atta's suitcase, an abandoned vehicle at Dulles airport, and the Pennsylvania crash site). Bearing the title "Instructions for the Last Night," the document offers the hijackers both spiritual ("purify your soul") and practical ("wear socks") advice. More importantly, "Instructions for the Last Night" reveals how al-Qaeda conceptualized the attack from within their own cultural framework (which proved to be very different from the United States). While the United States viewed the events of 9/11 as an act of terrorism, the hijackers conceived the attack as a *ghazwa* (raid),[19] which is a type of warfare practiced by many tribal societies. By referring to the attacks on 9/11 as a *ghazwa*, al-Qaeda linked their actions to "the prophet Muhammad's ghazwa at the time when the Islamic polity was established in Medina."[20] In so doing, they also revealed some of their intentions and objectives. Raids as practiced by tribal societies and in ancient Islam bear little resemblance to our Clausewitzian concepts of warfare.[21] Raids bring honor to the warrior and impose shame on the enemy, seeking neither annihilation nor capitulation of the enemy. Defeating an adversary with symbolic rather than political objectives arguably suggests a different strategy than "the war on terror."[22]

"Instructions for the Last Night" also revealed that al-Qaeda adopted a cultural pattern from the historical record of Islam called *taqiyya*, which had considerable implications for counterterrorism. Following the events of

9/11, investigators from the intelligence community and law enforcement scrutinized the actions and behavior of the hijackers prior to the attack to determine how and why indicators had been overlooked. While living in the United States and preparing for the attack, the hijackers behaved in a manner perceived by US observers to be *un-Islamic*, including drinking beer, visiting strip clubs, working out at the gym, and eating pizza. As Peter Bergen wrote, "These were not . . . impoverished suicide bombers of the type usually seen in the Palestinian intifada. Instead, they were generally well educated, technically savvy young men who blended all too well into various American communities in California, Florida, and Virginia."[23] But the intelligence community assumed that Islamic terrorists would conform to common expectations, and US counterterrorism policy was based on those assumptions.

Far from being un-Islamic, the behavior of the 9/11 hijackers corresponded *perfectly* with Islamic history and theology. Al-Qaeda operated according to a sixteenth-century theological doctrine called *taqiyya*, which permitted activities in the land of the enemy that would normally be prohibited or restricted. Abu Zubaydah, al-Qaeda's senior operational planner, instructed the hijackers "to cut their hair, to shave their beards and mustaches, and to always be polite. He told them what kinds of clothes to wear, what kinds of airline tickets to purchase, how to alter their appearances, and what to carry in order to avoid attracting suspicion from border authorities."[24] Similarly, "Instructions for the Last Night" instructed the hijackers to conceal their actions and intentions from unbelievers: "Do not show outward signs of embarrassment or nervousness, but be joyful and happy, open of heart and calm because you are going towards God's welcome and His favor."[25]

The practice of dissimulation (*taqiyya*) has a long history in Islamic theology. In the sixteenth century the Sunni majority persecuted the Shiite minority as heretics,[26] forcing the Shia to prove their adherence to Sunni Islam by cursing the House of Ali. No devout Shiite would normally disavow their beliefs, but "under compulsion or menace, a believer may be dispensed from fulfilling certain conditions of religion."[27] The principle of *taqiyya* was incorporated into the Islamic tradition of diplomacy and also incorporated into the contemporary practice of Islamic terrorism. In the context of modern statecraft, *taqiyya* permits tactical denial, deception, and other types of manipulation.

Had the United States understood the Islamic concept of *taqiyya*, perhaps fewer assumptions would have been made about the behavior and appearance of Islamic terrorists. But even after the fact, the intelligence community appeared uninterested in how the conceptual framework of Islam underpinned al-Qaeda's thinking and actions. The 9/11 report, for example, reconstructed the sequence of events that preceded the attack, but failed to mention "Instructions for the Last Night." As Hans G. Kippenberg noted, "Though the find was spectacular, the document had no major impact on the examination of the events and was widely ignored."[28]

The Real Touch Homework Hunch therefore reminds us that "doing our homework" means discovering the context of human thoughts and actions.

THE BETTER HUNCH

In the next illustration in *Hunches in Bunches*, a cheerful Better Hunch arrives clad in bedroom slippers and a large pink hat. With a gleeful smile, the Better Hunch yanks off the Real Tough Homework Hunch's large purple hat, stripping him of his authority and power. Then the Better Hunch tries to dissuade the boy from responsibly doing his homework by suggesting activities that sound much more fun. "We'll head downtown. / We'll pick up your good friend James / and together we'll trot to some real cool spot / and we'll play a few video games!"

Deciding to shirk one's responsibilities (whether that involves school homework or national security homework) might initially seem tempting. But a consideration of the inevitable negative consequences will usually dissuade most people from playing video games instead of doing their homework. In the realm of national security, the Better Hunch represents the imperative of considering the consequences of one's decisions. The scholarly literature on consequence theory generally distinguishes between first-order effects (or the immediate outcome) and second- and third-order effects (or subsequent cascading events).[29] While first-order effects can often be predicted, second- and third-order effects of any decision can be much more difficult to foresee.[30]

The US invasion of Iraq in March 2003, for example, had relatively predictable first-order consequences. Most armed conflicts produce causalities and cost money,[31] so invading Iraq would almost certainly have the same result. Here, even the second-order consequences could have been predicted. Even a cursory knowledge of the religious and ethnic fault lines in Iraq would have suggested the possibility of civil war in the wake of an overthrow of the Baathist regime.[32] Since most civil wars have regional consequences, the second-order effect of destabilization of the Middle East could have been predicted.[33] In this particular case, a civil war in Iraq had the second-order effect of empowering Iran, the geographically contiguous, historic enemy of Iraq. Since political destabilization creates opportunities for terrorist organizations, criminal enterprises, and other types of illegal activity that thrive in chaotic environments without the rule of law,[34] even the third-order consequence of empowering radical Islamic extremism could also have been predicted.

In cases where predicting the outcome might preclude the action being considered, however, denying foreseeability can be an effective strategy for carrying on as planned. In the immortal words of Donald Rumsfeld, the then US secretary of defense: "There are known knowns. These are things we know we know. We also know there are known unknowns. That is to say we know that there are some things we do not know. But there are also unknown unknowns, the ones we don't know we don't know."[35]

The Better Hunch encourages us to evaluate how decisions produce ripples with predictable patterns.

THE SOUR HUNCH

In the next illustration, the boy follows the Better Hunch downtown to play some video games. Suddenly a very bossy Sour Hunch wearing a fuzzy red sweater arrives. With his hands on his hips, Sour Hunch admonishes the boy "Your bicycle's rusting up! . . ./ Get yourself out back and oil it!" At this point, the boy becomes overwhelmed by contradictory impulses. "By now my mind was *so* mixed up / I really didn't know / if I wanted to go to the barber shop or/ to Boise, Idaho."

In the intelligence community, the inability to distinguish between important and meaningless information is referred to as the signal/noise problem. In her seminal book *Pearl Harbor: Warning and Decision*, Roberta Wohlstetter coined this phrase to explain why the Japanese attack at Pearl Harbor surprised the United States.[36] Although the United States had successfully decoded Japanese military and diplomatic traffic (including conversations, cables, and ship movements) indicating an attack was imminent, the US intelligence analysts failed to distinguish the signals from the noise. Because it was unimaginable that the Japanese might do something as "irrational" as attacking the headquarters of the US Pacific Fleet, intelligence analysts discounted the indicators of an attack, failing to see their meaning.[37]

Separating signal and noise is necessary but not sufficient. Intelligence analysts must also convince decision-makers that the signal actually exists and can be distinguished from the surrounding noise. In 2001, for example, very senior intelligence analysts reported multiple times that radical Islamic extremists were planning an attack on the United States using hijacked aircraft. George Tenet, the director of the CIA, later noted that "the system was blinking red."[38] More than forty intelligence articles in the *President's Daily Brief* (the daily summary of intelligence prepared for the president) mentioned Osama bin Laden between January 20 and September 10, 2001.[39] On August 6, 2001, the intelligence briefers warned the president that bin Laden was determined to strike the United States, exactly thirty-six days before the event.

In this case, intelligence analysts identified the signal and warned the president and his staff. But senior officials with sufficient power to take action to protect the United States did not take the threat seriously. "When these attacks occur, as they probably will," Richard Clarke, the senior director for counterterrorism on the National Security Council, wrote to National Security Advisor Condoleezza Rice: "We will wonder what more we could have done to stop them."[40]

The Sour Hunch reminds us to distinguish the signal from the noise, and that understanding the meaning of the signal depends on understanding the context from which it originated.

VERY ODD HUNCH

Next the boy encounters the Very Odd Hunch, who wears a blue polka dot ensemble with pink trim and matching accessories. Unlike the other normal-sized Hunches in the book, the huge size of the Very Odd Hunch enables him to balance the boy on one finger and the dog on the other finger. The enormous size of the Very Odd Hunch suggests tremendous power, concealed underneath attire suitable to a grandmother. We can therefore infer that the Very Odd Hunch is not only deceptive but also manipulative. Indeed, the Very Odd Hunch asks the boy a passive aggressive question under the guise of maternal concern: "Do you think it might be helpful / if you went to the bathroom, dear?"

The manipulative behavior of the Very Odd Hunch points to the problem of "politicization" in intelligence, when attempts are made to influence or change intelligence to reflect policy preferences.[41] Greg Treverton has identified five forms of politicization that may occur individually or simultaneously. First, senior policy officials may exert "direct pressure" on analysts to reach conclusions that accord with those officials' policy preferences. Second, a powerful faction in the intelligence community may have a "house line" on a particular subject, which suppresses alternative or heretical views. Third, senior officials may cherry pick their favorites out of a range of assessments. Fourth, senior officials may engage in "question asking," which implies a clear answer. Finally, intelligence and policy may have a "shared mindset" with strong presumptions and corroborating viewpoints.[42]

The potential for politicization is inherent in the structure of the relationship between intelligence producers and consumers. Because the intelligence producer supports the consumer, who is not required to use the material provided, this relationship is always asymmetrical. As Kent pointed out: "Intelligence is not the formulator of objectives; it is not the drafter of policy; it is not the maker of plans; it is not the carrier out of operations. Intelligence is ancillary to these; to use a dreadful cliché, it performs a service function."[43] The power imbalance creates a possibility for consumers to exert pressure on producers, albeit sometimes quite subtly.

Politicization can occur in both directions. "Policymakers are guilty of politicization if they directly or indirectly compel intelligence agencies to alter their conclusions in ways that are politically convenient or psychologically comforting."[44] As Paul Pillar noted during his assignment as the national intelligence officer for the Middle East, the Bush "administration used intelligence not to inform decision-making, but to justify a decision already made."[45] On the other hand, "intelligence officials are guilty of politicization if they shape their estimates to reflect their own beliefs and preferences."[46] For example, George Tenet chose to package intelligence to support the policy aims of the Bush administration once it was decided to go to war in Iraq.[47]

Politicization can occur when the relationship between decision-makers and intelligence analysts becomes too close, leading to a loss of objectivity. In the words of Sherman Kent, the storied originator of analytic methodology at the CIA, "Proper relationship between intelligence producers and consumers is one of utmost delicacy. Intelligence must be close enough to policy, plans, and operations to have the greatest amount of guidance, and must not be so close that it loses its objectivity and integrity of judgment."[48] In *Hunches in Bunches*, the Very Odd Hunch asks a question that presupposes agreement, which perfectly exemplifies the self-reinforcing "closed loop" of agreement that can occur between intelligence producers and consumers.

Thus, the Very Odd Hunch tells us to retain our objectivity and independence at all costs.

THE UNHELPFUL HUNCH

A disembodied orange arm wearing a blue glove is all we see of the next Hunch encountered by the protagonist. This Unhelpful Hunch rudely tells the boy that his mind "is frightfully ga-fluppted. / Your mind is murky-mooshy! Will you make it up?" Offended by the Unhelpful Hunch's insults, the boy scowls from a crouched, defensive position while his dog cowers in fear.

The Unhelpful Hunch brings to mind the occasional unwillingness of the intelligence community to understand the problem and/or provide answers that would actually be helpful. For example, George Tenet begins the chapter of his memoir dealing with the invasion of Iraq in March 2003 with the words, "One of the great mysteries to me is exactly when the war in Iraq became inevitable."[49] If the *director of the CIA* doesn't know the origins of the disaster, then outside observers have absolutely no chance of determining the cause. Tenet's disingenuous statement provides an example of intentional obstruction by the intelligence community, perfectly captured by the Unhelpful Hunch.

The Unhelpful Hunch confirms the well-known adage: 'In Washington, D.C., you always know who your friends are, because they stab you in the chest.'

THE SPOOKISH HUNCH

In the next illustration, the boy and his dog peer at a Spookish Hunch standing in a dark doorway. The Spookish Hunch wears an enormous purple hat, with gloves that point in opposite directions, forming a mirror image. In his hands, the Spookish Hunch holds a barbell with gloves on either end, which also point in opposite directions, forming a mirror image. Gazing from behind a wall in a room that resembles a fun house maze with distorting mirrors, the boy peers at the Spookish Hunch with trepidation. Mysteriously, the Spookish Hunch suggests something completely impossible: that the boy "*go four ways all at once!*" After his unpleasant encounters with so many Hunches, the boy has become skeptical of their propositions. He declares proudly, "I didn't fall for *that* one. / I am not that dumb a dunce." His experience has enabled him to anticipate the outcome of the Spookish Hunch's suggestion. "I knew where I would end up / if I tried a thing like *that*. . . . Most likely/ on some dead-end road / in West Gee-Hossa-Flat!" The accompanying illustration shows the boy and his dog looking over a precipice at the dead end of a road elevated by piers over a body of water which (presumably) shows his reflection.

The imagery associated with the Spookish Hunch involves mirrors and other reflective surfaces, calling to mind the analytical fallacy known as "mirror imaging." When intelligence analysts fall into the trap of mirror imaging, they falsely attribute their worldview (including decision-making processes, cultural norms, political objectives, and so on) to the adversary. This flaw in perception often leads to distorted conclusions, incorrect prediction, and ultimately strategic failure. For example, during the 1968 Tet Offensive US intelligence analysts assumed that the North Vietnamese operated according to the same strategic concepts as the US military, specifically that the North Vietnamese conceptualized military victory in the western sense. In North Vietnamese strategic theory, however, military victory cannot be attained through firepower. Rather, the psychological and diplomatic consequences of the use of firepower can lead to "decisive victory," such as when the French government lost the will to fight following Dien Bien Phu. "In short, intelligence knew all about the enemy, but failed to understand him. They knew the facts, but did not understand the meaning."[50]

The Spookish Hunch warns us to recognize the subjective reality of others and not to falsely impose or attribute our own.

THE NOWHERE HUNCH

As the boy's journey continues, he encounters a Nowhere Hunch wearing a blue-and-white body suit. With a determined expression on his face, the Nowhere Hunch marches purposefully in a circle. "I was following a Nowhere Hunch," says the boy. "A real dumb thing to do! / Everybody sometimes does it. / Even me. And even you. / I followed him in circles / till we wore the rug right through." As a result of his ceaseless marching, the Nowhere Hunch has worn a circle in the carpet, clearly an allegory for "chasing one's own tail" or persistent efforts in the utterly wrong direction.

Following a Nowhere Hunch has been a recurrent problem among intelligence and national security professionals, such as the US focus on the threat posed by al-Qaeda even after its demise. When bin Laden took credit for the 9/11 attacks, the United States demanded that the Taliban government hand over bin Laden and the other members of al-Qaeda in Afghanistan. The Taliban refused, and the United States invaded Afghanistan, toppling the Taliban government, scattering al-Qaeda, and all but eradicating the threat to the United States. But nobody in the intelligence community recognized al-Qaeda's demise. Although al-Qaeda inspired copycat attacks and former members joined other groups, the organization no longer posed a threat. Only the analysis of documents recovered from bin Laden's hiding place in Pakistan by US Navy SEALS on May 2, 2011, conclusively showed that American, Pakistani, and Iranian efforts to dismantle al-Qaeda had succeeded.[51]

Thus the Nowhere Hunch tells us to move forward along an azimuth rather than become entrapped in well-trod paths leading nowhere.

THE UP HUNCH

In the next illustration in *Hunches in Bunches*, the boy looks surprised by an Up Hunch flying past him with no obvious means of propulsion. Snapping his fingers, the Up Hunch tells the boy, "You are a stupid schlupp! / The way to go / is not *around*. / The way to go / is UP!" Convinced by the Up Hunch's enthusiastic pronouncement, the boy follows along. In the next illustration, the boy and his dog scramble up a steep, precarious staircase leading to an indeterminate destination. The boy explains that the Up Hunch's azimuth "seemed to make / a lot of sense. / I even took my chair. / I just knew / I'd make my mind up / if it had some high fresh air."

Carrying a chair up a staircase leading nowhere shows the boy's complete commitment to an irrational course of action. Since the Peace of Westphalia in 1648, the idea of rationality has been central to western strategic thinking. Rationality has been defined as "choosing to act in a manner which gives best promise of maximizing one's value position on the basis of a sober calculation of potential gains and losses, and probabilities of enemy actions."[52] The association of rationality with instrumentality reflects western norms.[53] But every society has its own frame of reference, historical constructs, and logical modes of thought. "To see the options faced by foreign leaders as these leaders see them," as Richard J. Heuer Jr. points out in the *Psychology of Intelligence Analysis*, "one must understand their values and assumptions and even their misperceptions and misunderstandings.... Too frequently, foreign behavior appears 'irrational' or 'not in their own best interest.' Such conclusions often indicate analysts have projected American values and conceptual frameworks onto the foreign leaders and societies, rather than understanding the logic of the situation as it appears to them."[54]

Iraq's possible possession of nuclear weapons offers a recent example of the rationality fallacy. For many years, CIA analysts had monitored development of weapons technology in Iraq. When Iraq began importing finely machined aluminum tubes with increasingly higher tolerances, intelligence analysts assumed that these tubes were part of a campaign to produce gas centrifuges for the production of weapons-grade nuclear fuel. Therefore, they viewed this as dispositive evidence of Saddam's nuclear weapons program. "An analyst looks for rational explanations and usually finds them in the technical realm they're used to," observed UN chief weapons inspector David Kay. "But Iraq was almost like a parallel universe. The explanations were driven not by technical reasons but by the moral and personal depravity engendered by the regime. A rational person would look at it one way, and it would be completely wrong, because in this parallel universe there was a different set of rules."[55] In fact, the increasing tolerances of the tubes did not indicate that Iraq was processing nuclear materials. Rather, they provided a pretext for purchasing more and more tubes, allowing the missile engineers to continue skimming money from the contracts.

In a different example from Iraq, we see the rationality fallacy creating havoc. The American decision to invade Iraq in 2003 was based on the belief that Iraq possessed nuclear weapons.[56] When Iraq refused to allow UN inspectors to verify that it had no stockpiles, the United States viewed this as confirmation of Iraq's weapons of mass destruction (WMD) program. But in fact, Saddam Hussein had effectively dismantled his WMD program after the United States pushed Iraqi forces out of Kuwait years earlier. Iraq's noncompliance served a different purpose, which was to convince Iran that nuclear devices were pointed at it, thereby deterring potential aggression.[57] This mistaken interpretation of Iraq's intentions provided the justification for the US invasion and therefore was probably the most costly intelligence failure in US history.[58] Of course, many members of the public, press, and US government believe that Iraq's possession of nuclear weapons was simply the bad-faith pretext for the Bush administration's decision to invade Iraq, which had already been decided on for other reasons.[59]

The Up Hunch reminds us that nothing can be truly irrational if you understand the context.

THE DOWN HUNCH

After following the Up Hunch hither and yon, the boy finally realizes that the Up Hunch "was a phony and a fake!" At the top of the staircase, the boy encounters an elderly man with a white beard wearing elaborate, long purple robes. Like a hermit on a mountaintop, the Down Hunch imparts his wisdom to the boy. "For goodness sake! / You should *never* trust an Up Hunch. / You have made a big mistake!"

In repudiating the fake, untrustworthy Up Hunch, the Down Hunch (who seems to embody the wisdom of experience) implies that some type of intentional deception has occurred. Deception has always been a key concern among intelligence professionals. In an ideal world, intelligence organizations could vet the credibility and reliability of information with complete accuracy. In practice, however, intelligence organizations (both "ours" and "theirs") can be fooled, tricked, and misled. "Practitioners of deception know this and prepare their disinformation, at the least, to confuse and, at the most, to mislead the intelligence services of their enemy."[60] When intelligence fails, deception succeeds.

The practice of deception can be enhanced through knowledge of the adversary's culture and worldview. "To be successful, the deceiver must recognize the target's perceptual context to know what (false) pictures of the world will appear plausible. History, culture, bureaucratic preferences, and the general economic and political milieu all influence the target's perceptions. False information should conform to the idiosyncrasies of strategic and even popular culture. Mistakes are easily detected and often appear comical to the target audience."[61] Because people often perceive what they expect to see, the most effective deception operations assist the adversary in their own self-deception by presenting false information that corroborates their expectations.[62]

Possibly the most successful military deception in history, Operation Fortitude, deceived Germany about the date and place of the 1944 Allied invasion of France. British officers executed the operation, but Col. William Harris, Maj. Ralph Ingersoll, and Capt. Wentworth Eldredge of the US Army designed the deception scenario. To fool the Germans, they devised a "ghost army" called the First Army Group and publicly announced that Lt. Gen. George S. Patton was the commander. To simulate a huge invasion force, they created fifty imaginary divisions with fictitious names and shoulder patches. For added credibility, they planted false stories that Patton and his white bull terrier had attended local town meetings to placate the civilian population. Corpses of British soldiers bearing false maps were dropped offshore, where they could be convincingly "discovered" by the Germans. Operation Fortitude pinned down the 15th German Army "with wholly imaginary assault forces dreamed up by the deception planners," thereby saving the lives of countless numbers of young soldiers.[63]

After the war, Eldredge reflected that in deception operations such as Operation Fortitude, the cover plan conceals "the real operational intention. It is aimed at the enemy commander through his own intelligence facilities of all sorts. It is a dangerous and subtle thing from start to finish because, if it is discovered by the enemy, it can point to the real plan." Operation Fortitude achieved remarkable success in part because it exploited the primary vulnerability in the German intelligence system, which was the German command's own hubris in their own invulnerability.

The Down Hunch in the Dr. Seuss book teaches us that deception succeeds when intelligence fails, and that self-deception facilitated by hubris is by far the most dangerous type of deception.

WILD HUNCHES IN BIG BUNCHES

The next illustration in the Dr. Seuss book shows at least nine different Hunches fighting with each other in a brutal melee. The boy reports that "things got really out of hand. / Wild hunches in big bunches / were scrapping all around me, / throwing crunchy hunchy punches."

Unfortunately this illustration accurately depicts the American intelligence community. According to the Office of the Director of National Intelligence (ODNI), at least eighteen different intelligence organizations exist in the

United States.[64] Each of these agencies, like each of the Wild Hunches, has its own bureaucratic position and interests to defend and will fight for resources, status, and continued organizational existence. The structure almost seems designed to produce internecine warfare and dysfunction.[65] Intelligence failures would appear almost inevitable, given the organizational fragmentation, procedural differences, division between foreign and domestic, information hoarding, legal constraints, mistrust, and so on.[66] In the wake of the 9/11 intelligence failures, the ODNI was created to referee the vehement fights among American intelligence agencies. But after five years of the ODNI's oversight, the Congressional Research Service concluded that the "coordinative mechanisms and authorities as currently established are inadequate to the goal of creating a more flexible and agile intelligence effort."[67]

Perhaps the Wild Hunches suggest that instead of attempting to control the chaos, we find a way to benefit from it?

SUPER HUNCH

Following the scuffle with the Wild Hunches, the boy encounters a yellow Hunch with a multiplicity of prehensile devices with gloves at the tip. Unlike all the other Hunches in the Dr. Seuss story, the Super Hunch moves his gloves mechanically by controlling a highly complex contraption. Operating the technology requires considerable effort, causing the Super Hunch to become increasingly exasperated. Poking the boy with a long yellow finger, the Super Hunch yells, "Make your mind up! Get it done! / Only *you* can make your mind up! You're the one and only one!"

The Super Hunch's advanced technology (unnecessarily complex, difficult to operate, and serving an unknown purpose) resembles a long-standing challenge faced by the US intelligence community: technocentrism. In the United States, the exponential pace of research and development in almost all scientific fields tends to confirm the view that technology solves every problem. Increasingly complex, precise, and accurate tools for intelligence collection and analysis are continually developed. Hundreds of photographs a day are beamed to the United States from surveillance satellites, yet "less than half of the pictures taken by our satellites ever get looked at by human eyes" or "by any sort of mechanized device or computerized device detecting change."[68]

All too often, information seems to be collected for its own sake. In the aftermath of the USS *Cole* bombing, former secretary of the navy John Lehman dismissed America's intelligence efforts as a "$30-billion jobs program that takes the most wondrous products of space and electronic technology and turns them into useless mush."[69]

To be anything more than "useless mush," technology must be suitable to the task at hand. Because most technical intelligence collection methods target a nation's infrastructure, they have only marginal utility against asymmetric or nonstate threats. For example, al-Qaeda's strict radio discipline effectively prevented the National Security Agency's efforts to monitor them using SIGINT. US imagery collection platforms can be defeated by moving underground or changing locations frequently. Many technical intelligence technologies developed during the Cold War to locate adversaries' armies prove useless when the intelligence objective concerns knowledge in the human mind. As Lt. Gen. Keith Alexander, former deputy chief of staff for Army intelligence, observed, "Now we're not looking for an armored division. We're looking for people—people who want to kill us, people who want to change things in their world and see us as the enemy. . . . That war and that problem set that we have is a far different intelligence problem set than what we had going into the Cold War with the Soviets."[70]

Technology can produce valuable intelligence about past events and about current movement but cannot predict the future. Following the bombing of the Khobar Towers in 1996, the Downing Task Force noted, "Precise warning of terrorist attacks depends on HUMINT to identify specific targets and the time and nature of the attack." The Task Force concluded: "Human Intelligence is probably the only source of information that can provide tactical details of a terrorist attack."[71] Nor can technology clarify the intentions of groups and individuals. "We have great sensors running around . . . that tell us what is happening," as Marine Lt. Gen. James Cartwright observed, but "they [are not] so good at telling us what is going to happen." In his view, the conflicts in Iraq and Afghanistan proved that "investment does not need to be in sensors. It probably needs to be in people."[72]

The Super Hunch confirms that the value of the technology depends on whether it actually solves problems or creates new ones.

IN CONCLUSION: MUNCH HUNCH

After the boy's encounter with the Super Hunch, he realizes his decision-making limitations. "One of me / could *never* do it." Making up his mind can only be achieved by arguing with himself. "To get a job like that done / would take more of me / . . . like two!" The boy's duplicate selves increase exponentially (as do his dog's) until a large group of boys and dogs are depicted standing on chairs "yelling" and "shoving." The situation appears to be dangerously out of control, but finally resolution is achieved. "We all talked the hunches over, / up and down and through and through. / We argued and we barg-ued! / We decided what to do." The boy and his dog finally decide to follow a Munch Hunch, who is "the best hunch of the bunch!" The Munch Hunch leads them into the kitchen, where they have "six hot dogs for lunch."

Seuss's optimism shines clearly in the final passage in *Hunches in Bunches*, where the boy's decision leads not to disaster but rather to an informal lunch of processed meat products. While *Hunches in Bunches* was never a bestseller in the Seuss oeuvre, the final lesson in the book should leave the national security community with a sense of hope. The Munch Hunch suggests that the "right" decision can be discovered, and that this decision will produce desirable results.

NOTES

1. Neil Sheehan, *A Bright Shining Lie: John Paul Vann and America in Vietnam* (New York: Random House, 1988); and John Keegan, *Intelligence in War: Knowledge of the Enemy from Napoleon to Al-Qaeda* (New York: Knopf, 2003).

2. David R. Mandel and Daniel Irwin, "Uncertainty, Intelligence and National Security Decisionmaking," *International Journal of Intelligence and CounterIntelligence* 34, no. 3 (2021): 558–82.

3. Thomas Fingar, "Analysis in the US Intelligence Community: Missions, Masters, and Methods," in *Intelligence Analysis: Behavioral and Social Scientific Foundations* National Research Council (Washington, DC: The National Academies Press, 2011), 4.

4. Charles Vandepeer, "Intelligence and Knowledge Development: What Are the Questions Intelligence Analysts Ask?, *Intelligence and National Security* 33, no. 6 (2018): 785–86.

5. James B. Bruce and Roger Z. George, "Intelligence Analysis—The Emergence of a Discipline" in *Analyzing Intelligence: Origins, Obstacles, and Innovations*, James Bruce and Roger George, eds.(Washington DC: Georgetown University Press, 2008): 1.

6. Stephen Marrin and Jonathan D. Clemente, "Improving Intelligence Analysis by Looking to the Medical Profession," *International Journal of Intelligence and CounterIntelligence* 18, no. 4 (2005): 709.

7. Marrin and Clemente, "Improving Intelligence Analysis," 708.

8. "It is Bateson's paper that the term 'frame' was proposed in roughly the sense in which I want to employ it." Erving Goffman, *Frame Analysis* (New York: Harper & Row, 1974), 7.

9. Gregory Bateson, *Steps to an Ecology of Mind: Collected Essays in Anthropology, Psychiatry, Evolution and Epistemology* (Northvale, NJ: Jason Aronson, 1987), 187.

10. David A. Snow et al., "Frame Alignment Processes, Micromobilization, and Movement Participation," *American Sociological Review* 51, no. 4 (August 1986): 464.

11. David A. Snow and Robert D. Benford, "Ideology, Frame Resonance and Participant Mobilization," *International Social Movement Research* 1 (1988): 197–217.

12. Robert D. Benford and David A. Snow, "Framing Processes and Social Movements: An Overview and Assessment," *Annual Review of Sociology* 26 (2000): 622.

13. Benford and Snow, "Framing Processes and Social Movements."

14. Amos Tversky and Daniel Kahneman, "The Framing of Decisions and the Psychology of Choice," *Science* 211, no. 4481 (January 30, 1981): 453.

15. Other sociologists, such as Todd Gitlin, adopted Goffman's approach and elaborated the concept: "Frames are principles of selection, emphasis and presentation composed of little tacit theories about what exists, what happens, and what matters." Gitlin, *The Whole World Is Watching: Mass Media in the Making and Unmaking of the New Left* (Berkeley: University of California Press, 1980), 6.

16. Suzanne Maloney, "1979: Iran and America," Brookings Institution, January 24, 2019, https://www.brookings.edu/articles/1979-iran-and-america/.

17. Sherman Kent, *Strategic Intelligence for American World Policy* (Princeton, NJ: Princeton University Press, 1966).

18. Adda Bozeman, *Strategic Intelligence and Statecraft* (New York: Brassey's, 1992), 79.

19. Yosri Fouda and Nick Fielding, *Masterminds of Terror: The Truth behind the Most Devastating Terrorist Attack the World Has Ever Seen* (London: Mainstream Publishing, 2003), 108, 121.

20. Hans G. Kippenberg, "'Consider That It Is a Raid on the Path of God': The Spiritual Manual of the Attackers of 9/11," *Numen* 52, no. 1 (2005): 31.

21. Montgomery McFate, "The 'Memory of War': Tribes and the Legitimate Use of Force in Iraq," in *Armed Groups*, ed. Jeffery Norwich (Newport, RI: Naval Institute Press, 2008).

22. Lee Harris, "*Al Qaeda's* Fantasy Ideology," *Policy Review* 114 (August–September 2002): 19–25.

23. Peter Bergen, *Holy War, Inc.: Inside the Secret World of Bin Laden* (New York: Free Press, 2001), 29.

24. Thomas R. Eldridge et al., *9/11 and Terrorist Travel: Staff Report of the National Commission on Terrorist Attacks Upon the United States*, August 21, 2004, 55, http://govinfo.library.unt.edu/911/staff_statements/911_TerrTrav_Monograph.pdf

25. "Instructions for the Last Night," *Frontline*, PBS, n.d., accessed February 13, 2024, https://www.pbs.org/wgbh/pages/frontline/shows/network/personal/instructions.html.

26. *Encyclopaedia of Islam*, vol. 10 (Leiden: E.J. Brill, 2000), s.v. "Taqiyya."

27. Bernard Lewis, *The Assassins* (New York: Basic Books, 1968), 25.

28. Kippenberg. "'Consider That It Is a Raid on the Path of God,'" 30.

29. Richard Vernon, "Unintended Consequences," *Political Theory* 7, no. 1 (1979): 57–73, http://www.jstor.org/stable/190824.

30. Samuel Barkin, "Realism, Prediction, and Foreign Policy," *Foreign Policy Analysis* 5, no. 3 (2009): 233–46, http://www.jstor.org/stable/24909777.

31. Peter H. Wilson, "The Causes of the Thirty Years War 1618–48," *English Historical Review* 123, no. 502 (2008): 554–86, http://www.jstor.org/stable/20108541.

32. Conrad Crane, *Cassandra in Oz: Counterinsurgency and Future War* (Annapolis, MD: Naval Institute Press, 2016).

33. Robert I. Rotberg, "Failed States, Collapsed States, Weak States: Causes And Indicators," in *State Failure and State Weakness in a Time of Terror*, ed. Robert Rotberg (Washington, DC: Brookings Institution Press, 2003), 1–2.

34. Patrick Stewart, "Civil Wars & Transnational Threats: Mapping the Terrain, Assessing the Links," *Daedalus* 146, no. 4 (2017): 45–58, https://www.jstor.org/stable/48563902.

35. US Department of Defense, "DoD News Briefing—Secretary Rumsfeld and Gen. Myers," September 12, 2002, https:// archive.defense.gov/transcripts/transcript.aspx?transcriptid=2636.

36. See Roberta Wohlstetter, *Pearl Harbor: Warning and Decision* (Stanford, CA: Stanford University Press, 1962).

37. Jeffrey Goldberg, "The Unknown: The C.I.A. and the Pentagon Take Another Look at Al Qaeda and Iraq," *New Yorker*, February 10, 2003, 40–47.

38. Thomas H. Kean and Lee H. Hamilton, eds., *The 9/11 Commission Report: Final Report of the National Commission on Terrorist Attacks Upon the United States* (New York: W. W. Norton, 2004), 259.

39. Kean and Hamilton, *9/11 Commission Report*, 254.

40. Kean and Hamilton, *9/11 Commission Report*, 262.

41. Joshua Rovner, "Is Politicization Ever a Good Thing?," *Intelligence and National Security* 28, no. 1 (2013): 55.

42. Gregory F. Treverton, "Intelligence Analysis: Between 'Politicization' and Irrelevance," in *Analyzing intelligence: Origins, Obstacles, and Innovation*, ed. Roger Z. George and James B. Bruce (Washington, DC: Georgetown University Press, 2008), 93.

43. Kent, *Strategic Intelligence*, 182.

44. Rovner, "Is Politicization Ever a Good Thing?," 55.

45. Paul R. Pillar, "Intelligence, Policy, and the War in Iraq." *Foreign Affairs* 85, no. 2 (2006): 15–27, https://doi.org/10.2307/20031908.

46. Rovner, "Is Politicization Ever a Good Thing?," 55.

47. Rovner, "Is Politicization Ever a Good Thing?," 61.

48. Kent, *Strategic Intelligence*, 180.

49. George Tenet, *At the Center of the Storm: My Years at the CIA* (New York: HarperCollins, 2007), 301.

50. Ronnie L. Ford, *Tet 1968: Understanding the Surprise* (London: Frank Cass, 1995), 4.

51. Nelly Lahoud, *The Bin Laden Papers: How the Abbottabad Raid Revealed the Truth about al-Qaeda, Its Leader, and His Family* (New Haven, CT: Yale University Press, 2022).

52. Glenn Snyder, *Deterrence and Defense* (Princeton, NJ: Princeton University Press, 1961), 25.

53. Ken Booth, *Strategy and Ethnocentrism* (New York: Holmes & Meier, 1979), 63–71.

54. Richard J. Heuer, Jr., *Psychology of Intelligence Analysis* (Washington, DC: Center for the Study of Intelligence, 1999), 33.

55. Michael Duffy, "So Much for the WMD," *Time*, February 9, 2004, http://www.time.com/time/archive/preview/0,10987,586175,00.html.

56. Pew Research Center, *Two Decades Later, the Enduring Legacy of 9/11*, September 2, 2021, https://www.pewresearch.org/politics/2021/09/02/two-decades-later-the-enduring-legacy-of-9-11/.

57. Tom Ricks, *Fiasco: The American Military Adventure in Iraq* (New York: Penguin, 2007), 3.

58. US Central Intelligence Agency, *Comprehensive Report of the Special Advisor to the DCI on Iraq's WMD, with Addendums* (Duelfer Report) (Washington, DC: Central Intelligence Agency, April 25, 2005).

59. Stefan Halper and Jonathan Clarke, *America Alone: The Neo-Conservatives and the Global Order* (New York: Cambridge University Press, 2004); and George Packer, *The Assassins' Gate: America in Iraq* (New York: Farrar, Straus and Giroux, 2005).

60. Joseph W. Caddell, *Deception 101—Primer On Deception* (Carlisle, PA: Strategic Studies Institute, US Army War College, 2004), http://www.jstor.com/stable/resrep11327.

61. Roy Godson and James J. Wirtz, "Strategic Denial and Deception," *International Journal of Intelligence and Counterintelligence* 13 (2000): 426.

62. Roy Godson, *Dirty Tricks or Trump Cards* (Washington, DC: Brassey's, 1995), 236.

63. H. Wentworth Eldredge, "Biggest Hoax of the War: Operation FORTITUDE; the Allied Deception Plan That Fooled the Germans about Normandy," *Air Power History* 37, no. 3 (1990): 15–22, http://www.jstor.org/stable/26271165.

64. This number includes two independent agencies, Office of the Director of National Intelligence (ODNI) and the Central Intelligence Agency (CIA); nine Department of Defense elements (Defense Intelligence Agency [DIA], National Security Agency [NSA], National Geospatial- Intelligence Agency [NGA], National Reconnaissance Office [NRO], Army Intelligence, Naval Intelligence, Marine Corps Intelligence, Air Force Intelligence, and Space Force Intelligence); seven elements of other departments and agencies (Department of Energy's Office of Intelligence and Counter-Intelligence, Department of Homeland Security's Office of Intelligence and Analysis and U.S. Coast Guard Intelligence, Department of Justice's Federal Bureau of Investigation and the Drug Enforcement Administration's Office of National Security Intelligence, Department of State's Bureau of Intelligence and Research, and Department of the Treasury's Office of Intelligence and Analysis).

65. Stephen J. Flanagan, "Managing the Intelligence Community," *International Security* 10, no. 1 (1985): 58–95, https://doi.org/10.2307/2538790.

66. Charles F. Parker and Eric K. Stern, "Bolt from the Blue or Avoidable Failure? Revisiting September 11 and the Origins of Strategic Surprise," *Foreign Policy Analysis* 1, no. 3 (November 2005): 301–31.

67. Richard A. Best, *Intelligence Reform after Five Years: The Role of the Director of National Intelligence* (Washington, DC: Congressional Research Service, 2010), 1.

68. Loch Johnson, "The CIA's Weakest Link: What Our Intelligence Agencies Need Are More Professors," *Washington Monthly*, July/August 2001.

69. Johnson, "CIA's Weakest Link."

70. Megan Scully, "'Social Intel' New Tool for U.S. Military," *Defense News*, April 26, 2004, 21.

71. *Force Protection Assessment of USCENTCOM AOR and Khobar Towers*, Report of the Downing Assessment Task Force (Washington, DC: US Government Printing Office, August 30, 1996).

72. Hunter Keeter, "Cartwright: Threat Location, Prediction Capability Should Be Priorities," *Defense Daily*, April 30, 2003, 5.

Part IV

THEORIES OF WARFARE

10

Horton and the Kwuggerbug and Deception in International Relations

Chris C. Demchak

> It happened last May, on a very nice day
> While the Elephant Horton was walking, they say,
> Just minding his business . . . just going his way . . .
> When a Kwuggerbug dropped from a tree with a plunk
> And landed on Horton the Elephant's trunk!
> The Kwuggerbug leaned toward the elephant's ear.
> "Perhaps you are wondering," he said, "why I'm here.
> "Well, I've got a secret!" he whispered. "I know
> "Of a Beezlenut tree where some Beezlenuts grow!"

So begins the leading story of Dr. Seuss's book *Horton and the Kwuggerbug and Other Lost Stories*. As the opening stanzas indicate, the stories that comprise the book focus on secrecy, intentional deception, and cascading consequences. These stories help underscore the rise in deceptive behaviors among states in the global system. Together they exemplify categories of deception, specifically willing credulity and false information. These stories also outline forums and forms in which deception emerges. In "Horton and the Kwuggerbug," for example, Horton is deceived (much like the complacent and largely undefended open democratic nations) into entering into a joint venture with a friendly acquaintance to obtain nuts from the rare Beezlenut tree. The seemingly benign and generous Kwuggerbug, much like a major peer adversary today, lures Horton to take potentially disastrous actions that are not in his best interest. The story involves manipulated elements akin to what we would today call lawfare, alliance management, phishing, emergence, and the surprise of rogue outcomes. Two other stories in the volume, "The Hoobub and the Grinch" and "Marco Comes Late," offer equally important considerations for deception, as will be seen later, while the story "How Officer Pat Saved the Whole Town" shows systemic resilience to be a combination of complex thinking and action.

Dr. Seuss published all four of these stories between 1950 and 1955. Having served in the military during World War II, Geisel returned to the continental United States to resume his career as a writer of children's books. However, the war had changed the balance of power in the international system of states, and a new type of conflict had emerged, namely the Cold War. During the latter years of the Cold War, conflict among large national systems or alliances was continuous, affected the whole of society, and was invisible. This form of conflict was confined by unspoken consensus to the spectrum between peace and war, yet its cumulative outcomes significantly changed the distribution of power internationally. Strangely enough, what conflict looked like in the latter years of the bipolar Cold War era bears great similarity to the type of conflict now emerging, refreshed and expanded, across all aspects of nations. In the modern form of "Great Systems Conflict,"[1] states prefer deception rather than overt military exchanges, with the goal of achieving strategic advantages, wealth, and the "avoidance of a greater cost" (e.g., plausible deniability) if operations are revealed publicly.[2]

The bulk of the international relations, economics, and even large-scale complex systems literature published over the past fifty years rarely mentions deception, even as a minor issue. Despite the healthy literature on the use of deception in war and an even larger discussion of intelligence and secrecy during the Cold War, why does deception *pur* remain absent from the main literatures of American international relations, economics, organization theory, and even national security studies? Deception is frequently embedded in the quintessential interactions often ignored by modern security, economic, and strategy scholars. Deception is to formal theories of international relations what domestic violence is to modern civil society: something everyone has but nobody wants to discuss.

The lack of attention to deception has a number of possible causes. For example, international relations scholars tend to presume objective rationality among key actors. Additionally, international relations scholars tend to come from neutral-speaking cultures in which a considerable amount of honesty in relations is normal. Especially prone to this blindness are American scholars and practitioners who (save for a select few such as Adda Bozeman) have absorbed the presumption of immunity and dominance associated with being a superpower.[3] Scholars of military history write about deception relatively frequently as a fact of the past.[4] Similarly, international relations scholars assume that the widespread presence and impact of deception does not exist in peacetime. Deception is thereby conceptualized as infrequently employed or trivial in overall effects—if it is even noted at all. The same indictment can (and should) be made against equally critical literatures of economics and large-scale complex adaptive systems.[5] While many military studies scholars in democracies will address deception in war, it is viewed widely as the slightly distasteful tool of the weaker powers, but it is not well studied by their academic peers.[6]

Unfortunately, adversaries of these westernized scholars are far less blind. The recent rise of "cybered conflict" on the spectrum between peace and war, which is increasingly being used by adversaries eager to avoid crossing the armed conflict threshold, demands that deception be taken seriously by international relations scholars and by states alike. In a global threat environment in which a shared but shoddy cybered substrate encourages and facilitates deception, the consequences of ignoring deception can be grave for the systemic resilience of a nation, an economy, or even an international system. Indeed, that which is assumed to be objectively true by scholars and practitioners alike is often being manipulated deceptively for an adversary's advantage. Indeed, the breakdown of the bipolar world, the decline of the liberal international system with the rise of China, and the spread of ubiquitous insecure global technologies means deception has a role in national security greater than merely being useful for the weak in times of war.[7] Practitioners and thinkers alike need to recognize that the new challenge for the states in the numerically small community of consolidated democracies is to remain democracies and prosperous while becoming strategically resilient to—and perhaps learning to employ—deception.[8]

The current crop of international relations scholars has difficulty recognizing and analyzing the implications of a rising use of technologically enabled deception as a national tool with strategic consequences. In the post-westernized global system, large peer states want to avoid war, but they still want to weaken their peer opponents in order to remain the dominant global player. The cybered world not only makes deception attractive, but it also offers five offensive advantages that enable the removal of resources and information from adversaries without overtly crossing the threshold into kinetic action. For the first time in history, virtually anyone with time and access to the global internet can afford to create any *scale of organization* desired; operate at any *proximity* from five to five thousand kilometers to gather high-value intelligence; and cheaply purchase any number of offensive tools at any level of *precision* in effect or targeting, all the while successfully maintaining *deception* in tools used and *opaqueness* in the person or sources who used them.[9] The reach of this now nearly ubiquitous cyberspace substrate has changed patterns in the global distribution of power among states. Adversaries are now able to remove wealth from consolidated democracies in vast quantities, as North Korea does to South Korea; coerce other nations' citizens directly, as Russia does to Ukraine; and collect and store critical technological innovations and intellectual property for future use, as China does to the United States. Moreover, they are able to obscure those campaigns through cybered deception and opaqueness. Strategically consequential and deception-replete events are now routinely occurring outside of war and changing the threat vectors around us, and international relations as a literature is silent.

Given this state of affairs, a campaign of intellectual shock and awe is needed to waken the international relations scholars to the new highly deceptive world order, which is dominated by authoritarian state actions and democratic defenses. In that light, the works of Dr. Seuss (aka Theodor Geisel, a World War II veteran and eminent children's' book author) may provide an antidote for this gap in perception. Dr. Seuss's works contain embedded lessons about deception in intersystem relations. In his world, deceivers are commonplace, creative, and persuasive, and for the most part they nearly always succeed. Using the directness of literature meant for a child's preparation for adulthood has the advantage of surprise and offers the opportunity for intellectual engagement. Dr. Seuss's imagery may help explain the unpleasant facts of deception to those scholars and practitioners who have been raised in the largely transparent, trust-based, neutral-speaking, well-established, and globally powerful democracies.

These "lost" stories of Dr. Seuss, which were written at the height of the Cold War, present a taxonomy of deception, ranging from willing credulity to false information. Additionally, these stories illuminate the forums and forms in which deception emerges. As mentioned earlier, in "Horton and the Kwuggerbug" the Kwuggerbug deceives Horton into entering into a joint economic venture under the guise of friendship and cooperation to obtain nuts from the rare Beezlenut tree. The Kwuggerbug manipulates Horton to act against his own self-interest. A similar theme emerges in "The Hoobub and the Grinch," in which a credulous Hoobub basking in the sun is approached by a deceptive Grinch, who promises something better than sunshine, but at what cost? At the end, Grinches are shown to repeatedly deceive Hoobubs for economic gain—profit or "rent" to be more precise—and the Hoobubs' credulity deprives them of wealth needed for better investments in a resilient future. Although short, this story is rich in deceptive spins involving the rational actor presumption, misperceptions of one's own interests, fragmented communications structures, and ultimately likely hollowing of the effective use of resources for the whole sociotechnical-economic system.

In "Marco Comes Late," the schoolboy Marco offers his teacher a wild, fabricated excuse for being late. Deception in this case is a strategy used to cover up poor or nonperformance in a rule-based normative system (namely school). Marco's efforts to deceive his teacher seemingly fail, and his innovative, creative thinking is not rewarded within the rules-based framework of traditional education. Similarly, scholars and practitioners who ignore deception because it is outside the acceptable behavior are also likely to miss the potential for failures due to creative, surprising operations on the part of adversaries who do not share those sticky, democratic values. In this vignette, Seuss makes the point that wrapping deceit in creativity often allows the deceiver (in this case Marco) to evade negative consequences, thus setting the foundations for future deception, likely even more effectively creative in fooling the target, whether a teacher or a defending state.

The final story in the volume, "How Officer Pat Saved the Whole Town," is not about deception per se but provides an example of how resilient societies ought to respond to deception in the international system. Namely, it is critical for resilient societies to engage in imagining potential consequences across complex systems, even scenarios in which deception plays a major role, in order to avert or prepare for catastrophic events downstream.

Sun Tsu famously said, "All war is based on deception." Yet in this present era in which conflict quietly rages between states,[10] too few international relations or national security scholars are addressing deception as a systematic strategy. The result may be democratic losses in power, resilience, and well-being over time. As Dr. Seuss noted about Hoobubs, cumulative losses may be the result because "Grinches sell Hoobubs such things every day." To ignore the rise of deception is, in effect, to become a Hoobub.

THE KWUGGERBUG AND THE DECEPTIVE FALSE FLAG STATE

In Seuss's "Horton and the Kwuggerbug," Horton the elephant is walking along when a Kwuggerbug drops onto his trunk with a proposal. The plan, according to the Kwuggerbug, is that Horton will carry the bug to a distant location—as yet undisclosed but too far away for the bug to travel—where there exists a prized Beezlenut tree. As a reward for serving as the transportation, Horton is promised half of the highly desirable nuts. The deception is multiplicative. The bug deliberately withholds critical information, focusing Horton on the lure of getting half the nuts from a whole tree. In his study of various forms of deception such as misinformation, Thomas J. Froelich said, "Deception is an act of . . . avoidance of a greater cost." In this case, the Kwuggerbug deceives Horton to avoid a greater cost to himself.[11] The deceiver leaves out the details about the difficulty and danger of the journey to the mountaintop neighboring the peak that holds the tree. The Kwuggerbug does not make it clear that only he would be able to get across to the tree, thereby leaving Horton at the final moment dependent on the Kwuggerbug's honesty in tossing half the nuts back across the gap between mountaintops to Horton. Further, when Horton expresses dissatisfaction, exhaustion, and even fear, the deceiving insect reminds him that he agreed to a deal, and Horton responds with begrudging but renewed effort, repeating to himself that a "deal is a deal." Finally, the Kwuggerbug also withholds his unique interpretation of "half," meaning that he would give Horton the shells of the nuts rather than equally dividing the pile. Thus, the lying bug keeps all the nut meat for himself.

There are a number of forms of deception in this story. Froehlich identifies four key misinterpretations of reality: misinformation, disinformation, missing information, and self-deception.[12] Self-deception is evident from the outset, spurred by not a small amount of greed. Horton quickly commits to a "deal," assuming that this stranger, a Kwuggerbug who is never named in the story, would be willing to share such delicacies merely for transport by an elephant. The question "why me" is never asked. Clearly a variety of other animals would have made the trip more rapidly than an elephant. The implication is that the Kwuggerbug had already done his research and knew Horton to be relatively credulous as well as stubbornly rule-oriented as an individual. Horton's rule-bound personality prevented him from reneging on an iniquitous bargain. The consequences are costly, damaging, and ultimately unrewarding for Horton.

In the current and foreseeable future, this form of deception will be common in both criminal and adversary operations due to the ease of acquiring high-quality intelligence on personal patterns of behavior. The opacity of the cybered world's technical substrate encourages the forms of deception identified by Don Fallis: misleading the targeted audience about the accuracy of the information, misleading the targeted audience about the identity of the source of that information, and misleading the audience about the implications of the information if it is accurate.[13] While the opportunity to exploit the credulous existed in personal letter scams and broadcast marketing prior to the advent of the internet,[14] only with the global spread of the cyberspace substrate can self-deception be exploited to such a pervasive extent.

Because distance is fundamentally irrelevant in modern cybered conflict, two types of wartime deception emphasized by Latimer have assumed increased importance: "the ambiguity-increasing variety (hinder identity of the true aim long as needed) or misleading (force someone to focus on one option or explanation over all others) variety."[15] First, ambiguity is easily achieved at great distance. At any proximity from five to twenty thousand miles away, attackers can use techniques to quickly, deceptively, and opaquely locate what once was high-value personal information about individuals, groups, organizations, or national governments.[16] Deception through the exploitation of openly available information is encouraged in this form of competition. In an open democracy, finding out that Horton loved Beezlenuts from his patterns of purchase, his texts (trunk-enabled special equipment), and his browsing behavior would have taken hardly any effort at all for the Kwuggerbug. Furthermore, the high grade of intelligence acquired from those indicators would enable the deceiving actor to conclude that Horton was stubborn and inclined to be obsessively attached to his decisions. Given his legalistic value system (which made Horton determined to follow through on a "deal" no matter what), the deception campaign hinged on exploiting Horton's credulity at the outset. However, this stratagem quickly developed into what is now called "lawfare,"[17] in which an aggressor uses the laws and law-abiding nature of the targeted victim against them. In this case, the Kwuggerbug holds his victim to his "word." Because honoring ownership is a key value legally protected in consolidated democracies, Horton's determination to honor his commitments proves useful as a strategic ploy to keep him from questioning the Kwuggerbug's veracity, despite Horton's increasing awareness of the difficulties.

Second, the bug used ambiguity and active misdirection throughout the operation. While stating that the tree was "far off," the Kwuggerbug did not identify how far away the tree was located. Nor, as they progressed on the journey, did the bug clarify how much farther they needed to travel. Only when physically arriving at the shore did Horton realize that they had to cross a lake "thirty miles long" and filled with crocodiles. His protests were met with reminders of his commitment to a "deal" in which he was only "the muscle." Having barely survived the treacherous lake, Horton was informed by the Kwuggerbug that the next step was to climb a tall mountain a "thousand feet high." When Horton protested that he could not climb such a colossal mountain, the bug reminded Horton of the delicious nuts and his commitment to their agreement. In modern states, this sort of lure-lawfare-lure cycle is common in phishing-fraud campaigns, online gambling, failed anti-corruption campaigns undertaken by the United States and its allies in Afghanistan, and in-game payment extractions from online game players.

Third, the deception operation used distorted information critical to the original commitment on Horton's part. The deceptive Kwuggerbug did not reveal his redefinition of the meaning of "half" the nuts. Instead of an equal division of the whole nuts, the Kwuggerbug was intending to claim his half in the form of nut meat, leaving Horton with nothing but shells. Of course the Kwuggerbug did not reveal this sneaky plan in advance. He waited until Horton and he arrived at the peak of the tall mountain, one elephant trunk's length away from the peak on which the Beezlenut tree sat. Horton then realized that only the bug could cross to the other peak, using Horton's extended trunk as a bridge. The bug clearly intended that Horton could not cross to the tree and thereby himself extract his half of the nuts. By inference, that fact was known to the bug undoubtedly through his extensive use of Google Earth, relatively granular satellite imagery, and research on the length of elephant trunks versus the gap between the two mountain peaks. Presumably he also knew elephants could neither jump that distance (or at all), nor reach far enough to touch the tree, nor squash a bug that had suddenly decided not to make good on a deal. As a result of all the data and analysis, the Kwuggerbug knew he only needed to continue the deception until he was across the gap between peaks and then no longer needed to pretend to share as promised.

Just as in interstate agreements, perpetrators need to integrate their deception with a longer-term plan to protect their advantage. Over time, the fifth offensive advantage—the ease of opaqueness—becomes less and less necessary. Although opaqueness may enable the use of a particular tool or the achievement of a particular objective, pure anonymity is unnecessary when the victims are constrained by their own legal system from striking back due to the passage of time or lack of certainty in data or assumptions. In this case, the duplicitous Kwuggerbug did not tell Horton that he had redefined the "half" until they reached the peak. At that point, deception was no longer necessary because the Kwuggerbug knew that Horton's rule-abiding nature would prevent him from objecting. The Kwuggerbug then declared the nuts could not be shared until they were shelled into two mounds, one with nut meats and one with shells. Once shelled, he then made good on his promise to give Horton half by giving him the half that was only shells. By the thinnest legalistic justifications, the Kwuggerbug thus fulfilled the "deal." As noted in the Dr. Seuss text, the fact that Horton received half of the nuts meant he "couldn't complain."

The Kwuggerbug's deception campaign ends with a stratagem similar to one used by states. Vague language is sometimes used in treaties and agreements to create ambiguity, thus preventing the aggrieved party from claiming that there has been a clear violation. In the emergent, post-western, and deeply digital world, this form of deceptive deal-making is facilitated by the general difficulty of observing what is occurring across the networks of the world. For example, the 2015 agreement—without enforcement mechanisms—between US president Barack Obama and Chinese president Xi Jinping stipulated that neither state would use military forces to steal economic intellectual property from the other.[18] As it happens, most democracies do not use their military cyber units for that purpose, making the agreement easy for the Americans to adhere to since it required no actual changes in behavior. On the other side, Chinese state officials knew that another civilian agency—the Ministry of State Security (MSS)—played a role in carrying out cyberattacks for economic gain, making the agreement easy for the Chinese to adhere to since it required no actual changes in overall behavior.[19]

Applying the lessons of Dr. Seuss's story to the current world of international security, one could argue that deceptive manipulation is just under the threshold of armed conflict of adversaries such as China and Russia and has increased the hesitation of the United States and its allies with respect to engagements and economic ties with these countries. Duplicitous treaty agreements are well-established fare in the annals of diplomatic history, but the state-sponsored link to massive, digitized transfers of wealth from democracies to adversary states or proxies (estimated to total 1 to 2 percent of annual GDP of consolidated democracies) has meant deception via the cyberspace substrate is changing the power and wealth distribution globally. Just as the Kwuggerbug's cheating prevented Horton from obtaining the nuts for which he risked so much, these deceptive diplomatic efforts have real costs on the ground. They fail to curtail bad state cybered behavior and fail to provide the democracies with the stability or security of the digital world that their future well-being requires. With no consensus regarding the indicators of either deceit or enforcement of agreements, the victim states must turn to their own cybered defense and offense. In short, the ease of deception across the ubiquitous and shoddy cybered substrate, along with the other four offense advantages, has induced the "unwelcome militarization of the internet,"[20] and despite the early dismissals, the greater possibility of escalation toward cyber-physical conflict.

THE GRINCH DECEIVES THE HOOBUBS . . . AGAIN

Deception has been a component of the operation of economic markets throughout history. However, economists view deception as an ugly concept, and (just like international security scholars) prefer not to discuss it. As George A. Akerlof and Robert J. Shiller note in their seminal 2015 book, economists simply do not want to deal with the messiness of recognizing that people, institutions, and tools lie to themselves and others.[21] Lying—aka marketing—influences whole populations to act in ways contrary to their interests, thereby making the clean predictions of a multitude of economic and statistical models inaccurate at best.

In "The Hoobub and the Grinch," Dr. Seuss captures the beauty of a persuasive use of self-deception and deceptive misinterpretation of reality in a recurring campaign for economic gain. Like Horton, the Hoobub is credulous and apparently from a community prone to open acceptance of deceptive appeals.

The Grinch persuades the Hoobub, who is lounging in the sun (which of course can be read as an indictment of those states that neglect their resilience to deception) that a piece of green string—which the Grinch is generously willing to sell—offers more value than the sun. By reverse logic, the Grinch notes that the string has more versatility, more endurance year-round, and better strength for personal uses, and poses less danger to skin and health than the sun. The catch of course is that the Hoobub must buy it off the Grinch, who feigns some reluctance to part with the colossal, immense green string (in capital letters!) for merely "ninety-eight cents." And the Hoobub bites . . . apparently repeatedly. As Dr. Seuss notes, "Grinches sell Hoobubs such things every day."

There are two classic elements of deception in the story. First, the Grinch deceives the Hoobub by manipulating his perception of reality. Barton Whaley's 1969 report on deception and war defines "deception as information . . . intended to manipulate the behavior of others by inducing them to accept a false or distorted perception of reality— . . . physical, social, or political environment."[22] Indeed, nothing the Grinch said in his sales pitch for string (aside from the bombast at the end) was inaccurate.[23] If one puts a string in one's pocket and leaves it there, then it will indeed be there every day of the year; be as strong as it was when one put it in there; and never cause skin cancer or freckles, or come close to broiling a person "like fat." The Grinch did not lie to the Hoobub. Rather, the Grinch's rhetorical presentation of the information (e.g., using declarative statements, making false comparisons to the sunlight of a distant star, focusing on the 98 cent bargain) indirectly creates a false belief. Robert M. Clark and William L. Mitchell observe that "deception is a process intended to advantageously impose the false story on a target's perception of reality."[24] In short, the Grinch has clouded the Hoobub's perception of reality, resulting in a willing suspension of disbelief and a corresponding shift in behavior that benefits the Grinch. In modern terms, some similarities to the deception campaigns conducted particularly well through open social media outlets by Russian operatives against democratic elections resonate here.

The second classic element of deception found in Dr. Seuss's story is that the Grinch's deception causes the Hoobub to act against his own self-interest. Akerlof and Shiller, in their study of modern economics, noted that deception was getting someone to act in the interest of the deceiver rather than in their own interest.[25] The Grinch is able to deprive the Hoobub of some of his wealth (to be precise, 98 cents) for an item that has no identifiable value in any market, not even to the Hoobub at the outset of the sale. Yet by the end of the discussion the Grinch is able to change the Hoobub's sense of his own self-interest long enough to affect the transfer of some wealth to the Grinch. Systemic economic analysis shows that the Hoobub's interests are not sustained in this transaction, but the Grinch's interests are well-served. There is no mention of paybacks, return policies, negotiated changes in the terms of the transactions, or even a use that the Hoobub later discovers for the otherwise valueless (within his sociotechnical-economic system at least) piece of string. Rather, the picture suggests the Grinch simply nets a transfer of wealth with no other expenditure than a well-spun, deceptive tale. Contemporary Russian use of social engineering campaigns to trick individuals in democratic nations into clicking on poisoned hyperlinks that allow access and removal of wealth from victims has considerable resonance here. The comparison between the resources spent by the deception campaign's attackers and the losses in resources by the victims are similar to the value asymmetry in the exchange of the Grinch and the Hoobub. Deception pays, and it pays systemically in the global order as well as in the world of Dr. Seuss.

The "charm" campaigns of various nation-states provide another example of how deception can be used to overcome a victim's natural suspicions and cause them to act against their own self-interest.[26] For example, both Russian and Chinese state-sponsored and non-state-sponsored actors have engaged in extensive campaigns to persuade corporate and developing nation victims to accept distorted information about potential, production, or future access to key markets.[27] For example, Huawei has publicly declared for years that it has no connection to the Chinese government and that it abides by all the laws of every state in which it operates. In recent years, persuasive reports have shown that neither assertion is true. Huawei, like all Chinese major corporations, has long had a Chinese Communist Party cell in its headquarters and offers its employees bonuses for extracting intellectual property from westernized firms with whom it partners or for whom it provides raw materials.[28] The difficulty for democratic states, leaders, and even the corporations bound by the established democracies' rule of law is that this kind of systemic deception at scale by a peer adversary throughout the cyberspace substrate is exceptionally difficult to withstand without a national strategy to counter and employ deception as well.

MARCO COMES LATE AND SPINS A TALE . . .

In the third story of the collected volume, "Marco Comes Late," Dr. Seuss presents deception as an attempt that fails to deceive but achieves its aims nonetheless by impressing its intended audience. Arriving two hours late to school, the boy Marco weaves a fantastic tale of how a bird laid an egg on the arithmetic book balanced on his head. He claims that to save the baby bird's life, he had to wait for the egg to hatch. To break that dear egg would be terribly cruel. / An egg's more important than going to school. / That egg is that mother bird's pride and her joy. / If he smashes that egg, he's the world's meanest boy!" Interspersed throughout the story are characters such as a worm couple and a cat couple who argue about the relative ethical value of being on time to school or saving a bird's life. Marco's own sense of his innate goodness as a human being depends on his allowing the egg to rest undisturbed on his head despite the rule-breaking consequence. "So I stayed where I was / With the egg on my head, and my heart full of fears / And the shouting of cats and of worms in my ears." Marco's story brings to mind Hank Pruncken's work on intelligence and counterintelligence, in which he notes that deception "present[s] a situation that resembles what one might expect to exist in a particular setting, but in fact is a pure fiction."[29] In the Dr. Seuss story, the irked teacher does not believe the false retelling of events but calls it a "very good tale" in recognition of its creativity. She asks Marco if it is true, and he admits "not quite all." He notes that he did actually see a worm, and the story ends with no negative consequences for either being late or lying about the circumstances.

Deception in this latter case succeeds not because Marco's teacher believes it but because it is done with such considerable novelty and flair that those authorities with the power to punish the perpetrator stay their hand in admiration or confusion, or a bit of both. The available inference is that lying—if creative enough—is mildly admirable, even if not believable, and that the ambiguity of the reception achieves at least a temporary reprieve for the deceiver. As recently as 2020, the condemnation of a Russian cyber group's elaborate deceptive heist of SolarWinds was suffused with grudging admiration, though not forgiveness, by American and European cyber defenders for the cleverness of the perpetrators.[30] In contrast, the Chinese attack on a broad swath of Microsoft Exchange Servers across the United States and its allies a few months later was perceived by the same democratic cybersecurity community as indiscriminate and inelegantly executed. The Chinese campaign exposed a great number of organizations not only to data loss but also to devastating ransomware attacks by a multitude of bad actors beyond the Chinese perpetrators.[31] Both the Chinese and Russian campaigns were highly deceptive, yet the inelegance of the Chinese campaign caused a greater sense of outrage in the victims than the Russian campaign. More than one commentator urged US president Joseph Biden to respond more harshly to China than Russia despite the very likely much greater targeted loss of data from the Russian exploitation.[32] The Biden administration has indeed deepened sanctions against Russia as a response to the SolarWinds attack, but the publicly stated reason for the increase in sanctions is past-due debt payments and election manipulation campaigns through social media.[33]

RESILIENCE IN "HOW OFFICER PAT SAVED THE WHOLE TOWN"

Horton and the Kwuggerbug contains a fourth story that is not about deception per se but rather about resilience. Resilience comes from thinking beyond first- and second-order effects to fifth- and sixth-order consequences, or (as one might say in complex adaptive system studies) following the potential cascade or the "deviant amplitudes."[34] "How Officer Pat Saved the Whole Town" illustrates just such forward imagining of consequences across complex systems in order to act appropriately and at what point to avert much worse rippling events. In the story, Officer Pat sees a gnat about to land on and bite a sleeping cat, who will jump up with a loud yowl, startling passing babies, whose cries will scare a flock of birds into sudden flight, shocking a fishmonger to toss his fish onto a passing horse, whose wagon will tip and cause a maintenance man to wrench open a huge water main, and so on. The "trouble with trouble [is that] it grows and grows," in this case through fourteen cascading evenings until, finally, a sudden horn distracts a passing dynamite truck driver who hits a tree and blows up the town "to small bits." Being alert on his beat and knowing his resilience lessons well, Officer Pat can think through the consequences of allowing a gnat to bite a sleeping cat. He therefore swats the gnat at just the right moment, the cat never wakes, and a long series of disastrous events never occurs.

"How Officer Pat Saved the Whole Town" shows us how modern democratic states need to develop a long-linked, complex view of likely consequences. Democracies must think through how to strategically employ deception in their national defense strategies, in large part because it is a central and growing feature of those of their adversaries.

CONCLUSION: DECEPTION IS A STRATEGY THAT PAYS

In the stories of Dr. Seuss, the effectiveness of deception is time dependent. That is, the longer the victims remain unaware of the falseness of the imposed reality, the more likely the success of the operation. For the Kwuggerbug, deception needed to last long enough for the bug to get to the nuts and move out of reach of Horton's wrath, should the beleaguered elephant strike back. For the Grinch, the deception only needed to last until the Hoobub paid, but it was in the Grinch's interests to keep things pleasant so he could return often to the same victim for more wealth transfers. For Marco, deception needed to last until the teacher indicated through her smile that she was disarmed. In the international system (as in the tales of Dr. Seuss), the longer the deception continues, the more drastic the consequences. The 2020 SolarWinds cyber campaign conducted by the Russian-state-sponsored offensive cyber group against firms and agencies in the United States deceived its victims about its exploitations for more than nine months and was thereby able to do considerable damage.[35]

In the past no single spy could have the kind of reach and retrieval demonstrated by Russia in the SolarWinds operation, which delivered enormous stocks of intelligence and marketable knowledge to be stored for future use. As this chapter has argued, the inherent structural characteristics of cyberspace enable manipulation on a wide scale; at any distance; and across myriad tools, targets, and objectives. Like the Russian SolarWinds operation, cyber campaigns are predominantly sub-rosa, rarely reaching the level of overt conflict. Just as in the era of the Cold War, states now eschew overt military exchanges. Instead, states now use deception to achieve strategic advantages. Cyberspace, as the universal medium of interaction, encourages the rise of deception to levels of strategic consequence.[36]

Despite the clear importance of deception as a political-military tool in the present era, westernized democracies have not paid enough attention to the role of deception in the cybered world. For example, allied democracies missed an opportunity to employ deception in a variety of circumstances between 2010 and 2020 to cause their adversaries to pause, unsure of the consequences of their malicious operations. Likewise, the importance of deception has been ignored in international relations and security studies. In 2010, for example, senior public scholars were claiming that the warnings about cybered conflict were merely threat inflation.[37] The past decade has been wasted persuading key scholars of the importance of cyberspace and deception. Had both senior political leaders and senior scholars been able to grasp the implications earlier, the intervening ten years might not have been spent futilely trying to reach agreement on cyber deterrence.

Time is of the essence for the development of a national strategy to counter and employ deception. Consolidated democracies are ill-advised to spend another ten years thinking about how the world may have changed with the advent of cyber and the rise of China. The neglect, bias, complacency, and fear of technology must change in order for the consolidated democracies to face the world as it is, not as it was during the post–Cold War period. "As deterrence became foundational in the nuclear era, deception should rise in prominence in a world that increasingly depends on technology to mediate interaction."[38] While some critics might argue that it is unseemly for a superpower to engage in vulgar acts of deception, the refusal to employ deception on moral grounds despite its tangible strategic benefits will certainly have grave consequences for the systemic resilience of democracies.[39]

Today a critical question must guide the development of any national strategy for deception in the cybered world. How should the consolidated democracies defend themselves across their collective sociotechnical-economic systems in the face of an authoritarian-leaning international community strongly influenced by one enormous central state well-skilled in deception in political, economic, and technological strategies? The first step these democracies must take in their own defense is understanding the fundamental components of deception. The Dr. Seuss stories discussed in this chapter do much more than demonstrate to children how commonplace deception is in everyday life. Rather, they serve as a warning about how adversaries can distort reality in the digitized world for strategic effect. They contain pertinent examples of deception in the emerging, deeply digitized international system, including wishful self-deception, distorted information, and deliberate efforts at creating ambiguity. Current and future generations of practitioners and scholars must recognize deception as an element of both national power and vulnerability in the emerging era of great systems conflict.[40]

NOTES

The ideas contained in this piece are solely those of the author and do not reflect the views or positions of any element of the US government.

1. Chris C. Demchak, "Achieving Systemic Resilience in a Great Systems Conflict Era:Coalescing against Cyber, Pandemic, and Adversary Threats," *Cyber Defense Review* 6, no. 2 (Spring 2021), https://cyberdefensereview.army.mil/Portals/6/Documents/2021_spring_cdr/COVID_CDR_V6N2_Spring_2021_r4.pdf.

2. Thomas J. Froehlich, "A Not-So-Brief Account of Current Information Ethics: The Ethics of Ignorance, Missing Information, Misinformation, Disinformation and Other Forms of Deception or Incompetence," *BiD*, no. 39 (December 2017), https://bid.ub.edu/en/39/froehlich.htm.

3. Adda Bruemmer Bozeman, *The Future of Law in a Multicultural World* (1971; Princeton, NJ: Princeton University Press, 2015).

4. Jon Latimer, *Deception in War: The Art of the Bluff, the Value of Deceit, and the Most Thrilling Episodes of Cunning in Military History from the Trojan Horse to the Gulf War* (New York: Abrams Press, 2003).

5. George A. Akerlof and Robert J. Shiller, *Phishing for Phools: The Economics of Manipulation and Deception* (Princeton, NJ: Princeton University Press, 2015); and J. L. Casti, *Complexification: Explaining a Paradoxical World through the Science of Surprise* (New York: Abacus, 1994).

6. Robert M. Clark and William L. Mitchell, *Deception: Counterdeception and Counterintelligence* (Washington, DC: CQ Press, 2018). See the introduction discussion, especially p. 6.

7. William Hutchinson, "Information Warfare and Deception," *Informing Science* 9, no. 20 (2006), https://doi.org/10.28945/480.

8. Clark and Mitchell, *Deception*.

9. Chris C. Demchak, "Resilience, Disruption, and a 'Cyber Westphalia': Options for National Security in a Cybered Conflict World," in *Securing Cyberspace: A New Domain for National Security*, ed. Nicholas Burns and Jonathon Price (Washington, DC: Aspen Institute, 2012).

10. Chris C. Demchak and Peter J. Dombrowski, "Rise of a Cybered Westphalian Age," *Strategic Studies Quarterly* 5, no. 1 (March 1, 2011): 31–62.

11. Froehlich, "Not-So-Brief Account."

12. Froehlich, "Not-So-Brief Account," 142.

13. Don Fallis, "The Varieties of Disinformation," *The Philosophy of Information Quality*, ed. Luciano Floridi and Phyllis Illari (Cham: Springer International Publishing, 2014).

14. Akerlof and Shiller, *Phishing for Phools*.

15. Latimer, *Deception in War*.

16. Chris Demchak, "Defending Democracies in a Cybered World," *Brown Journal of World Affairs* 24, no. 1 (2017): 139–58, https://www.jstor.org/stable/27119084.

17. Wouter G. Werner, "The Curious Career of Lawfare," *Case Western Reserve Journal of International Law* 43 (2010): 61–72.

18. Joseph S. Nye Jr., "Deterrence and Dissuasion in Cyberspace," *International Security* 41, no. 3 (2016): 44–71.

19. Robert Williams, *The "China, Inc." Challenge to Cyberspace Norms* (Stanford, CA: Hoover Institution, 2018).

20. Jonathan Zittrain, "'Netwar': The Unwelcome Militarization of the Internet Has Arrived," *Bulletin of the Atomic Scientists* 73, no. 5 (2017): 300–304.

21. Akerlof and Shiller, *Phishing for Phools*.

22. Clark and Mitchell, *Deception*.

23. Roderick M. Chisholm and Thomas D. Feehan, "The Intent to Deceive," *Journal of Philosophy* 74, no. 3 (1977): 143–59.

24. Clark and Mitchell, *Deception*, 9.

25. Akerlof and Shiller, *Phishing for Phools*.

26. Robin Emmott and Ben Blanchard, "Wary of Trump, China Launches EU Charm Offensive: Diplomats," *Reuters*, March 28, 2017, http://www.reuters.com/article/us-eu-china-idUSKBN16Z22S.

27. Andrew Scobell et al., *China's Grand Strategy: Trends, Trajectories, and Long-Term Competition* (Washington, DC: RAND, 2020); and Thomas Graham, "Let Russia Be Russia," *Foreign Affairs* 98 (2019): 134.

28. Gordon Corera, "Huawei: MPs Claim 'Clear Evidence of Collusion' with Chinese Communist Party—the Inquiry Focused on Huawei's Telecoms Kit Rather Than Its Consumer Handset Business," *BBC*, October 7 2020, https://www.bbc.com/news/technology-54455112; Cho Mu-Hyun, "Huawei Regularly Tried to Steal Apple Trade Secrets: Report," *ZDNet Online*, February 18, 2019. https://www.zdnet.com/article/huawei-regularly-tried-to-steal-apple-trade-secrets/; and Nick Bastone, "Chinese Electronics Giant Huawei Allegedly Offered Bonuses to Any Employee Who Stole Trade Secrets," *Business Insider*, January 28, 2019. https://www.businessinsider.com/huawei-indictment-trade-secrets-2019-1.

29. Hank Prunckun, *Counterintelligence Theory and Practice* (Rowman & Littlefield, 2019), 187.

30. Liam Tung, "Microsoft: This Is How the Sneaky Solarwinds Hackers Hid Their Onward Attacks for So Long—the Solarwinds Hackers Put in 'Painstaking Planning' to Avoid Being Detected on the Networks of Hand-Picked Targets," *ZDNet*, January 21, 2021, https://www.zdnet.com/article/microsoft-this-is-how-the-sneaky-solarwinds-hackers-hid-their-onward-attacks-for-so-long/.

31. Dan Goodin, "There's a Vexing Mystery Surrounding the 0-Day Attacks on Exchange Servers—a Half-Dozen Groups [Including Hafnium] Exploiting the Same 0-Days Is Unusual, If Not Unprecedented," *Ars Technica*, March 11, 2021, https://arstechnica.com/gadgets/2021/03/security-unicorn-exchange-server-0-days-were-exploited-by-6-apts/.

32. Dmitri Alperovitch and Ian Ward, "How Should the U.S. Respond to the Solarwinds and Microsoft Exchange Hacks?," *Lawfare Blog*, March 12, 2021, https://www.lawfareblog.com/how-should-us-respond-solarwinds-and-microsoft-exchange-hacks.

33. Alex Scroxton, "Biden Sanctions Russia over Solarwinds Cyber Attacks—US President Imposes New Sanctions on Russia Following Malicious Cyber Attacks against the US and Allies," *Computer Weekly*, April 15, 2021, https://www.computerweekly.com/news/366551317/UK-and-US-slap-fresh-sanctions-on-Conti-ransomware-crew.

34. L. S. Sproull, S. B. Kiesler, and S. Kiesler, *Connections: New Ways of Working in the Networked Organization* (Cambridge, MA: MIT Press, 1992).

35. Steven J. Vaughan-Nichols, "Solarwinds: The More We Learn, the Worse It Looks," *ZDNET*, January 4, 2021, https://www.zdnet.com/article/solarwinds-the-more-we-learn-the-worse-it-looks/.

36. Emily O. Goldman, "The Cyber Paradigm Shift," in *Ten Years In: Implementing New Strategic Approaches to Cyberspace*, ed. Emily Goldman, Jacqueline Schneider, and Michael Warner (Newport, RI: US Naval War College Press, 2020).

37. Stephen M. Walt, "Is the Cyber Threat Overblown?," *Foreign Policy*, March 30, 2010.

38. Erik Gartzke and Jon R. Lindsay, "Weaving Tangled Webs: Offense, Defense, and Deception in Cyberspace," *Security Studies* 24, no. 2 (2015): 316–48.

39. A more complex ground for objection is that the use of deception in the context of an open democracy, no matter what the aim, may unintentionally result in the very outcome one is trying to avoid: the destruction of democracy itself. However, using deception as needed in the course of strategic actions for national security and resilience does not inevitably mean becoming deceptive as a people or as a nation.

40. Demchak, "Achieving Systemic Resilience."

11

Did I Ever Tell You How Lucky You Are? and Luck in Warfare

Erich Henry Wagner and Montgomery McFate

When Maj. Theodor Geisel went to the front lines of the European theater to screen a film for the approval of US Army top brass in December 1944, he did not expect to end up trapped for three days in the largest battle fought on the Western Front during World War II, the aptly Seussian-termed Battle of the Bulge. Indeed, this would become the third deadliest campaign in American history.[1] Major Geisel, the commander of the animation department of the 1st Motion Picture Unit, arrived in Bastogne only a few hours before the onslaught of Hitler's Ardennes offensive. The soldiers of the 106th Infantry Division called this quiet and uneventful sector the 'Ghost Front.' Within days, the 106th would be destroyed, and ten thousand men of the division would surrender, in the largest battlefield capitulation of American troops in the European War.[2] Geisel credited his survival to the fact that "I got there in the early morning and the Germans didn't arrive until that night." "Nobody came along and put up a sign saying, 'This is the Battle of the Bulge,'" he would later reminisce. "How was I supposed to know? I thought the fact that we didn't seem to be able to find any friendly troops in any direction was just one of the normal occurrences of combat."[3] Finding himself ten miles beyond friendly lines, cut off, and in the freezing rain, Geisel overheard two griping GIs bellyaching: "Rain, always rain! Why can't we have something different for a change?"[4]

When British forces rescued him three days later, Geisel recalled to his biographers with characteristic flare: "The retreat we beat was accomplished with a speed that will never be beaten."[5] Perhaps Geisel thought back to these cold and scary days when, using the pen name Dr. Seuss, he published *Did I Ever Tell You How Lucky You Are?* in 1973. The book is an admonition to children (and perhaps a reminder to adults) that no matter how bad a day you might think you're having, there's always someone who has it worse. We can just imagine the Old Man in the Desert of Drize admonishing:

> Just tell yourself, Duckie, you're really quite lucky . . .
> oh, ever so much more lucky . . .
> Not to be in alone. . . . Surrounded by Nazis
> In the freezing rain of Bastogne!

It is safe to assume Dr. Seuss would agree he was pretty lucky to have emerged unscathed from the Wehrmacht's last major offensive in the West.

Geisel firmly believed in luck and attributed one of the most significant events in his life to this force. Geisel credited fate for the chance encounter with his old Dartmouth classmate Mike McClintock. His friend "picked [him] off Madison Ave" and asked him what he had under his arm:

> I said, "A book that no one will publish. I'm lugging it home to burn." Then I asked Mike, "What are you doing?" He said, "This morning I was appointed juvenile editor of Vanguard Press, and we happen to be standing in front of my office; would you like to come inside?" So, we went inside, and he looked at the book and took me to the president of Vanguard Press. Twenty minutes later we were signing contracts. That's one of the reasons I believe in luck. If I'd been going down the other side of Madison Avenue, I would be in the dry-cleaning business today![6]

Thanks to this chance encounter with his friend, Geisel's first work *And to Think That I Saw It on Mulberry Street* was published in 1937 under the pen name Dr. Seuss. The book earned strong reviews as a "highly original and entertaining" work, thus beginning Dr. Seuss's influential and prolific career as a children's books author-illustrator.[7]

Luck was important in Dr. Seuss's civilian career and also during his military service. For someone who had once sworn that if he ever saw action he would "grab [his gun] by the barrel and throw it," Geisel was certainly lucky to have emerged unscathed from the Battle of the Bulge.[8] Indeed, the concept of luck in war is one of the most constant and enduring themes echoing through the memoirs, letters, and reminisces of those who have survived the experience of combat. The concept of luck in warfare is culturally agnostic, spanning from the battlefields of antiquity to the present day. Political philosopher Hannah Arendt maintained that "nowhere else does Fortune, good or ill luck, play a more fateful role in human affairs than on the battlefield."[9] Regardless of the label used (whether luck, fortune, chance, fate, divine intervention, karma, or kismet), thousands of warriors who have avoided death, sidestepped a disaster, achieved victory, or gained glory attributed their success to the intervention of the fickle Fates. For many practitioners, the outcome of violence on the field resulted in part from the interplay of luck and chance.

Luck may be important to soldiers, but it is apparently not important to the US Army. The US Army's current leadership doctrine, Field Manual 6-22, *Developing Leaders*, does not mention luck.[10] On the other hand, the earlier, superseded Field Manual 22-100, *Army Leadership* (1999), simply stated: "Failing through want of experience or luck is forgivable."[11] However, the manual did not recognize luck as a value, attribute, skill, or action officially associated with a professional officer. This lack of recognition stems from the Army's (and everyone else's) inability to define or quantify luck.[12]

Even the grand master of strategy, Carl von Clausewitz, struggled with the notion of luck. In *On War*, Clausewitz described war as a trinity consisting of "primordial violence, hatred, and enmity, which are to be regarded as a blind natural force; *of the play of chance and probability within which the creative spirit is free to roam*; and of its element of subordination, as an instrument of policy, which makes it subject to reason alone."[13] When combatants engage each other in the battlespace, an infinity of petty circumstances accumulate to make even the most simple thing difficult. "This tremendous friction," the Prussian observed, "which cannot, as in mechanics, be reduced to a few points, is everywhere in contact with chance, and brings about effects that cannot be measured, just because they are largely due to chance."[14] Clausewitz offered a few examples of how chance and friction intersect. Fog may prevent the enemy from being discovered or hinder a report from reaching the general. Rain may prevent a battalion from arriving on time or stop the cavalry from charging because it is stuck in the mud. Any random event might be the proximate cause of victory or defeat, yet it cannot be measured, quantified, or controlled. "No other human activity is so continuously or universally bound up with chance," wrote the general. "And through the element of chance, guesswork and luck come to play a great part in war."[15]

The theory of luck inherent in *Did I Ever Tell You How Lucky You Are?* aligns well with Clausewitz's conception of luck. Dr. Seuss's book begins with the voice of the narrator: "When I was quite young / and quite small for my size, / I met an old man in the Desert of Drize." Perched upon a "terrible, prickly" cactus, the old man in the desert wears a long blue robe and a chunky necklace of beads. This wise oracle tells the narrator (who is depicted a young boy): "When you think things are bad, / when you feel sour and blue, / when you start to get mad . . . / you should do what *I* do! / Just tell yourself, Duckie, / you're really quite lucky! / Some people are much more . . . unlucky than you!"

Presumably, many children who read *Did I Ever Tell You How Lucky You Are?* learn that many events in one's life cannot be predicted or controlled and that all you can control is your attitude toward the situation. However, if we interpret the book from the perspective of strategic theory, we see that Dr. Seuss understood the role of friction in human affairs (e.g., "things are bad"), the "primordial violence" in Clausewitz's trinity (e.g., "you start to get mad"), and the "element of subordination" that makes warfare "subject to reason [law, rationality]" (e.g., "you should do what I do!"). Beginning with this framework, the author of *Did I Ever Tell You How Lucky You Are?* presents a variety of scenarios that represent how luck, emotion, and constraint intersect in warfare. In other words, Dr. Seuss's text is a highly sophisticated analysis of Clausewitz's theory of warfare.

THE BUNGLEBUNG BRIDGE AND THE DANGERS OF WAR TECHNOLOGY

At the beginning of *Did I Ever Tell You How Lucky You Are?* the reader is thrust into a bustling world of workhands on Bunglebung Bridge, a precarious structure spanning a waterway overflowing with hazards. The narrator is told, "Be glad you don't work on the Bunglebung Bridge / that they're building across Boober Bay at Bumm Ridge. / It's a troublesome world. All the people who're in it / are troubled with troubles almost every minute." Danger is ubiquitous on Bunglebung Bridge, and safety standards are incredibly lax: a man stands on the very edge of a crane with no safety harness, a worker wheels a cartload of bricks that have not been secured, a boat captain has tied up his boat to a structurally unsound piling under the bridge, and so on. Chaos abounds.

Let us recall that Theodor Geisel was trapped at Bastogne during the Battle of the Bulge in 1944. Bastogne lies at the confluence or origin of a number of rivers, including the Meuse, the Clert, the Our, the Wiltz, and the Sure. Naturally with so many rivers in the vicinity of Bastogne, control of the bridges became a critical objective of both the Allies and the Nazis. During the battle, German engineers built bridges over the rivers in a matter of hours, and US engineers blew them up. Seen in the context of Dr. Seuss's personal experience, the meaning of Bunglebung Bridge in *Did I Ever Tell You How Lucky You Are?* swiftly becomes apparent.

Bunglebung Bridge can be read as a metaphor for the role of technology in warfare. Technology has not improved safety conditions for laborers on the bridge; it has only made their work more perilous. Similarly, technology has also made the military profession more perilous and made accidents more lethal. Although our ancient warrior ancestors faced dangers in the profession of arms, any mishaps occurred from relatively primitive weapons. Catapults could misfire; hot oil could fall on friendly troops. Because these weapons only had a limited lethality, accidents involving them killed or injured correspondingly fewer people. Since World War I, however, military technology (such as weapons, explosives, and platforms) has become increasingly lethal. Correspondingly, accidents have become more common and deadlier. "In the post-bourgeois world [following World War I]," historian Stephan Fritz asserted, "where the worker and the soldier were becoming identical and where technology resulted in the penetration of the dangerous into everyday life, combat seemed merely a factory job: the machines of destruction were fed and let loose, and the tight web of the machine and its routine determined a man's life."[16]

Why have military accidents become more common? First, the weaponry of contemporary militaries is so complex that a high degree of specialization is necessary. Artificial intelligence, weaponry, swarms of drones, and hypersonic arsenals require sizable crews of soldier-scientists to operate and manipulate them. According to then chairman of the Joint Chiefs of Staff, Gen. Mark Milley, at least forty to fifty highly complex new technologies (including unmanned systems and artificial intelligence) must be mastered by US personnel to prepare them for war.[17]

Second, this complexity leads to "normal accidents." Recent American military operations, for example, have been plagued by a wide variety of normal accidents, including explosions, collisions, and fratricide. For example, during the Gulf War, allied airplanes, tanks, and helicopters fired missiles and guns at friendly troops, inflicting 35 of 145 total American combat deaths.[18] Today, a disturbing rise in deadly military training accidents among armor and aviation units has lawmakers calling for public hearings on Capitol Hill. As sociologist Louis Hicks noted: Modern combat operations score so highly on complexity and interdependence, they can be expected to trigger structurally induced normal accidents. As military operations become more tightly coupled and more complicated, system accidents may become more frequent. "Some military technologies are sufficiently complex and tightly coupled for system accidents to be 'normal,' that is, inevitable."[19]

There are two basic types of military accidents: industrial accidents that occur in a military setting (such as nuclear reactor accidents that destroy submarines) and combat accidents whose character is peculiarly military, such as "friendly fire" accidents. While both types of these military accidents involve luck, Dr. Seuss seems to emphasize what could be categorized as combat accidents in *Did I Ever Tell You How Lucky You Are?*, and so we will retain that focus here.

The death and injury resulting from negligent discharges and fratricide frequently appear to be the result of bad luck simply because there is no other way to explain causation. For example, in 1800 "George Elers was in his colonel's tent when a random shot whizzed by, whereupon inquiry, it was found one of the Regiment's armorers was repairing a pistol when it discharged striking a regimental tailor and blinding him in both eyes."[20] In another instance, a soldier who lit his pipe after the fierce street fighting of Lucknow in 1857 ignited loose gunpowder, fatally wounding five other comrades, who died in great pain day later.[21]

While such military accidents involving negligent discharge and friendly fire are horrifyingly random, in aggregate they are a normal occurrence. According to a French general writing in 1921, French artillery caused 75,000 French casualties during World War I.[22] In 1968, the 1st Marine Division committed 323 accidental discharges, killing 40 and wounding another 309 men, which was more than twice the number of casualties inflicted in 1967 (156 wounded and 16 killed).[23] Even in Iraq, deaths from accidental gunshots in the Marine Corps increased every year between 2004 and 2008, including those involving "more senior, highly trained personnel."[24]

Combatants faced with the inevitability of random accidents involving highly complex and lethal machines often conceptualize bad luck not as a characteristic of themselves, but as something external to the self. For example, air crew in the RAF Bomber Command and the US 8th Air Force over Germany (which was plagued by accidents) often spoke of "gremlins," meaning the inexplicable, invisible forces that caused misfortune.[25] In 1943, for example, twenty-two-year-old RAF pilot Sidney Cohen was flying a Fairy Swordfish plane on a rescue mission when the compass "had a fit of the gremlins." Low on fuel, Cohen spotted the small Italian island of Lampedusa and landed his plane. Two Italian officers approached waving white sheets, surrendering the island to his crew. This is certainly the only historical example of an island surrendering to a plane, made even more extraordinary because of the role of luck.[26]

Similarly, airmen over Germany during World War II attributed luck not to themselves but to the machines they were flying. Some aircraft were referred to as "hangar queens." Certainly the poor maintenance records of these aircraft caused breakdowns, but perhaps luck played a role as well.[27] Complex systems, such as aircraft, tend to suffer more sudden failures that are not readily comprehensible to the system's users and operators, and which also thwart recovery and repair efforts.[28] Because they are incomprehensible in their complexity, these failures are often perceived by users as "luck."[29]

The plethora of accidents occurring on Bunglebung Bridge are the result of a host of complex, interdependent factors. While user error certainly plays a role, the technology on the bridge is incredibly complex and appears to be outside of human control. If we were able to ask the workers on Bunglebung Bridge the cause of their mishaps, no doubt they would attribute them to luck rather than to "normal accidents."

ZAYT HIGHWAY EIGHT AND THE DANGERS OF ROADS

In *Did I Ever Tell You How Lucky You Are?* the narrator is reminded that he would be very unlucky to live in Ga-Zayt and be "caught in that traffic / on Zayt Highway Eight!" The illustration shows a crossroad with heavy traffic coming from all four directions, with no traffic lights to control the flow. At the very center stands a young boy with red hair, wailing because his blue bicycle has been crushed by a camel pulling a load of barrels. One can see a pair of anxious elephants stuck in traffic with a decorated howdah; a man balancing fish tanks on his head; three yaks pulling a steamroller; and so on. To avoid the hazards of broken machines, falling watermelons, and the lash of a whip on Zayt Highway Eight, one requires luck.

Luck on the road is also critical for military organizations throughout history. Armies, like water, most often take the path of least resistance to move efficiently. These avenues of approach depend mainly on man-made roads and trails. Control of roadways can give a combatant a tactical advantage, but the concentration of forces along the road also creates an opportunity for the adversary. In recent conflicts, some roads have been particularly dangerous. Route Irish, for example, was the 7.5-mile section from Baghdad's Green Zone to the airport. Because there was no alternative route, coalition forces had to use this road and the surrounding areas. Route Irish, as a result, became a magnet for improvised explosive devices (IEDs). During the American-led occupation of Iraq it was labeled "the world's most dangerous highway."[30]

Few, if any, threats weigh more on a soldier's mind when on patrol than encountering an IED or landmine. Almost universally soldiers attribute luck and chance to their ability to survive encounters with IEDs. Stories about soldiers walking over IEDs and escaping because they were duds or misfired are commonplace. Company E, 2nd Battalion, 1st Marines encountered two dud IEDs on patrol in Afghanistan within a matter of days in 2010. "It was just by pure luck that [our IED] was made by one bad terrorist," one of the fortunate Marines testified. Finding the insurgents who placed the IEDs was also commonly attributed to luck. Marines of 2nd Battalion, 8th Regiment were searching for roadside bombs in 2009, when they intercepted a Taliban radio transmission saying the terrorists were "ready for the guests." "Guests" was the Taliban codeword for Marines. A relieved Marine noted, "We're lucky if we find them. Better than when they find us, I guess."[31] Military personnel in Iraq and Afghanistan often carried lucky charms (such as rabbit's feet) with them, just as soldiers of past eras did. Afghanistan veteran Jose De Matos, for example, was wearing his lucky charm wristband when he walked over two IEDs in Afghanistan in 2010–11, both of which failed to detonate.[32]

Mines and IEDs engender great terror, partly because they are invisible. Marine 1st Lt. Justin Gray said, "We can see the guys shooting at us, but we can't see the stuff in the ground. We have dogs and metal detectors, both great assets. Sometimes it's luck."[33] Mines and IEDs also cause fear among military personnel because they are an impersonal, inhuman threat (much in the way World War II pilots feared the impersonal flak more than the personal but more deadly fighters).[34] One armor officer in Vietnam believed "nothing is more terrifying in the early stages of combat than mines, booby traps, and unexpected rockets or mortars. It is much easier to face a known enemy and his fire than the unknown enemy."[35] IEDs provoke anxiety in part because there are no direct, affirmative actions that soldiers could take to avoid the weapon. Raleigh Trevelyan captures this feeling of inability to take action when he notes, "I don't mind a fighting chance, but I have a dread of mines."[36] Mines and IEDs pose an additional fear for soldiers because of the nature of the injuries they produce. As Philip Caputo notes, "The infantryman knows that any moment the ground he is walking on can erupt and kill him; [k]ill him if he's lucky. If he's unlucky, he will be turned into a blind, death, emasculated, legless shell."[37]

When dealing with IEDs, some military personnel describe luck as a finite resource. "Luck is like a sum of gold, to be spent," according to Gen. Edmund Allenby.[38] In a recent example, Lt. Col. Colin Whitworth trained explosive ordnance disposal (EDO) teams in Helmand and lost an arm to a poorly made IED when "luck was not on [his] side." In his view, luck was a limited resource. "You can be the best trained operator, and your luck may just run out. The more things you do by hand, and the more devices you deal with, your luck gets shorter and shorter and shorter."[39] The view of luck as a finite resource is ubiquitous among soldiers.

Not unlike the combatants in the Ardennes or coalition forces in Iraq, the protagonist in Dr. Seuss's book discovered that while roads may make travel easier, they can also pose significant hazards, including IEDs and terrible traffic. Because of the intractability of these thoroughfare hazards, travelers must rely on their luck to avoid the many snares at the crossroads of Zayt Highway Eight.

THE SCHLOTTZ AND THE WOUNDS OF WAR

Dr. Seuss invites the reader to "consider the Schlottz, / the Crumple-horn, Web-footed, Green-bearded Schlottz, / whose tail is entailed with un-solvable knots." The sad-looking creature has webbed feet like a duck and a long green beard that looks like moss. Judging by his crumpled horns and his knotted tail, it appears that the Schlottz has sustained considerable physical damage. The wounded Schlottz represents the experience of many fighters who have survived combat but sustained grievous physical and psychological wounds that persist throughout the veteran's life. These wounds are "un-solvable knots" because healing presents a significant challenge.

Generally civilians view wounded soldiers as unlucky. For many soldiers, however, certain types of wounds were actually considered to be fortunate. In an environment where an ambulance or a coffin seemed like the only escape from the fighting, a man with a serious but not dangerous wound was viewed as a "lucky bastard."[40] For the British in World War I, as Siegfried Sassoon recalled, the goal was a "blighty" wound, or one just serious enough to get sent back to England to recuperate. Dick Winters (of *Band of Brothers* fame) recalled that, "When a man was wounded, we felt glad for him, we felt happy for him. He had a ticket to get out of there, and maybe a ticket to get home."[41]

In Dr. Seuss's drawing, only the Schlottz's tail and horns have sustained injury. The specific locations of the Schlottz's wounds remind the reader that soldiers have distinct preferences for the type and location of combat wounds. The best location for a wound is the legs and arms, since the human body has two of each and therefore some functional redundancy. Audie Murphy told of a comrade whose luck held up when a shell fragment "simply clipped off half his right hand [but would] . . . not greatly affect his ability to hoist a bottle."[42] Hans Woltersdorf, similarly disenchanted, prayed "that if my leg was hit, it would be the left one, which was not of much use to me anyway." Indeed, he added laconically, "My prayer was answered."[43] Veterans often felt lucky when eyes, brain, abdomen, and genitals were not damaged. Marine Corps Col. R. Mark Wood recalled the collective expressions of relief from the staff at the nightly Commander's Update Briefs in Helmand Province, Afghanistan, when the day's casualty reports indicated only limbs on the unlucky wounded Marine had been lost; the genitals had remained intact.[44]

The crumpled horn and knotted tail of the Schlottz are a powerful metaphor for the wounds that combatants experience in combat, some lucky and some not.

HAWTCH-HAWTCHER BEE-WATCHER AND MILITARY INTELLIGENCE

In an illustration, the Hawtch-Hawtcher Bee-Watcher (an elderly man with glasses) stands in a field of scraggly plants with yellow blossoms watching a bee as it collects nectar. The job of the Hawtch-Hawtcher Bee-Watcher "is to keep both his eyes on the lazy town bee. / A bee that is watched will work harder, you see." Apparently the Hawtch-Hawtcher Bee-Watcher fails to successfully observe the insects gathering nectar, because "in spite of his watch, / that bee didn't work any harder. Not mawtch." In order to solve the problem of the lazy bee, a by-stander suggests, "Our old bee-watching man / just isn't bee-watching as hard as he can. / *He* ought to be watched by *another* Hawtch-Hawtcher! / The thing that we need / is a Bee-Watcher-Watcher!" The illustration on the next page shows the price of failure: additional watchers (with identical glasses, bald heads, and long red robes) have accumulated to watch the first bee watcher. "And today all the Hawtchers who live in Hawtch-Hawtch /are watching on Watch-Watcher-Watchering-Watch /Watch-Watching the Watcher who's watching that bee. / *You're* not a Hawtch-Watcher. You're lucky, you see!"

This tale of misfortune and misadventure might bring to readers' minds the Latin phrase *quis custodiet Ipsos custodes?* (who will guard the guards themselves?), commonly rendered as "who watches the watchmen?" Sometimes this phrase is used in the context of marital fidelity and other times to refer to political accountability. However, if one considers the wartime experience of Theodor Geisel, it becomes clear that the Bee-Watchers may very well refer to military intelligence. It was, after all, military intelligence that had *assured* Geisel "the 'safest' place would be the 106th Division front—a very calm sector," only hours before two hundred thousand Germans broke through and entrapped him.[45]

Military intelligence is the collection and analysis of information concerning the operational environment and the forces deployed therein (whether hostile, friendly, or neutral) in order to assist commanders in their decision-making processes. Here too, luck plays a powerful role. In Clausewitz's view, the collection of precise, relevant, and complete data on the battlefield lies within "the realm of uncertainty . . .[and] the realm of chance." "Whatever is hidden from full view in this feeble light," he surmised, "has to be guessed at by talent, or simply left to chance. So once again for the lack of objective knowledge, one has to trust to talent or to luck."[46]

Of the plethora of historical examples of how luck affects military intelligence, September 11, 2001, is undoubtedly the best-known to the general public. The story begins on July 10, 2001, when a Federal Bureau of Investigation (FBI) agent in the Phoenix Field Office identified a coordinated effort by Osama bin Laden to send "students" to the United States for civil aviation-related training. In his subsequent "Phoenix Memo" he expressed his suspicion that these men were learning to fly planes to conduct a terrorist attack. Unfortunately, neither the FBI's Bin Laden Unit nor the FBI's Radical Fundamentalist Unit saw the memo until after September 11. Although the FBI field office in New York received the memo, it took no action.[47] The Phoenix Memo exemplifies how bad luck can lead to missed opportunities even when correct intelligence is possessed.

As noted earlier in this chapter, one man's luck in warfare may be another's misfortune. One of the witnesses at the 2003 9/11 congressional hearings, Mindy Kleinberg, testified: "It has been said that the intelligence agencies have to be right 100 percent of the time and the terrorists only have to get lucky once. This explanation for the devastating attacks of September 11th, simple on its face, is wrong in its value." In her view, the 9/11 terrorists were lucky over and over again: "Is it luck when 15 visas are awarded based on incomplete forms? Is it luck when Airline Security screenings allow hijackers to board planes with box cutters and pepper spray? Is it luck when Emergency FAA and NORAD protocols are not followed? Is it luck when a national emergency is not reported to top government officials on a timely basis?"[48]

Of course luck is not the sine qua non of successful military operations. Preparation, persistence, courage, accurate information, and hard work build the foundation for a successful military outcome; luck merely increases the odds of a positive outcome. Indeed, the 1996 version of the US Army War College's *Battalion Commander's Handbook* noted, "Luck favors the soldier with better tactical intelligence."[49] This adage is exemplified by the hunt for Osama bin Laden. As US intelligence community officer Paul Pillar acknowledged:

> The manhunt for Bin Laden was not a straightforward matter of trying hard enough, and using good enough skill, to collect good enough intelligence. Given the nature and impenetrability of the target, it always depended on a lucky break or a slip-up by the quarry, which from our point of view is another form of luck. It was impossible to predict when such a break might occur. . . . Ultimately the identification of a courier was what brought the United States to Bin Laden's hide-out, but that also required luck, as well as several years of trying to develop further information.[50]

In the hunt for Bin Laden, several years of intelligence collection and analysis took place before a "lucky" break enabled US forces to identify the courier who led them to Bin Laden's lair in Pakistan. The difficult (and sometimes tedious) work of military intelligence professionals—which mostly involves *watching*—creates the preconditions for the intercession of luck. Like those military professionals in the intelligence field, the Hawtch-Hawtcher Bee-Watchers spend days upon end doing nothing but watching, with their hands folded patiently in front of them and no other distractions. Hawtch-Hawtch clearly illustrates that no matter how intently our intelligence professionals execute their mission, "time and chance happeneth to us all."[51]

POOGLE-HORN PLAYERS AND THE DANGER OF BROKEN TOOLS

In *Did I Ever Tell You How Lucky You Are?* Dr. Seuss reminds us of the bad luck of "the poor puffing Poogle-Horn Players, / who have to parade / down the Poogle-Horn Stairs / every morning to wake up / the Prince of Poo-Boken. / It's awful how often / their poogles get broken!" The illustration shows three horn players in blue uniforms riding very tall unicycles while playing giant horns that resemble tubas.

Like Geisel's Poogle-Horn players, militaries often go to war with the tools they have, not the tools they want. And often those tools simply do not work. The annals of warfare are replete with examples of unlucky soldiers and sailors who were provided tools that were antiquated, broken, or in other ways ineffective. For example, during the first eighteen months of World War II, the lives and success of US Navy submariners depended on the Mark XIV torpedoes, which proved to be at least 70 percent unreliable. In one instance, a sub commander fired fifteen torpedoes at a Japanese ship, scoring twelve hits, but with only one detonating.[52] The single torpedo that detonated caused only minimal damage to the target. Even worse, a faulty gyro system on these torpedoes caused the weapon to boomerang, sinking the submarine that fired it. During World War II, out of twenty-four such incidents submarines managed to evade the torpedo in twenty-two cases. The USS *Tang* and USS *Tullibee* were not so lucky, however, both succumbing to fishtailing torpedoes.[53]

Like such historical examples, the Poogle-Horn Players' musical instruments are defective and subject to breakage, and so they must rely on their luck while riding down stairs on a unicycle while playing large horns. Proper maintenance and training, of course, might improve performance in both scenarios.

HARRY HADDOW'S SHADOW: CHANGING ONE'S LUCK

In *Did I Ever Tell You How Lucky You Are?* the reader is invited to imagine the bad luck of "poor Harry Haddow. / Try as he will / he can't make any shadow!" In the illustration of Harry Haddow's predicament, a group of men wearing identical one-piece pajamas face the setting sun. The light from the sun causes their shadows to stretch out behind them. But when he looks over his shoulder, Harry cannot see his own shadow. In Harry Haddow's case, "He thinks that, perhaps, something's wrong with his Gizz. / And I think that, by golly, there probably is." Here, Dr. Seuss reminds the reader that the victims of bad luck frequently try to identify a reason for their misfortune.

Like Harry Haddow (who identified the malfunctioning Gizz as the source of his mischance), soldiers often attribute their bad luck to something concrete and external to the self. Psychologist Harvey Irwin conjectured that by carrying, wearing, or displaying amulets or talismans, combatants intend to influence the future, ensuring protection, good fortune, or personal gain.[54] Like Harry Haddow's Gizz, these talismans and amulets are concrete objects believed to be imbued with protective power that will shield soldiers from harm.

The belief in the lucky power of objects transcends cultures; soldiers from many different societies have carried talismans into battle. Japanese troops, for example, wore good-luck *sennin-bari*, or "belts with a thousand-stitches," handsewn by Japanese women as Shinto amulets.[55] In 1942 *The American Soldier* observed that many US GIs carried protective amulets or cherished articles of clothing that they had worn in previous dangerous battles.[56] Even those who did not necessarily believe in the protective power of objects adopted them nevertheless—*just in case*. Some American combat airmen convinced themselves that wearing their lucky socks, carrying a fortunate silver dollar, or settling into their compartment in the very same way at the exact same moment before each mission would ensure success.[57] In 1944, *Yank* magazine spoke informally with bomber crewmen in England who acknowledged that they all carried some sort of lucky charm on missions. One man believed in his lucky flight jacket and refused to fly without it. Others put faith in a ball cap or a sword; yet another never washed his underwear until the end of his combat tour.[58] Confederate general A. P. Hill carried a lucky hambone, bestowed on him by his mother when he entered West Point, and was rumored to be without it the day he died at Petersburg.[59]

The belief in the power of talismans also transcends rank. General Dwight Eisenhower, for example, carried a five-guinea gold piece for good luck, and General George Kenney carried a pair of dice that had been blessed by a priest in World War II.[60] The perception of a talisman's protective power (just as Harry Haddow believed in the power of his Gizz to make a shadow) is not just a historical artifact but occurs in the present era. Thai soldiers serving in the multinational forces in Iraq took more than six thousand Buddha amulets, pieces of blessed cloth, and sacred phalluses with them to bring good luck.[61] In some campaigns, soldiers and police officers were ordered to wear the amulets at all times or face a punishment of imprisonment. In the ongoing conflict in southern Thailand, the number of talismans worn by the members of Thai battalions sometimes far exceeds the number of actual troops, with some soldiers having over one hundred charms each. "This amulet makes me feel safe even if I don't have a bulletproof vest," one soldier said.[62]

The question naturally arises: Why do soldiers believe a talisman will protect them from injury or death in combat? One explanation is that professional warfighters believe that they can make their own luck. Carrying a material object imbued with perceived powers is a means of changing their luck. John C. McManus, in his work on American combat airmen in World War II, summarized what was true for every airman, indeed, every combatant, in the conflict: "Would today's mission be the day that luck ran out? Or would it be tomorrow? Or would the luck ever run out? Could the verdict be influenced by bravery, competence, or some kind of personal talisman? Every combat airman lived with these questions and, either consciously or unconsciously, struggled to answer them."[63] Soldiers do not passively accept the misfortune of bad luck. Rather, they tend to believe that they can take affirmative action to change their luck. In his memoirs, Adm. William Halsey offers an example of how he changed his luck:

> When we received orders for our next operation, we were appalled to find that not only had we been designated Task Force 13, but our sortie had been set for February 13, a Friday! Miles Browning and my Intelligence Officer, Col. Julian P. Brown of the Marines, immediately went to CINCPAC's headquarters and asked his chief of staff, Capt. Charles H. McMorris, 'What goes on here? Have you got it in for us, or what?' 'Sock' Morris agreed that no sane sailorman would dare buck such a combination of ill auspices and changed our designation to TF 16 and our sortie to the fourteenth.[64]

Another more mundane explanation for soldiers' faith in the power of talismans is that it alleviates anxiety. Approaching battle, a soldier's confidence in their own luck plays a significant role in overcoming stress and fear. A belief in luck has a positive psychological effect: it emboldens combatants with the courage to hold on, as Gen. George S. Patton would say, "a minute longer."[65] In this sense, magic and luck (like prayer) represent alternate approaches to dealing with anxiety. "Individuals fall back on ritual," wrote Richard Holmes, "to which they sometimes attribute magical properties as a means of defending the ego against anxiety."[66] Even if it is illusory, a belief in luck provides a measure of psychological comfort. "Soldiers find some measure of comfort behind the flimsiest of cover if it makes them feel less exposed," wrote cultural historian William Miller, "even though in fact it provides no protection whatsoever."[67]

In addition to soothing anxiety, soldiers' belief in luck also prepares them psychologically for battle. Fetishes and precombat rituals fortify the warrior by generating feelings of invulnerability. This prebattle mindset (whether fostered by religious rituals, amulets, prayers, incantations, drugs, or other consumables) creates personal and collective energy toward the desired outcome. During the 1900 Boxer Rebellion in China, for example, members of a secret society known as the Fists of Righteous Harmony believed they could become possessed by the gods through martial arts, incantations, and charms. Under the influence of the spirits, the Boxers claimed they could fly, become invisible, multiply their bodies, kill at a distance, and bring the dead back to life.[68] Mastering this commingling concoction made them invulnerable, so they believed.

The best way to reduce physical vulnerability in combat is to become invisible. Perhaps Harry lacks a shadow because he has become invisible? Because invisibility is the ultimate camouflage, soldiers and militaries throughout time have pursued it with great effort and zeal. Boko Haram warlord Ibrahim Tada Ngalyike (notorious for kidnapping the Chibok girls in 2014) was known for sporting charms to ward off bad spirits and make him invisible. In Sierra Leone's devastating eleven-year armed conflict (1991–2002), both rebels and government Christian soldiers relied on magical talismans and invincibility inducing charms, coupled with drugs provided by local holy men, to make them invisible to their enemy.[69] In Côte d'Ivoire, a five-thousand-strong loyalist militia calling itself the Invisible Army sported amulets while attacking the former president's stronghold. Foreign Legion Command sergeant major Patrick Nevin, who participated in the United Nations peacekeeping force in Abidjan, recalled local indigenous forces crediting the rapid success of French intervention forces to "superior French magic enabling them to become invisible," rather than superior tactics and modernized weapons.[70] In Cambodia, soldiers wear magic tattoos that they believe make them invisible to enemies or repel bullets.[71] At the Thai-Burma border, a guerrilla movement known as God's Army flourished from the early 1990s until 2006. This ragtag insurgent group was led by twelve-year-old, illiterate messianic twins who claimed to be invisible to the enemy. They commanded about two hundred guerrillas who believed government bullets would "just bounce off."[72]

While losing one's shadow would be unfortunate in a civilian context, losing one's shadow in a military context would not necessarily be a tragic event. In fact, soldiers perpetually seek talismans (such as Harry Haddow's Gizz) that will change their luck, reduce their anxiety, and keep them safe from harm.

THE BROTHERS BA-ZOO: LUCKY COMRADES IN WAR

In Dr. Seuss's *Did I Ever Tell You How Lucky You Are?* the reader is invited to imagine the bad luck of "the poor Brothers Ba-zoo! / Suppose *your* hair grew / like *theirs* happened to do!" In the illustration, in an endless line unfortunate old men wearing identical red robes and red slippers walk calmly, holding their hands clasped in front of them. The beard of the man in front is connected to the hair of the man behind, creating an endless link between brothers. Like the Brothers Ba-Zoo, the connection between soldiers in warfare sometimes coheres into a "band of brothers."

Becoming a member of a band of brothers is not a sure thing but depends to a certain degree on luck. Obtaining quality "battle buddies" has always been of vital importance to the collective experience, motivation, and effectiveness of soldiers, both in and out of combat. They complement strengths and weaknesses and develop enduring bonds. John McNamara remembered of his Marine platoon mates: "Luckily, we were like the three Musketeers, just as close as buddies could be, inseparable throughout our training."[73] Finding a compatible battle buddy is lucky, and conversely, working with an incompatible colleague is unlucky. One soldier considered himself favored not to be "paired up with people they didn't get along with or were sub-standard soldiers, which could present a problem. Nobody wants to be stationed with somebody they just spent 14 weeks with who they do not like."[74] Sometimes, however, being paired with unlucky individuals might actually be a different sort of good fortune. Accessing the air campaign over Germany, one scholar concluded that "even those, who, because of lack of skill or bad luck were shot down, may have prevented others—more capable of hitting the target—from becoming casualties themselves."[75]

Another form of luck involving military units concerns the competence of the individuals within a unit. Heraclitus, the philosopher tactician, allegedly uttered that "out of every one hundred men, ten shouldn't even be there, eighty are just targets, nine are the real fighters, and we are lucky to have them, for they make the battle. Ah, but the one, one is a warrior, and he will bring the others back."[76] Similarly, the nineteenth-century French strategist Ardant du Picq claimed an army only had a "few truly brave" or courageous soldiers. Gideon, du Picq asserted, "was lucky to find three hundred in thirty thousand."[77] With many chances left to fate in a battle, he continued, "the greater the necessity for the best troops who know best their trade, who are most dependable and of greatest fortitude."[78] One can only wonder if there were true warriors among the Brothers Ba-Zoo.

THAT FOREST IN FRANCE:
LUCK AND THE ENVIRONMENT IN WARFARE

In one of the darkest drawings in *Did I Ever Tell You How Lucky You Are?*, a young boy is alone in a forest of spooky green and brown trees. The branches and twigs reach toward the boy menacingly. One of the branches snatches off his yellow pants, leaving him bare bottomed in a dark forest in France. The boy appears to be surprised by rather than terrified of the creepy trees, so perhaps he expected to encounter this particular misfortune. In the accompanying text, Dr. Seuss invites the reader to "suppose that you lived in that forest in France, / where the average young person just hasn't a chance / to escape from the perilous pants-eating-plants!"

Geisel's "forest in France" depicts the natural environment as a hazardous place, full of terrible predicaments that might befall a person or an army.[79] The young boy's apparent situation in a forest lacking light and full of dangerous roots is evocative of Geisel's own experience in the Ardennes forest years earlier. The significance of forests to the German psyche—and especially the Nazis' rhetoric—in terms of culture and mythology is noteworthy. Historian Peter Caddick-Adams outlined Hitler's preoccupation with Wagnerian-themed dark woods where humans are tested, as well as the emphasis both historically and culturally the Führer placed on the famous and important events of history, like the AD 9 Roman disaster in the Teutoburg Forest at the hands of Germanic tribes. This ancient battle occurred in similar dark, wet, lonesome woods and was a defining moment for the Teutonic people. Even the chosen German name for the 1944 Ardennes operation, *Herbstnebel* ("Autumn Mist") connoted Norse woodland. Caddick-Adams argued that it was no accident the Bulge offensive was launched from large forests, in heavy fog.[80]

Indeed, throughout history, the physical environment has been interconnected with warfare. "Despite the evolving technology in warfare," wrote climatologist and historian Harold Winters, "physical geography has a continuous, powerful, and profound effect on the nature and course of combat."[81] Battles have been won or lost due to environmental factors affecting terrain, such as the dust and wind at Cannae (216 BC), the freezing cold in Saxony (1708), the setting sun at Coronel (1914), and the bitter cold on the Eastern Front (1941–42).[82]

Physical topography, terrain, and climate have influenced the outcomes of many wars and campaigns, benefiting the winner or hampering the loser.[83] Thus, soldiers need luck not just against the enemy but also within the natural environment. "In combat," according to Harold Winters's view, "an environmental advantage for one side always means some degree of misfortune for the other, and that situation can easily reverse itself on the next battlefield."[84] For example, at the Battle of Waterloo in 1815, the rain, while mutually miserable, fortuitously benefited the Duke of Wellington and debilitated Napoleon, whose artillery emplacement was delayed. As Victor Hugo penned, "If it had not rained in the night between the 17th and 18th of June, 1815, the fate of Europe would have been different. A few drops of water, more or less, made Napoleon waver. All that Providence required in order to make Waterloo the end of Austerlitz was a little more rain, and a cloud crossing the sky out of season sufficed to overthrow the world."[85]

Because of its power to determine the fate of nations, predicting and adapting to the physical environment better than one's enemy has always been important to military strategists.[86] Yet the tactical situation in any battlespace is frequently ambiguous, uncertain, and shifting. "With chance at work everywhere," Clausewitz wrote, "the commander continually finds that things are not as he expected." Some leaders can overcome this "relentless struggle with the unforeseen" through what is known as *coup d'oeil*. Literally translated as a "stroke of the eye," this phrase refers to a commander's innate ability to appraise terrain, the disposition of forces, and other factors and to rapidly decide the best course of action.[87] Whether one possesses this intuitive ability is a matter of fate, a "gift," according to Napoleon, "of being able to see at a glance the possibilities offered by the terrain." Napoleon believed that the *"coup d'œil militaire . . .* is inborn in great generals," and of course he counted himself among those elite gifted with *coup d'oeil*.

Unlike such great commanders, the pantless boy in the French forest lacked coup d'oeil. He did not intuitively grasp the hazards of the environment in which he found himself, and therefore he hadn't "a chance / to escape from the perilous pants-eating-plants." Napoleon could claim he walked "accompanied by the god of war and the god of luck."[88] Seuss's boy, on the other hand, lacked such deities.

THE SEVENTEENTH RADISH:
LUCK AND THE RANDOMNESS OF SURVIVAL

In the illustration of Farmer Falkenberg's seventeenth radish, a farmer snoozes on his hammock after a hard day's work cultivating radishes. Meanwhile, under the soil the unlucky seventeenth edible root is about to be devoured by a hungry worm. "You should be greatly glad-ish / you're not Farmer Falkenberg's / seventeenth radish," Seuss admonishes. Having spent time in Italy, Geisel's assignment of "seventeenth" to this vegetable's location should not be overlooked, as the belief that the number 17 is unlucky dates backs to ancient Roman times.[89] This unlucky radish serves as an allegorical comment on the randomness of luck in war, which is a theme that resonates through recollections and experiences of so many combatants. In the words of former aide to Patton Col. Richard Stillman, survival "in battle . . . is always an element of good fortune or just plain luck."[90] Almost every combat veteran can relate a "lucky" combat survival tale. Sergeant Charles Twiggs, for example, attributed his absence from the disastrous Battle of iSandlwana during the Zulu War of 1879 to serendipity. "It was lucky I was out; I was going to make a pair of trousers for an officer, but luckily I went out at eight o'clock so I have saved my life, but lost all my things, books and all, except what I stand in."[91]

Out of all the radishes in the field, Farmer Falkenberg's seventeenth radish has the bad luck to be selected by the worm. Nothing indicated to the worm that this particular radish would make a better meal than any other radish. The selection was simply arbitrary luck. Dr. Seuss's portrayal of the random misfortune of a single individual within a larger numbered group resembles the practice of decimation. In ancient Rome, if legionnaires fled the battlefield, the unit was broken into groups of ten troops, lots were cast, and the "short stick" was beaten to death by his comrades. Two thousand years later, Nazi forces randomly wiped out whole villages in reprisal for partisan attacks against German troops. Such examples recall to posterity Winston Churchill's assertion that "countless and inestimable are the chances of war."[92]

IN CONCLUSION: A GAME OF CARDS

In 2020, author and illustrator Caitlin Fitz Gerald published the *Children's Illustrated Clausewitz*. In the book, the wise rabbit Hare Clausewitz imparts his warrior expertise to a forest-based class that includes a bear, fox, badger, boar, and otter, among other woodland creatures found in Clausewitz's Prussia. Children "live in the same world we live in and need ways of interpreting it," Fitz Gerald said. "There's a long history in children's literature of books that help children understand and process difficult realities and complicated politics."[93] Like the *Children's Illustrated Clausewitz*, Dr. Seuss makes difficult concepts accessible to and appropriate for children and conveys meaningful messages about the human condition. Through reading Dr. Seuss, both children and warriors may learn that "from the very start, there is an interplay of possibilities, probabilities, good luck and bad, that weaves its way throughout the length and breadth of the tapestry. In the whole range of human activities, war most closely resembles a game of cards."[94]

War, however, is a game of chance not intended for children . . . no matter how charmed their birth star.

NOTES

1. "Battle of the Bulge: The Greatest American Battle of the War," National Veteran's Memorial and Museum, December 16, 2020, https://nationalvmm.org/battle-of-the-bulge-the-greatest-american-battle-of-the-war/.

2. David Eisenhower, *Eisenhower: At War, 1943–1945* (New York: Random House, 1986), 570.

3. E. J. Kahn Jr., "Children's Friend," *New Yorker,* December 17, 1960, 47–93.

4. Brian Jay Jones, *Becoming Dr. Seuss: Theodor Geisel and the Making of an American Imagination* (New York: Dutton, 2019), 212.

5. Rick Atkinson, *The Guns at Last Light* (New York: Henry Holt, 2013), 416.

6. Alison Flood, "Dr. Seuss's Thank-you Letter to Man Who Saved His First Book," *Guardian UK*, January 30, 2019.

7. Brian Jay Jones, *Becoming Dr. Seuss: Theodor Geisel and the Making of an American Imagination* (New York: Dutton, 2019), 114.

8. Jones, *Becoming Dr. Seuss*, 173.

9. Richard Holmes, *Acts of War: The Behavior of Men in Battle* (New York: Free Press, 1989), 193.

10. US Army, Field Manual 6-22, *Developing Leaders* (Washington, DC: US Government Printing Office, November 2022).

11. US Army, Field Manual 22-100, *Army Leadership* (Washington, DC: US Government Printing Office, August 31, 1999), 6–16.

12. In statistical principle, chance, as distinguished from luck, should influence both sides equally, or in other words, favor no one. By any definition, luck is not something quantifiable.

13. Carl von Clausewitz, *On War*, trans. and ed. Michael Howard and Peter Paret (Princeton, NJ: Princeton University Press, 1976), 89.

14. Clausewitz, *On War*, 120.

15. Clausewitz, *On War*, 85.

16. Stephen Fritz, *Frontsoldaten: The German Soldier in World War II* (Lexington: University Press of Kentucky, 1995), 190.

17. Abraham Mahshie, "As China's Military Might Rises, the US Must Master Change to Prepare for Next War, Milley Says," *AirForceMag.com*, August 2, 2021, https://www.airandspaceforces.com/as-chinas-military-might-rises-the-us-must-master-change-to-prepare-for-next-war-milley-says/.

18. Louis Hicks, "Normal Accidents in Military Operations," *Sociological Perspectives* 36, no. 4 (1993): 377–91.

19. Hicks, "Normal Accidents in Military Operations," 381, 388.

20. Richard Holmes, *Sahib: The British Soldier in India 1750–1914* (London: HarperCollins, 2005), 411.

21. Holmes, *Sahib*, 411–12.

22. Alexandre Percin, *Le Massacre de notre Infanterie 1914–1918* (Paris: Albin Michel, 1921), 10, 13–14, 217–18.

23. Jack Shulimson et al., *U.S. Marines in Vietnam: The Defining Year, 1968* (Washington, DC: US Government Printing Office, 1997), 565.

24. Bryan Bender, "More U.S. Troops Dying Outside of Combat," *Boston Globe*, May 3, 2009.

25. Mark Wells, "Aviators and Air Combat: A Study of the U.S. Eighth Air Force and R.A.F. Bomber Command" (PhD dissertation, Kings College London, July, 1992), 60–63.

26. Laddie Lucas, *Out of the Blue: The Role of Luck in Air Warfare, 1917–1966* (London: Hutchinson, 1985), 184–85.

27. Wells, "Aviators and Air Combat," 250–51.

28. Charles Perrow, *Normal Accidents* (New York: Basic Books, 1984).

29. In light of the inevitable failures resulting from complex, lethal technological systems, some observers have proposed that the way to defeat technology is not with

more technology, but with no technology. Today, at the US Naval Academy, midshipmen are again learning how to use sextants. While the United States needs technologies, the military should expect these systems to fail and have a fallback, low-tech "Plan B" solution. Mary Louise, "Years of Military Service Helped Inform '2034': A Novel of the Next World War'," *NPR*, March 18, 2021.

30. W. Thomas Smith, "The Roads Are Hell," *National Review*, April 16, 2007.

31. Tom Bowman, "On the Hunt for Roadside Bombs in Afghanistan," *NPR*, October 27, 2009.

32. National Army Museum, "Luck and Superstition," accessed February 13, 2024, www.nam.ac.uk/explore/luck-and-superstition.

33. Soryaya Nelson, " Taliban Flees Marjah, Threat Remains for Marines," *NPR*, March 1, 2010.

34. Holmes, *Sahib*, 211–12.

35. Holmes, *Sahib*, 211–12.

36. Holmes, *Sahib*, 211.

37. Philip Caputo, *A Rumor of War* (New York: Ballantine, 1978), 273.

38. Robert Heinl, *The Dictionary of Military and Naval Quotations* (Annapolis, MD: Naval Institute Press, 1966), 177.

39. Mark Townsend, "Being Shot at While Trying to Disarm a Bomb—Just Another Day at Work," *UK Guardian*, November 7, 2009.

40. Holmes, *Sahib*, 183.

41. Dick Winters, interviewed in *We Stand Alone Together*, directed by Mark Cowen, aired on HBO in 2001.

42. Audie Murphy, *To Hell and Back* (New York: Henry Holt, 2002), 202.

43. Fritz, *Frontsoldaten*, 63.

44. Marine Corps Colonel R. Mark Wood, conversation with author, September 9, 2021.

45. Rick Beyer, "Seuss On the Loose," *Dartmouth Alumni Magazine*, November–December 2014, https://dartmouthalumnimagazine.com/articles/seuss-loose.

46. Clausewitz, *On War*, 140.

47. Thomas H. Kean and Lee H. Hamilton, eds., *The 9/11 Commission Report: Final Report of the National Commission on Terrorist Attacks Upon the United States* (New York: W. W. Norton, 2004), 272.

48. Mindy Kleinberg, *Statement to the First Public Hearing of the National Commission on Terrorist Attacks Upon the United States*, March 31, 2003, https://govinfo.library.unt.edu/911/hearings/hearing1/witness_kleinberg.htm. After bombing the Brighton hotel where Margaret Thatcher was staying in the 1980s and failing to kill her, the IRA issued a statement: "Today you have been lucky. But you have to be lucky every time. We only have to be lucky once."

49. US Army, *The Battalion Commander's Handbook* (Carlisle, PA: US Army War College, 1996), 42, https://apps.dtic.mil/sti/pdfs/ADA309539.pdf.

50. Paul R. Pillar, "Where Luck Comes In," *New York Times*, May 3, 2011.

51. Eccles. 9:11 (King James Version).

52. Chester W. Nimitz, *Command Summary of Fleet Admiral Chester W. Nimitz, USN: Nimitz "Graybook" 7 December 1941–31 August 1945* (Newport, RI: US Naval War College, 2013), 1625.

53. Roger Branfill-Cook, *Torpedo: The Complete History of the World's Most Revolutionary Naval Weapon*, (Annapolis, MD: Naval Institute Press, 2014), 220–21.

54. H. J. Irwin, "The Measurement of Superstitiousness as a Component of Paranormal Belief—Some Critical Reflections," *European Journal of Parapsychology* 22, no. 2 (2007): 95–120.

55. "Good Luck Belts," *New York Times Magazine*, January 11, 1942.

56. Samuel A. Stouffer et al., *The American Soldier* (Princeton, NJ: Princeton University Press, 1949), 2:188–91.

57. John McManus, *Deadly Sky: The American Combat Airman in World War II* (New York: Penguin, 2016), 343.

58. McManus, *Deadly Sky*, 345–47.

59. James Robertson, *General A. P. Hill: The Story of a Confederate Warrior* (New York: Vintage, 1992), 25.

60. "Soldiers' Superstitions," *New York Times Magazine*, April 2, 1944.

61. Wassana Nanuam, "Soldiers Keep Faith with Sacred Charms," *Bangkok Post*, November 28, 2003.

62. Wassana Nanuam, "Troops Get Amulet Protection," *Bangkok Post*, January 1, 2008, https://bike-a-way.com/thailand-january-2008/.

63. McManus, *Deadly Sky*, 2.

64. William Halsey, *Admiral Halsey's Story* (New York: Wittlesley House, 1947), 98.

65. Gary Bloomfield, *George S. Patton: On Guts, Glory, and Winning* (Guilford, CT: Lyons Press, 2017), 174.

66. Holmes, *Sahib*, 238.

67. William Ian Miller, *The Mystery of Courage* (Cambridge, MA: Harvard University Press, 2000), 215.

68. Mark Elvin, "Mandarins and Millenarians: Reflections on the Boxer Uprising of 1899–1900," *Journal of the Anthropological Society of Oxford* 10, no. 3 (1979): 122.

69. Barak Salmoni and Paula Holmes-Eber, *Operational Culture for the Warfighter: Principles and Applications* (Quantico, VA: Marine Corps University Press, 2011), 61.

70. Command Sergeant Major Patrick Nevin, French Foreign Legion, interview with author, Washington, DC, October 24, 2012.

71. Kounila Keo, "Modernity Counteracts Magic Tattoos in Cambodia," *Daily Telegraph UK*, November 18, 2009.

72. "Is the News Media Making Too Much of Karen twins, Johnny and Luther?," *The Straits Times Interactive*, August 11, 2000, www.hartford-hwp.com/archives/54/170.html.

73. John McNamara, *Millville's Mac—The Life Story of a World War II Combat Marine* (n.p.: Changing Attitudes LLC, 2014), 106.

74. Peter Ramsberger, Peter Legree, and Lisa Mills, *Evaluation of the Buddy Team Assignment Program* (Arlington, VA: US Army Research Institute for the Behavioral and Social Sciences, October 2002), B- 62, https://apps.dtic.mil/sti/pdfs/ADA408486.pdf.

75. Wells, "Aviators and Air Combat," 110–11.

76. Peter Tsouras, *Warriors' Words: A Dictionary of Military Quotations* (London: Cassell Arms and Armour, 1994).

77. Ardant du Picq, *Battle Studies: Ancient and Modern Battle* (New York: Macmillan, 1921), 120.

78. Du Picq, *Battle Studies*, 117.

79. As the Roman poet Cicero wrote two thousand years ago, "Woods are an ornament in peace and a fortification in war." It is of note that Geisel chose the woods as a setting for this lesson, as his experience of luck in war occurred against the dark backdrop of a Wagnerian Teutonic cataclysm, in one of the most famous woods in Europe, the Ardennes, where Hitler launched his last great offensive in the West, and where Geisel found himself cold and trapped and presumed dead. See Beyer, "Seuss on the Loose."

80. Peter Caddick-Adams, *Snow & Steel: The Battle of the Bulge, 1944–45* (Oxford: Oxford University Press, 2015), 125–31.

81. Harold Winters, *Battling the Elements: Weather and Terrain in the Conduct of War* (Baltimore, MD: Johns Hopkins University Press, 1998),4.

82. The winter of 1942–43 was the coldest European winter of the twentieth century, and around Moscow, the coldest in approximately 250 years. See Russel Stolfi, "Chance in History: The Russian Winter of 1941–1942," *History* 65, no. 214 (1980): 222.

83. For readers who wish to pursue further the intercorrelation between military geography and warfare, there have been many fine studies written. See Patrick O'Sullivan and Jesse Miller, *The Geography of Warfare* (London: Croom Helm, 1983); and Winters, *Battling the Elements*.

84. Winters, *Battling the Elements*, 1.

85. Victor Hugo, "Waterloo," quoted in *Men at War*, ed. Ernest Hemingway (New York: Brammel House, 1979), 675.

86. Sun Tzu, who famously distinguished six kinds of terrain in chapter 10 of his famous work, *The Art of War*.

87. John J. O'Brien, "Coup d'Oeil: Military Geography and the Operational Level of War" (master's thesis, School of Advanced Military Studies, US Army Command and General Staff College, May 16, 1991).

88. Frank McLynn, *Napoleon: A Biography* (New York: Arcade Publishing, 2002), 198–99.

89. Nick Harris, "Bad Omen for Italy as Their Unlucky Number Comes Up," *Independent UK*, November 15, 2007.

90. Richard Joseph Stillman, *The U.S. Infantry: Queen of Battle* (New York: F. Watts, 1965), 4.

91. *Canterbury Journal and Farmers' Gazette*, March 15, 1879, quoted in Frank Emery, *The Red Soldier* (Johannesburg: Jonathan Ball, 1977), 96.

92. Winston Churchill, *The River War: An Account of the Reconquest of the Sudan* (Mineola, NY: Dover, 2006), 138.

93. J. D. Simkins, "'On War' Is Now a Children's Book Featuring 'Hare' Clausewitz Teaching Woodland Critters," *Military Times*, July 8, 2020.

94. Clausewitz, *On War*, 85–86.

12

The Butter Battle Book and Deterrence and Escalation

Sam J. Tangredi

> Some critics claim that *The Butter Battle Book*
> Is only a moral relativist's look,
> That makes two systems equal to blame
> For what to Seuss seemed a great power game.
> Yet if one dives deep beneath this conclusion,
> You find its usefulness not an illusion.
> It cites a reason for war's non-occurrence
> And a game-theoretical view of deterrence.
> But there are other observations to make,
> That perhaps Seuss did not consciously take.
> For example, it took only one renegade Zook
> To provoke the other sides' caution forsook.
> And technology diffusion routinely resulted
> In scientists by military leaders consulted.
> Moral questions throughout can certainly be phrased,
> The nature of nationalism a big issue raised.
> My use of rhyme is so readers don't dose;
> But let me continue with practical prose.

The Butter Battle Book (1984) tells the story of the Yooks and the Zooks, whose disagreement about whether bread should be eaten buttered side up or down quickly escalates to a crisis point. "So you can't trust a Zook who spreads bread underneath," warns one of the central characters in the book. "Every Zook must be watched! / He has kinks in his soul!" As hostilities mount, in their efforts to defend themselves against their neighbor, the Yooks and the Zooks develop increasingly lethal and complex weaponry.

When *The Butter Battle Book* was published, the Cold War seemed never-ending, and some questioned the policies of the Reagan administration's forceful relationship with the Soviet Union. However, in the years immediately after publication the Strategic Arms Reduction Treaty (START) and Intermediate-Range Nuclear Forces (INF) Treaty were signed by President Ronald Reagan and Soviet general secretary Mikhail Gorbachev. The number of superpower nuclear weapons was being reduced. By the time animator Ralph Bakshi put the story to film in 1989, the Berlin Wall was crumbling. Following the end of the Cold War, the international system appeared to have been transformed and the dilemma of nuclear deterrence—with the dire image of the doomsday clock approaching midnight—seemed much less pressing.

Is *The Butter Battle Book* an artifact of the Cold War or an illustration of principles still relevant for twenty-first-century international relations? Though written during the Cold War, *The Butter Battle Book* continues to illustrate the principles of deterrence, compellence (defined later), and escalation in international relations and provides an intriguing starting point for a discussion of weapons development, arms races, deterrence, escalatory policies, miscalculation, and the effects of individual actions, particularly in the context of great power competition.

The possibility of major war between technologically advanced nations now appears more likely than at any time since 1991, which was the year of Geisel's death. By 2010 the Russian Federation had already embarked on a major recapitalization of its nuclear forces. Meanwhile, the People's Republic of China, unconstrained by effective arms control treaties, substantially increased its nuclear weapons and ballistic missile inventory, including missiles that could alternately carry conventional (nonnuclear) or nuclear warheads.[1] In 2019 both the United States and Russia withdrew from the INF treaty. Yet a caveat must also be added; "more likely" does not mean certain. In January 2021 Russian president Vladimir Putin signed an extension of the 2010 US-Russian New START treaty, which originally had a ten-year lifespan, retaining existing numerical limits on strategic nuclear weapons.[2]

Nevertheless, such agreements have little effect without credible deterrent capabilities able to hold signatories to their promises. As a result of the Budapest Agreement of 1994, Ukraine (along with Belarus and Kazakhstan) gave up the nuclear weapons left within its territory following the collapse of the Soviet Union in exchange for a permanent guarantee of Russian respect for its sovereignty. Obviously the sovereignty of a nuclear-disarmed Ukraine was not respected. Given the second Russian invasion of Ukraine, understanding of the principles of deterrence and escalation has taken on an even greater sense of urgency.

CONTROVERSY AND REALITY

The Butter Battle Book has frequently been portrayed as Geisel's most controversial book. The uncertain, cliffhanger ending—a realistic portrayal of the nuclear standoff of the time—came under criticism specifically for being inconclusive, thereby causing fear and worry among children. Others countered that it was an idea to which children needed to be exposed and was a useful teaching tool for elementary school students.[3]

Certain conservative media, such as *National Review*, argued that the book portrayed an attitude of moral equivalence between the United States and the Soviet Union, which indeed may have contributed to its having lower sales than most of the other Seuss books. There is clear evidence that the publisher, Random House, recognized this possibility and attempted to stoke the controversy to increase sales rather than diffuse it. Random House widely publicized a complaint letter sent to it by one reader's mother, along with their rather bombastic reply concerning the publisher's commitment to children's critical thinking.[4] There is no indication that the complainant ever intended the letter to be public.

An internet myth that *The Butter Battle Book* was banned by libraries—always useful in spurring interest—has persisted despite the fact that a University of Saskatchewan study identified only one public library in North America (out of over forty thousand) that ever banned it.[5]

Supposedly Geisel thought *The Butter Battle Book* was one of his most important books, defending it by saying, "I'm not antimilitary. . . . I'm just anti crazy."[6] However, intellectuals and Seuss buffs (perhaps we should call them Seussies) have frequently detected—with valid reason—support for many political causes (sometimes anti-military) in Geisel's later books, many of which were subsequently viewed as rebukes to US government policies of the time. Some Seussies speculate that Geisel attempted to atone for his work on propaganda and instructional animated films for the US war effort in World War II, some of which portrayed considerable hostility (naturally enough) toward the Japanese.[7] Others maintain that publication of *The Butter Battle Book* made Geisel an "enemy of the right."[8]

The reality is a bit more prosaic. *The Butter Battle Book* was one of Geisel's least popular books. Since the success of *The Cat in the Hat* in the 1950s, books associated with Dr. Seuss (even later dictionaries and board books in which he had little involvement) have sold tremendously, making him the world's best-selling children's author. Yet despite releasing precise sales numbers (with fanfare) for Geisel's most popular books, Random House has never released that information on *The Butter Battle Book*, probably because sales were relatively low for a Seuss book. Nor did the book receive much critical attention when published, since there were many more-controversial books at the time. Yet *The Butter Battle Book* still contains many lessons about warfare with continued relevance today.

ARMS RACE

At the beginning of *The Butter Battle Book*, the situation seems relatively stable. "The Wall wasn't so high," recalls Grandpa "and I could look any Zook / square in the eye." The arms race between the Yooks and Zooks begins when "a very rude Zook by the name of VanItch / snuck up and slingshotted by Snick-Berry Switch!" Every time the Yooks develop a new innovative weapon (such as the Triple-Sling Jigger, the Kick-a-Poo Spaniel, the Utterly Sputter, and the Bitsy Big-Boy Boomaroo), the Zooks respond with a countering weapon (the Jigger-Rock Snatchem, Eight-Nozzled, Elephant-Toted Boom-Blitz). Eventually the weapons become so complex that the users can no longer understand the mechanism. The giant self-propelled Utterly Sputter was "*so* modern, *so* frightfully new, / no one knew quite exactly just *what* it would do." The Yooks continue until both they and the Zooks possess nuclear-like devices in an action-reaction type process that Paul Warnke, the chief nuclear arms negotiator and director of the US Arms Control and Disarmament Agency in President Jimmy Carter's administration, referred to in real life as "apes on a treadmill."[9]

Critics of the use of nuclear weapons generally see the development or improvement of such weapons as constituting an "unwinnable" arms race, a term that came into vogue in the 1960s to describe what seems like constant competitive development by the superpowers in increasing their nuclear weapons stockpile and delivery systems.[10] However, it has also been applied to other competitive military buildups in history, particularly in naval fleets. Whether or not arms races actually result in wars has been long and inconclusively debated, although proponents of arms control have fiercely clung to the belief that arms races have inevitably led to hostilities, even if other factors were the proximate causes of conflict.[11]

Such a belief had its roots in the years immediately after World War I, which was popularly blamed on "merchants of death," companies that sold the armaments used by the combatants. The result was a series of international conferences attempting to place formulaic limits on types of naval vessels during the 1920s and 1930s.[12] Arms control proponents look at these treaties as a success, but critics point out that they had no effect on the actual outbreak of war.

Having grown up during World War I and served in World War II, Geisel is not likely to have viewed arms control as creating a legacy for future success.[13] But Geisel clearly recognized the process of "technology diffusion" in which advanced technology is replicated by other nations through publication, scholarly exchange, reverse engineering, purchase of components, and industrial espionage.[14] The Soviet Union tested its own fission atomic weapon in 1949, only four years after one was developed by the United States.[15] Russian physicists had kept abreast of the prewar scientific discoveries; however, Soviet focus during the war was naturally enough on immediate improvements to existing weapons (as well as survival of the state following the German invasion). However, the Soviets did possess the underlying knowledge as well as receiving detailed information from espionage within the Manhattan Project.[16] Having a comparable weapons development infrastructure, the Soviets were able to replicate the more powerful hydrogen bomb thermonuclear fusion device in 1955, four years after the successful US test in 1951.

TECH PACE

On the next page of *The Butter Battle Book* we find an illustration that shows Grandpa "head hung in shame" visiting the leader of the Yooks, Chief Yookeroo, who assures him, "You're not to blame. / And those Zooks will be sorry they started this game." Behind the leader's desk, a group of nerdy Yooks with small tufts of hair and eyeglasses appear to be doing calculations on a drafting board. Chief Yookeroo tells Grandpa, "We'll dress you right up in a fancier suit! / We'll give you a fancier slingshot to shoot." Chief Yookeroo then orders the "Boys in the Back Room to figger / how to build me some sort of a triple-sling jigger."

Any arms race requires a scientific, engineering, and manufacturing infrastructure that can produce increasingly sophisticated weapons (or at least greater numbers of weapons), a combination that is disparagingly known as the military-industrial complex. During World War II many American scientists volunteered their research for the war effort, which became evident particularly in the atomic bomb effort. With the introduction of nuclear weapons, scientific advisers became a permanent feature of American national security. The label Boys in the Back Room suggests both the scientific advisers and civilian strategists who publicly expressed (if not actually determined) the US nuclear strategy in the 1960s and 1970s, such as the game theorists advising Secretary of Defense Robert S. McNamara (1961–68).

Part of the role of scientific advisers in the Department of Defense (DoD) consists of monitoring the pace of overall technological development to avoid surprise as well as benefit from advancements. The common perception is that the pace of technological advances in the late twentieth and early twenty-first centuries has increased exponentially at a rate dwarfing past history, primarily through the development of enabling information technology. This in turn has increased the importance of the scientific community in national security. However, some contend that their involvement biases policy toward technological solutions for problems of strategy and diplomacy. *The Butter Battle Book*'s portrayal links science and engineering to the Yook/Zook arms race, giving the nerdish Boys in the Back Room a dominant role. Since no other military leaders or diplomats are portrayed in the book, perhaps their roles have been overshadowed by the evident pace of technological development in the need to keep up with or surpass the opponent's weapons technology.

ESCALATION AND HESITATION

After visiting the Boys in the Back Room, Grandpa now steers a three-wheeled platform carrying a much larger slingshot, with three rubber slings that can accommodate much larger rocks. "With my Triple-Sling Jigger / I sure felt much bigger." But now the Zooks must develop an even larger and more lethal weapon to counter the Yook threat. "Now," says the Zook combatant, "I have *my* hand on a trigger!" His weapon, the Jigger-Rock Snatchem, now has a fully automatic electric firing system, which "will fling 'em right back just as fast as we catch 'em."

Thus begins the Yook/Zook arms race. Arms races are often seen as a form of escalation, in which both sides become more deadly. Analysts at the federally funded research and development center RAND—an organization that frequently provides "Boys in the Back Room" support for DoD—define escalation "as an increase in the intensity or scope of conflict that crosses threshold(s) considered significant by one or more of the participants."[17] From this perspective, escalation requires a psychological component; one of the participants must perceive an action as increasing the danger to itself, no matter the specifics of the action. As the RAND report continues, "Not every increase or expansion of violence is escalatory: Escalation occurs only when at least one of the parties involved believes that there has been a significant qualitative change in the conflict as a result of the new development."[18] In RAND's view, escalation as a deliberate choice of action can "be considered both as a strategic tool to be wielded or deterred and as a potential problem to be managed."[19]

No proof exists that arms races themselves ever caused a particular war, just as there is no firm evidence that deterrence ever prevented a particular war. Rather, arms control proponents argue that such agreements reduce the tensions between potential opponents and encourage a habit of compromise, thereby preventing continuing arms races that could escalate into war. Deterrence itself is seen as a preventative to escalation to outright war.

RATIONAL OR NATIONAL

The next illustration in *The Butter Battle Book* portrays Grandpa reporting to Chief Yookeroo at headquarters. The Chief does not blame Grandpa for the failure of the slingshot, recognizing that the fault lies with the technology: "That was old-fashioned stuff. / Slingshots, dear boy, / are not modern enough." The Boys in the Back Room have been tasked with developing a "newfangled kind of a gun." Looking very determined, Grandpa marches along with the new weapon, "a gun called the Kick-a-Poo Kid," which uses ammunition made of "ants' eggs and bees' legs / and dried-fried clam chowder." What's more, the Boys in the Back Room have "carefully trained a real smart dog named Daniel / to serve as our country's first gun-toting spaniel." With an anxious expression on his face, the dog carries a large black weapon on his back. In the background of the illustration, a group of Yooks cheers for Grandpa, the spaniel, and the new weapon, crying, "Fight! Fight for the Butter Side Up! / Do or die!"

The cheering Yooks in this *Butter Battle Book* illustration seem highly emotional, suggesting that highly charged, emotive topics such as tribalism and nationalism caused the dispute. By implication, Geisel seems to suggest that nationalism can lead to irrational acts, such as nuclear weapons and a stalemate.[20] In Geisel's view the entire scenario was irrational because the cause—buttering one's bread on different sides—was irrational. None of the characters or scenes in *The Butter Battle Book* appear to be rational.

In suggesting that nuclear escalation results from emotion rather than calculation, *The Butter Battle Book* raises the issue of human rationality in a deterrence situation. Since deterrence is a psychological state whose results can never be definitively demonstrated or quantified, theorists naturally focus much of their attention on the elements and characteristics that make up the collective psychologies of the leaders of the contestants. Many of these academic explorations have been in the field of game theory, invented in the 1920s by mathematician John von Neumann and further developed in the 1940s in collaboration with Oskar Morgenstern.[21] A formal definition of "game theory" is that it is "the study of the ways in which *interacting choices* of *economic agents produce outcomes* with respect to the *preferences* (or *utilities*) of those agents, where the outcomes in question might have been intended by none of the agents."[22] A less formal description is that the actions of participants in a contest or negotiations are observed to see how the results were determined. Presumably the contestants are all trying their best to win the contests and achieve the desired outcome; however, the nature of their interactions (and the structure of the contest itself) might create consequences that none of them intended.

Both von Neumann and Morgenstern had influence in the political decisions behind the American nuclear posture. Even more influential (particularly on McNamara) was the game theorist Thomas Shelling and his strategy of conflict, which popularized the prisoner's dilemma experiment (among other 2 x 2 matrix games) as the quintessential example of game theory.[23] In the prisoner's dilemma scenario, two criminal accomplices are separated and questioned. Each is offered the same deal if they confess, but there are three different possible outcomes:

1. If one confesses and turns state's evidence and the other does not, the one who confesses goes free and the other receives a fifteen-year prison sentence.
2. However, if both confess, both will get a ten-year prison sentence.
3. If neither confesses, they will both receive five-year sentences based on a lesser charge.

The prisoner's dilemma is intended to illustrate a situation in which individuals need to make critical choices without information about the opponent's intentions. The question is whether to gamble on getting off by confessing while the accomplice does not or to "cooperate" with the accomplice by assuming they also do not confess. As in a situation in which a nuclear-armed state might perceive an opportunity to conduct a decapitating strike on an opponent's nuclear force to eliminate the threat of itself being attacked, a prisoner has great temptation to turn state's evidence and get off freely, condemning the accomplice to take the rap. Being criminals, trust would be an uncertain commodity. For a nation-state protecting its security in an international environment or anarchy, trusting an enemy to "play nice" also appears an uncertain prospect. In structuring the game to fit a certain situation, game theorists attempt to determine what factors might elicit cooperation (such as neither confessing) or what might cause a player to defect (assuming the other won't confess).

Contrary to Geisel's view that the war of the butter was irrational because of the cause, game theorists tend to assume that the players are "rational"— or, to avoid debates over ideology, more "economically rational."[24] An economically rational actor is assumed to be able to "assess outcomes, in the sense of rank-ordering them with respect to their contributions to his or her welfare; calculate paths to outcomes, in the sense of recognizing which sequences of actions are probabilistically associated with which outcomes; and select actions from sets of alternatives . . . that yield the most-preferred outcomes, given the actions of the other players."[25] From a game theoretical situation, an actor can be rational in threatening nuclear war if that is the optimal method of achieving a desired outcome such as strategic stability or to prevent an opponent's invasion. To be rational requires a calculation of other players' responses and taking those probable responses into consideration before acting. Most game theorists assume that the more knowledge a player has about other players' capabilities and characteristics, the more "rational" the action being taken will be.

In assessing two nuclear armed superpowers, game theorists view a strategic stalemate as being "cooperation." It is from this perspective that the idea of mutually assured destruction (MAD) was derived; if both sides are capable of destroying the other no matter who attacks first, it makes no rational sense for either side to attack. Both sides cooperate in the sense that they have the capability to attack but they do not because they are mutually deterred. This would correspond in the above prisoner's dilemma game as both sides not confessing and therefore getting a five-year sentence—living under the threat of nuclear war being the five-year sentence instead of gambling by actually starting a war.

RENEGADES AND PROVOCATEURS

Deviation from the stalemate of deterrence is represented in *The Butter Battle Book* by the action of an individual Zook named VanItch. At the beginning of the book, as one might recall, Grandpa marches along the wall in his own territory, swishing his Switch at any Zook that approaches too close. Being branch-like, the switch has limited range and (based on the drawings) can only reach a few inches over the wall. Grandpa states that he could give any Zook near the wall "a twitch." It is hard to determine if that means a physical strike or merely a shake in their direction. In any event, the Zook VanItch physically begins the spiral of escalation by "slingshott[ing Grandpa's] Snick-Berry Switch!"

Although VanItch subsequently becomes the primary Zook antagonist, appearing with all of the Zook weapons that counter those of the Yooks, it is unclear whether his actions are officially sanctioned by the Zook government or if he is a renegade who takes actions into his own hands. In the latter case, the Zook government part of the escalation cycle would seem inadvertent. If VanItch's action is indeed unauthorized, Geisel added an apparent historical element to his cautionary tale, whether intentional or not. Born in 1904, Geisel was ten years old at the start of World War I, ostensibly set in motion when a Bosnian Serb terrorist, Gavrilo Princip, assassinated Archduke Franz Ferdinand, crown prince of the Austro-Hungarian Empire. Princip was a member of the group Young Bosnia, which advocated the independence of Bosnia from the empire and uniting it with the independent Kingdom of Serbia. Princip himself made it clear that his act was a provocation, stating, "I am a Yugoslav nationalist, aiming for the unification of all Yugoslavs, and I do not care what form of state, but it must be freed from Austria."[26]

Since Princip was a member of groups linked to the Kingdom of Serbia, the Austro-Hungarian Empire blamed the Serbian government, making demands concerning issues of sovereignty. Despite last-minute Serbian acquiescence to most of the demands, Austria-Hungary's declaration of war on Serbia subsequently resulted in the mobilization of contending alliance networks, leading to the "war to end all wars." As historians have long argued, there were many, many other causes for the war, but Princip's act (along with those of his associates) is popularly seen as the spark.

Geisel himself gave no indication that he had Princip and World War I in mind when writing the book in the 1980s. However, the anti-German prejudice he experienced in his hometown of Springfield, Massachusetts, the occurrence of World War I during his early teens, and his years at Oxford in the mid-1920s (where memories of the war and antiwar reaction were presumably still intense) provide circumstantial evidence that Princip's act had a subconscious effect on Geisel. Conscious, subconscious, or not present at all, what Geisel presents might be considered an inadvertent (third-party) action that triggers the spiral of escalation leading to a war between the two "superpowers." The fact that the initial action was small (a stone fired from a slingshot) might also be part of the cautionary tale.

The concept of a provocateur or a third party causing an escalation between two superpowers leading to the use of nuclear weapons has never been a staple of deterrence literature. Fidel Castro may have had the potential for doing so by commandeering the Soviet nuclear-armed missiles stationed in Cuba during the Cuban Missile Crisis in 1962, but that now seems improbable. However, the deterrence literature does express concern that extending nuclear umbrellas over allies might cause an "inadvertent" war between nuclear powers, similar to the alliances of World War I being sparked by Serbian intrigue and the resulting assassination.

DETERRENCE AND NONOCCURRENCE

Perhaps the most important moment in *The Butter Battle Book* happens when the Yooks and the Zooks have invented an identical weapons platform, the Blue Gooer. These giant machines have a bubble on the front (like a World War II aircraft) in which the pilot sits. Also reminiscent of World War II aerial bombardment, the blue machine has a tube extending from the top that can shoot blue goo onto the target. The pilots make funny faces at each other, both aware of their firepower equality. "I heard a voice yell," says Grandpa Yook. "'If you sprinkle us Zooks, / you'll get sprinkled as well!'"

In *The Butter Battle Book*, escalation as part of an arms race eventually leads to a potential act of great destruction. Both sides have chosen to escalate and react to the opponent's escalation, but they hesitate to actually start the catastrophic war. Geisel does not explain the hesitation of the Yooks and the Zooks. Neither participant fears for their personal safety and, by that point, they seem to be determined military professionals. Nor do they seem to have a sudden revelation concerning their common humanity ("-ookness"?) Perhaps neither can foretell the consequences for their respective nations, or perhaps they know the consequence to be mutual (assured) destruction?

In any case, both the Yooks and the Zooks demonstrate successful deterrence, in that both hesitate to actually start the war. This state of "strategic stability," albeit in a hair-trigger posture, could be indefinitely maintained on a lower level of alert, which the rival superpowers achieved during the Cold War.

As a work of satire, Geisel crafted *The Butter Battle Book* to illustrate the ridiculous irony of the nuclear weapons conundrum of MAD. In a more general sense, however, *The Butter Battle Book* satirizes war in general. Humanity has always been ambivalent about the legitimacy of war. Views have ranged from absolute horror of its occurrence, to the justification of its use for self-defense, to its glorification as building the character of nations. However, as a general rule, most individuals seek to avoid its physical harm, if not its dangers (which can themselves be exciting and intoxicating). Most nations—even vile dictatorships seeking aggressive gains—prefer to achieve their objectives using less-than-violent means, whether through peaceful relations or in the hope that their victims of aggression do not resist. In their model of the relations between sovereign states (generally rendered as nations or countries), international relations scholars postulate the international system as being "anarchic," in which no overriding authority can resolve the disputes between individual states.[27] International organizations designed to promote cooperation do exist, such as the United Nations. However, cooperation between independent, sovereign entities (presumably equal under international law) always remains to some degree voluntary and subject to the use of force. *Anarchy* remains the principal characteristic.

Some theorists might indeed argue that cooperation is the more natural state of human action, and there are constant appeals to the "better angels" of human nature.[28] Nevertheless, the historical record inclines most international relations scholars to concede the validity of the observations of ancient Greek strategist and writer Thucydides (often referred to as the father of the study of international relations) concerning a dialogue in 416–415 BCE between representatives of the Greek city-states of Athens and Melos. In convincing the Melians to surrender to their wishes, the more powerful Athenians point to what they see as the natural order: "The strong do as they will, the weak suffer what they must."[29] Thus, every state (be it strong or weak) is responsible for its own security. In other words, international politics remains a "self-help" system in which those with more "help" from themselves gain their objectives more often.

How therefore can war be prevented in a self-help system in which the strong go about doing what they can? Moreover, if the strong go about doing what they can and the weak necessarily suffer what they must, why aren't there more wars or perhaps a continuing state of war or aggression? Most international relations scholars would argue that rather than morality, the reason is "deterrence"—the concept that, by taking a particular military posture and related actions, one state can dissuade another from taking hostile action without resorting to actual violence. A definition of deterrence that is frequently used in academic literature is the "persuasion of one's opponent that the costs and/or risks of a given course of action he might take outweigh its benefits."[30]

The practical outcome of deterrence in foreign policy is supposed to be the nonoccurrence of war, or at least a limit on organized violence between armed groups. Theories of deterrence generally break down the concept into two distinct types: deterrence by *punishment* and deterrence by *denial*. Additionally, deterrence is linked to the concept of "compellence," which some consider the opposite of deterrence. In fact, their relationship is more familial.[31] Compellence is the act or attempt to compel another to take an action that is in the interest of the compelling state without that state resorting to violence. Another term for compellence is "coercive diplomacy," but the term "influence" is often used since it seems less forceful and thereby gentler than "compel" or "coerce." There are those, however, who argue that the latent threat of force is an actual use of force in itself (similar to how assault and battery are always linked—assault being the threat of harm and battery being the act of harm). Yet the act of deterrence—providing incentive for an opponent *not* to do something they otherwise might do—does indeed appear very much related to compellence, which involves providing incentive for an opponent to do something they might otherwise *not* do.

As logical as deterrence might seem, however, the theory is difficult to demonstrate in practice since there are many other extraneous factors ("variables" is the academic term) that might prevent undesired activity such as war from occurring.[32] Being a psychological state, deterrence is certainly difficult to quantify, and theorists end up having to prove a negative. But how does one prove what actually caused a "nonoccurrence"?

PUNISHMENT OR DENIAL

The next weapon the Boys in the Back Room invent in *The Butter Battle Book* is the Big-Boy Boomeroo. Apparently this weapon poses such danger that the Chief orders the Yook population "to stay safe underground." In the next illustration, the Yook population files passively into an underground shelter for protection, "obeying our Chief Yookeroo's grim command."

This section of *The Butter Battle Book* harkens back to the 1950s–1960s, when the US government, in the face of MAD, did make modest efforts to promote civil defense measures such as fallout shelters. Although this effort largely atrophied in the 1970s–1980s when Geisel was writing the book, the fear that such protection would be ineffective in the event of a tit-for-tat nuclear exchange remained in the public consciousness.

"Deterrence by punishment" is what most people view as the prime characteristic of nuclear war: a tit-for-tat response in which you nuke them because they nuked you.[33] Deterrence by punishment is understandable in a criminal justice situation because it is logical in theory to assume that the fear of a prison sentence is sufficient to deter at least some people from committing crime. (Whether it has that effect in practice is hotly debated.[34]) However, in a criminal justice situation, the apparatus of the legal system is likely to be much more powerful than the resources of the individual—even when armed with clever lawyers. Thus, there is the omnipresent possibility of a punishment that only affects the individual and not the prosecution.[35]

In the situation of two nuclear-armed nations with relatively equal capabilities, punishment smacks of mere revenge since the act deserving punishment—a nuclear attack—has already been carried out. Nevertheless, the logic of deterrence under relatively equal terms is that both sides would be deterred from striking first. This situation gained its own terminology as "assured destruction" or, to emphasize the relative equality in capabilities, "mutually assured destruction." Naturally, mutually assured destruction quickly gained the acronym MAD, since many critics saw it as exactly that. Geisel seems to have agreed with this understanding of the nature of the nuclear stalemate between the United States and the Soviet Union. In the final scene in the book, both Grandpa and VanItch stand on the wall holding weapons that will blow the other (and potentially all other people) into nonexistence.[36]

To achieve such a stalemate requires that some of the weapons capabilities of both sides be able to survive an initial attack (particularly a surprise attack) by the opponent. It was Robert McNamara, the notable secretary of defense in the Kennedy and Johnson administrations, who accepted the further theoretical premise that stability was best achieved if both sides accepted mutual vulnerability by forgoing any effort to defend themselves or increase their own capabilities so as to be superior to their opponent.[37]

Obviously, to suggest that the United States should make no attempt to defend itself or use its economic strength to develop weapons superior to those of the Soviet Union was a difficult sell, particularly given the assumption that America's Department of Defense was supposed to *defend* America. Nevertheless, theorists applauded and stability was presumed until the Soviet Union's activities and its opaque decision-making caused many Americans of the late 1970s to yearn for a "defense that defends"—a phrase that President Reagan used to justify a general increase in military spending and his desire to construct a strategic defense initiative, popularly known as his Star Wars program.[38]

Deterrence by punishment was the original strategy of the Cold War, particularly in its earliest period (1950s/early 1960s), when the United States had a monopoly, then a near-monopoly, and then an overwhelming superiority in "atomic weapons" (now considered a quaint term). Fearing that Soviet forces would be successful in attacking and capturing Western Europe during the USSR's period of expansion, the United States directly threatened to attack Russian means of production—and the cities in which they were contained—with nuclear bombs and missile warheads, referred to as the "massive retaliation" strategy. Holding this over the Soviet population (and perhaps leadership, although the leadership would naturally have greater access to the means of protection) was conceived to be a sufficient threat to deter a Soviet invasion. This supposedly led to "strategic stability" between the superpowers.[39] Western officials preferred to refer to their target sets as comprising military forces and war-related industries as opposed to population centers, although obviously civilians would suffer in any deterrence by punishment scenario. McNamara even developed a formula for what he conceived of as sufficient deterrence: nuclear weapons capability that could destroy between 20 to 33 percent of the Soviet population and 50 to 70 percent of their industrial capacity.[40] In strategic parlance this was referred to as "countervalue" targeting.

By representing the Yooks and Zooks as having weapons directed at each other with the capacity to destroy their population, *The Butter Battle Book* captures the concept of countervalue targeting, a deterrence posture based on ensuring the death of civilians. The discomfort that countervalue targeting caused many Americans—a horror at the moral turpitude of ensuring the death of civilians—along with concern that the Soviet dictatorial regime (recently emerging from Stalinism) might not value human life enough to be deterred, prompted McNamara and the Kennedy and Johnson administrations to somewhat back away from exclusive reliance on "city busting." Various terms such as "city avoidance" and "flexible response" (which could include conventional military operations in the context of the nuclear threat) were used to describe the idea that nuclear deterrence might move back toward the traditional military objective of destroying an enemy's military forces. This led to the concept of "counterforce," in which a nuclear attack would be primarily directed at nuclear and other military targets. Within the Soviet Union may of these targets were located away from population centers. Many of the US land-based nuclear-armed intercontinental ballistic missile (ICBM) sites were located in Montana, North Dakota, and Wyoming, and the United States relied more heavily on its sea-based ICBMs housed in patrolling submarines.

By the time of *The Butter Battle Book*'s publication, the Carter and Reagan administrations had actually moved further away from deterrence by punishment and countervalue targeting toward counterforce targeting, constituting a shift from targeting industries and cities to targeting the opponent's nuclear (and some conventional) forces. Counterforce targeting was the primary method for the second form of strategic nuclear deterrence, deterrence by denial. In contrast to focusing on revenge or retribution, the nuclear posture was intended to "deny" the opponent's (perceived) objectives, such as conquest of another's territory. This capability would require instilling a perception in the adversary that any aggressive gains could be neutralized, and thus the attempt would not be worth the cost. Exactly what level of latent force would be required would depend on the particular circumstances and location of the threat. However, Soviet decision-makers appeared to take this concept seriously, continuously calculating what they referred to as the "correlation of forces": the relative balance between their military strength and that of the West.[41] The closest American term for this was (and remains) "net assessment," although the latter term includes economic and other political factors.

The greatest fear of the western alliance was that deterrence might break down if the Soviets ever calculated a correlation that indicated they held a significant force advantage. If indeed they were committed to communist expansion into Western Europe, a positive correlation might result in a "window of vulnerability" in which there would be an incentive to "act now," before the correlation indicated less favorable conditions.[42] It remained unclear throughout the Cold War how the Soviet leadership calculated this correlation (or even believed it), but very fortunately they did not act in Europe.

Whether or not this nonoccurrence can be attributed to deterrence remains a mystery. However, the United States and its allies certainly tried—within changeable political support for higher levels of defense spending—to look comparable in force to the Soviets.

Within the confines of the nuclear weapons posture, a window of vulnerability is seen as existing when a nuclear-armed state perceives itself to be in a "use or lose" situation. In this circumstance, a state that fears its nuclear forces would be at greater risk in the future because its opponent is making significant and unanswerable improvements in its nuclear war fighting capabilities might be tempted to attack first in order to reduce this risk of future vulnerability. While this would seem a very dangerous gamble, some theorists point to the conjectured results of the prisoner's dilemma scenario used as a staple of game theory as justifying their concern over this possibility. Others view this as a flaw in deterrence theory itself.[43]

The fear of vulnerability is also seen as a cause of "unintended" arms races, in which potential opponents continuously increase and improve their stock of weapons without examining the far-off consequences of such mutual activity. This is one of *The Butter Battle Book*'s underlying premises: no one is thinking clearly about the consequences of war.

WAR TERMINATION

As the other Yooks march off to their underground shelter, Grandpa proceeds with the Chief's order to "run to the wall like a nice little man. / Drop this bomb on the Zooks just as fast as you can." As he heads to the wall, Grandpa discovers the young narrator has not joined the Yook civilians in the fallout shelter but remains on the surface. Impromptu, Grandpa takes him to the wall to "see me make history! / RIGHT HERE AND RIGHT NOW!" Instead of obeying the Chief's orders and dropping the bomb, when Grandpa reaches the wall, he encounters VanItch, who is also holding a Bitsy Big-Boy Boomeroo. Both Grandpa and VanItch stand poised to drop their weapons but remain motionless on top of the wall.

Until the Chief Yookeroo's last order to Grandpa, neither side seems to have sought to annex the other's territory or conduct a war for the sheer pleasure of destruction. A desire for security and status seems to be behind each successive wave of weapons development. As much as the Yooks/Zooks might want to persuade the others to butter their bread the right way, there seems to be no desire to convert by the sword. Thus, *The Butter Battle Book* does not capture the ideology of a Vladimir Lenin in outwardly spreading world revolution, nor of a Joseph Stalin intent on imposing a Soviet-style regime in every country reached by the Red Army, nor of an Adolf Hitler (which was the threat that Geisel well recognized). The conquest of territory or blatant genocide did not seem to be Yook objectives.

What then made the Chief Yookeroo order Grandpa to drop the bomb? That decision is never explained. Perhaps Geisel saw Americans' hostility toward and fear of communism (on display in the denunciations and hearings led by Senator Joseph McCarthy in the 1950s) as an ideological motive that could have prompted such a decision? By the time of the writing of *The Butter Battle Book*, however, American administrations had largely agreed to a policy of détente with the Soviet Union, more of a cold peace than the earlier periods of the Cold War. Or perhaps Geisel was portraying the military buildup of the Reagan administration. However, Reagan routinely portrayed the Soviet and Warsaw Pact population not as heretics but as prisoners of brutal authoritarian regimes. The most logical conclusion is that Geisel saw the arms race and the concept of MAD as eventually leading to a disastrous decision, whether caused by fear or by the corruption of power.

Despite the adoption of a nuclear weapons–based deterrence strategy, almost all American leaders viewed the possibility of nuclear war as a tragic mistake. In adopting deterrence, the goal was to prevent the other side (whose actions were out of US control) from making that tragic mistake. While American strategists theorized how such a war could begin and built plans to conduct it, very few formally examined how it could be terminated. Many Americans—including political leaders—adopted a fatalistic attitude: if such a war occurred, life on the planet would be largely doomed by nuclear winter, if not from the direct destruction. This was reflected in a series of movies and television films such as *The Day After* (1983) in the United States and *Threads* (1984) in the United Kingdom, released roughly at the same time as *The Butter Battle Book*. There is no evidence that these movies in themselves influenced Geisel's thinking, but rather that they tapped into what was a common theme in creative thought.

Nevertheless, a number of advisers and officials of the Reagan administration had previously examined the concept of interwar deterrence and the possibility that wars could deescalate due to a number of factors, including the resilience of national populations, a remaining relative balance of power, or perhaps mutual exhaustion. A noted study by sociologist Fred Ikle entitled *Every War Must End* was one of the first to raise war termination as a field of inquiry for academic strategists.[44] Ikle, like Warnke, served as director of the Arms Control and Disarmament Agency (1973–77) and then as undersecretary of defense for policy under President Reagan. He observed that the termination of a (nuclear) war deserves as much study and planning as the determination to go to war.[45] Critics replied that the very contemplation of the possibility of "terminating" a nuclear war somehow made its occurrence more likely. The logic of examining it seemed to them illogical. Fortunately, the Cold War ended in the exact same posture as Grandpa and VanItch, with both sides retaining the bomb but neither side using it. Perhaps *The Butter Battle Book* is as much a prescient prediction as a cautionary fable?

Although the end of the Cold War curtailed this debate, interest in the study of war termination has recently revived in the wake of US intervention in Afghanistan and Iraq and their apparently inconclusive endings.[46] Most scholarly interest in war termination assumes that future great power conflicts would be conventional rather than nuclear.

IN CONCLUSION: GUNS OR BUTTER

Does *The Butter Battle Book* remain relevant? Previously, Geisel's writing seemed always to be published in exactly the right place at the right time. Young readers and their parents were bored of Dick and Jane, and the Cat in the Hat stood ready. But *The Butter Battle Book* appeared in 1984, shortly before the 1989 fall of the Berlin Wall, making it seem now like a work of history rather than controversy. The arms races, nuclear standoffs, and Boys in the Back seem out of date. Even though they are more than seventy-five years old, nuclear weapons still may appear frightening, but the world no longer exists in a tense nuclear standoff. Whether that perception will continue in the coming rounds of great power competition between China, Russia, and the United States remains to be seen.

Like *The Sneetches*, *The Butter Battle Book* concerns the topic of ethnocentrism, which remains highly relevant in the era of great power competition. The Yooks and Zooks lack any evident distinction in appearance but still despise each other because of customary behavior. With the exception of an inconsequential difference in eating habits, Yook and Zook society appear as exact copies, something it would be difficult to argue if one looked from atop the Berlin Wall rather than Grandpa's wall. Of course, deep differences existed between the United States and its allies and the Soviet Union. The conflict was not about buttering bread, but about how humans should be governed. Throughout its history, the Soviet Union actively used force and intrigue to spread communist-style governance to other nations under the "scientific Socialist" theory that communism was historically inevitable. Both the United States and the Soviet Union had been allies against Hitler's Germany, a war Geisel had supported by the creation of military propaganda and training films. While not an apologist for authoritarianism, Geisel evidently did not see the ideological competition between the United States and the Soviets as meriting hostile nuclear postures.

Because the Yooks and the Zooks had no discernible ideological differences, critics argued that by implication no distinction existed between the United States and the Soviets. Critics argued that the fable was compromised by its moral relativism: "the view that moral judgments are true or false only relative to some particular standpoint (for instance, that of a culture or a historical period) and that no standpoint is uniquely privileged over all others."[47] Clearly *The Butter Battle Book* did not adhere to the general consensus of the time, that democracy had stark economic and social differences from communism. Of course, no satire could fully capture the seriousness of the political and moral concerns about the potential for continuing Soviet expansionism during the Cold War.

Likewise, the book seems to denigrate military service and nationalism. Yet Geisel saw himself as not anti-military but anti crazy. *The Butter Battle Book* makes it clear that Geisel did not view ideological differences as a rational cause for war. We can also infer that Geisel did not view nuclear strategists as rational actors, nor the resulting nuclear weapons standoff as rational. Therefore, nothing that nuclear strategists postulated, promoted, or modeled could possibly be rational.

Of course, *The Butter Battle Book* adopts a satirical approach to the potential for conflict propelled by ideology, which most Americans in the 1980s saw as legitimate and serious issues. Like Jonathan Swift's *Gulliver's Travels* (1726), the conflict in *The Butter Battle Book* derived from what the authors (and prospectively, their readers) view as inconsequential differences. Geisel studied English literature as a graduate student at Oxford (1926–27), so his familiarity with Swift's works and use of political satire would seem inevitable. Geisel never acknowledged that his inspiration came from *Gulliver's Travels*, but in Swift's book blood is to be shed over boiled eggs instead of buttered bread. In *Gulliver's Travels* the very tiny Lilliputians of a previously undiscovered South Seas island are at war with the nearby Blefuscudians regarding the proper way to break a hard-boiled egg.[48]

Even though *The Butter Battle Book* is satire, it nevertheless illustrates the progression of arms races, the role of science and technology, and some types of deterrence. *The Butter Battle Book* also touches on the more complex elements of deterrence (e.g., denial, compellence, game theory, extended deterrence, and tactical nuclear weapons). As a concept, deterrence remains relevant in an anarchic system that retains nuclear weapons, is prone to conflict, and can be affected by individual actions.

Or, to continue the poem:

> Overall, BBB teaches much more
> than just that war is right to abhor.

NOTES

1. David C. Logan, "Making Sense of China's Missile Forces," in *Chairman Xi Remakes the PLA*, ed. Phillip C. Saunders, Arthur S. Ding, Andrew Scobell, Andrew N. D. Yang, and Joel Wuthnow (Washington, DC, NDU Press, 2019), 393–435, https://ndupress.ndu.edu/Portals/68/Documents/Books/Chairman-Xi/Chairman-Xi _Chapter-11.pdf?ver=2019-02-08-112005-803; P. W. Singer and Ma Xiu, "China's Ambiguous Missile Strategy is Risky," *Popular Science*, May 11, 2020, https://www .popsci.com/story/blog-network/eastern-arsenal/china-nuclear-conventional-missiles/.

2. Kingston Reif and Shannon Bugos, "U.S., Russia Extend New START for Five Years," *Arms Control Today*, March 2021, https://www.armscontrol.org/act /2021-03/news/us-russia-extend-new-start-five-years.

3. Hayley Wood, "Interview with Filmmaker Ron Lamothe about *The Political Doctor Seuss*," *Massachusetts Foundation for the Humanities—News & Events*, September 2007, https://web.archive.org/web/200709160421/http://www.mfh.org /lamotheinterview/.

4. Jack St. Rebor, "Butter Battle Book," *Seussblog*, December 30, 2013, https:// seussblog.wordpress.com/2013/12/30/the-butter-battle-book/.

5. Wikipedia, s.v. "*The Butter Battle Book*," accessed August 15, 2020, https:// en.wikipedia.org/wiki/The_Butter_Battle_Book.

6. St. Rebor, "Butter Battle Book." See also M. Croft, "Hidden Meanings in Classic Dr. Seuss Books," *Best of New Jersey*, February 25, 2016, https://bestofnj.com /features/entertainment/hidden-meanings-dr-seusss-classic-books/.

7. Wood, "Interview with Ron Lamothe."

8. John Wilmes, "Dr. Seuss's Forgotten Anti-War Book Made Him an Enemy of the Right," *Outline*, July 30, 2018, https://theoutline.com/post/5601/dr-seuss-the -butter-battle-book-history?zd=1&zi=h5webi5x.

9. Paul C. Warnke, "Apes on a Treadmill," *Foreign Policy*, no. 18 (Spring 1975): 12–29, https://doi.org/10.2307/1147960.

10. Lawrence Freedman, *Deterrence* (Malden, MA: Polity Press, 2004), 240–42.

11. Greg Cashman, *What Causes War?* (Lanham, MD: Lexington Books, 1993), 177–84.

12. These meetings produced the Washington Naval Treaty of 1922, London Naval Treaty of 1930, and Second London Naval Treaty of 1936. Despite the fact that imperial Japan and Nazi Germany eventually denounced these treaties in the late 1930s, arms control proponents look at them—or at least the Washington Naval Conference (1921–22), and despite some cheating by Imperial Japan—as a success in arms limitation. Others point to the fact that naval limits seemed to have no effect on the actual outbreak of war.

13. Of note is that the terminology for such agreements shifted from disarmament—the rhetorical goal of the international conferences—to arms control by the 1960s, reflecting the fact that total disarmament remained an illusion in a still anarchic international environment.

14. On technology diffusion see Paul Stoneman and Giuliana Battisti, "The Diffusion of New Technology," in *Handbook of the Economics of Innovation*, ed. Bronwyn H. Hall and Nathan Rosenberg (Amsterdam: Elsevier, 2010), 2:733–60.

15. Freedman, *Deterrence*, 60–64.

16. There are literally thousands of sources and numerous books on Soviet spies in the Manhattan Project. Perhaps the first two were Oliver Pilat, *The Atom Spies* (New York: Putnam, 1952) and Bernard Newman, *Soviet Atomic Spies* (London: Robert Hale, 1952).

17. Forrest E. Morgan et al., *Dangerous Thresholds: Managing Escalation in the 21st Century* (Santa Monica, CA: RAND, 2008), 8, https://www.jstor.org/stable/pdf/10.7249/mg614af.9.pdf.

18. Morgan et al., *Dangerous Thresholds*.

19. Morgan et al., *Dangerous Thresholds*, 9.

20. A discussion of both the criticisms and logic of nationalism is found in Gustavo de la Casas, "Is Nationalism Good for You?," *Foreign Policy*, October 8, 2009, https://foreignpolicy.com/2009/10/08/is-nationalism-good-for-you/.

21. John von Neumann and Oskar Morgenstern, *Theory of Games and Economic Behavior* (Princeton, NJ: Princeton University Press, 1944).

22. Emphasis in original. *Stanford Encyclopedia of Philosophy*, s.v. "Game Theory," March 8, 2019, https://plato.stanford.edu/entries/game-theory/.

23. Thomas Schelling, *Arms and Influence* (New Haven, CT: Yale University Press, 1966). A concise description of the prisoner's dilemma (along with game theory in general) is found in Cashman, *What Causes War?*, 198–202.

24. Cashman, *What Causes War?*

25. "Game Theory."

26. Vladimir Dedijer, *The Road to Sarajevo* (New York: Simon & Schuster, 1966), 341.

27. Kenneth Waltz, *Theory of International Politics* (Reading, MA: Addison-Wesley, 1979); John Baylis et al., eds., *Strategy in the Contemporary World* (Oxford: Oxford University Press, 2002), 6–7, 67, 73; and Cashman, *What Causes War?*, 224–32.

28. Steven Pinker, *The Better Angels of Our Nature: Why Violence Has Declined* (New York: Penguin Books, 2012); and John Garnett, "The Causes of War and Conditions of Peace," in *Strategy in the Contemporary World*, ed. John Baylis, James Wirtz, Eliot Cohen and Colin S. Gray (Oxford: Oxford University Press, 2002), 75–76.

29. Robert B. Strassler, ed., *The Landmark Thucydides: A Comprehensive Guide to the Peloponnesian War* (New York: Free Press, 1996), 352 (book 5, section 89 in the standard English translation by Richard Crawley).

30. Alexander L. George and Richard Smoke, *Deterrence in American Foreign Policy: Theory and Practice* (New York: Columbia University Press, 1974), 11.

31. Lawrence Freedman, *Strategy: A History* (Oxford: Oxford University Press, 2013), 163, 190–191.

32. Colin S. Gray, *National Security Dilemmas: Challenges & Opportunities* (Washington, DC: Potomac Books, 2009), 58, 62–66.

33. Robert Axelrod, *Evolution of Cooperation* (New York: Basic Books, 1984), 53–84; and Cashman, *What Causes War?*, 202–6.

34. See official refutation in National Institute of Justice, *Five Things about Deterrence*, US Department of Justice, May 2016, https://www.ojp.gov/pdffiles1/nij/247350.pdf.

35. Freedman, *Deterrence*, 60–71.

36. According to the strict definition, however, this would really *not* be a MAD situation since—with only one weapon apiece—whichever side struck first would destroy the other side's weapon, thereby "winning" the struggle. However, MAD was never popularly viewed by its strict definition, and it was generally assumed by nonstrategists that its failure would result in mutual destruction—hardly a victory for the side receiving a lesser amount of destruction. In Geisel's portrayal, both weaponeers (Grandpa and VanItch) stand side by side; therefore neither would be likely to survive (nor would the narrator).

37. Robert Jervis, *The Illogic of American Nuclear Strategy* (Ithaca, NY: Cornell University Press, 1984), 26–27.

38. Lawrence Freedman, *Evolution of Nuclear Strategy*, 3rd ed. (New York: Palgrave Macmillan, 2003, 394–97.

39. See discussion in Colin S. Gray, "Strategic Stability Reconsidered," *Daedalus* 109, no. 4 (Fall 1980): 135–54.

40. Freedman, *Evolution of Nuclear Strategy*, 233.

41. See, for example, James K. Womack, "Soviet Correlation of Forces and Means: Quantifying Modem Operations" (master's thesis, US Army Command and General Staff College, 1990). Perhaps the definitive work is a partially declassified (sanitized) report by Science Applications International Corporation for the Defense Nuclear Agency: Judith K. Grange et al., *Soviet Measurements of Strategic Balance and Arms Control*, DNA-001-84-C-0307 (Washington, DC: Defense Nuclear Agency, April 21, 1986), https://apps.dtic.mil/sti/pdfs/ADA221221.pdf.

42. Freedman, *Evolution of Nuclear Strategy*, 72–85.

43. Robert H. Johnson, "Periods of Peril: The Window of Vulnerability and Other Myths," *Foreign Affairs* 61, no. 4 (Spring 1983), https://doi.org/10.2307/20041562.

44. Fred C. Ikle, *Every War Must End* (New York: Columbia University Press, 1991).

45. Ikle, *Every War Must End*, 1–2.

46. E. Stanley and J. Sawyer, "The Equifinality of War Termination," *Journal of Conflict Resolution* 53, no. 5 (October 2009): 651–76; Dominic J. Caraccilo, *Beyond Guns and Steel: A War Termination Strategy* (Santa Barbara, CA: Praeger, 2011); and Gideon Rose, *How Wars End: Why We Always Fight the Last Battle* (New York: Simon & Schuster, 2011).

47. Emrys Westacott, "Moral Relativism," *Internet Encyclopedia of Philosophy* (ISSN 2161-0002), https://iep.utm.edu/moral-re/ (accessed February 2, 2021).

48. Jonathan Swift, *Gulliver's Travels into Several Remote Nations of the World* (Paris: A. and W. Gaglignani, 1826), 1:74–79. Interpreters generally equate Lilliput with Great Britain and Blefuscu with France. Although Lilliputians traditionally broke their hard-boiled eggs on the larger end, an emperor of Lilliput, whose son was cut by

an egg shard while breaking the larger side, decreed that henceforth all Lilliputians would break eggs on the smaller end. This decision inspired wars both foreign and domestic, pitting—within Lilliput itself—the Big-Endians versus the Little-Endians and the emperor. Interpreters see the initial emperor as representing Henry VIII and his royal descendants, the Little-Endians as English Protestants, and the Big-Endians as English Catholics. Sometimes the emperor reverts to being a Big-Endian (Charles I and James II) and ends up (bad pun) losing his crown or his head. For later Lilliputian generations, the Big-Endians appear to represent the "high church" adherents of the Anglican Church/Church of England (supported by the emperor/king), with the Little-Endians being the "low church" members of the dissenting (other) Protestant sects (supported by the commoners/Parliament). Since the Lilliputian religious text states that an egg should be broken on "the convenient end," there is no divine sanction to resolve the dispute. Meanwhile, Blefuscu (France) remains a staunchly Big-Endian (Catholic) empire, readily supporting Big-Endian pretenders to the Lilliputian throne (Charles II/James Stuart) and willing to hazard a continuing series of wars against Lilliput. Both sides attempt to utilize Gulliver's size in their cause.

Part V

CONSEQUENCES OF WARFARE

13

How the Grinch Stole Christmas and Traumatic Stress

Montgomery McFate

Hailed as one of Dr. Seuss's most successful books, *How the Grinch Stole Christmas* (1957) tells the tale of the cantankerous, forsaken Grinch. Living alone in an isolated ice cave on top of a mountain, his heart brims with resentment toward his neighbors, the Whos. "He stood there on Christmas Eve, hating the Whos," observes the narrator. "Staring down from his cave with a sour, Grinchy frown / At the warm lighted windows below in their town." Not only does the isolated, grumpy, cave-dwelling Grinch growl and sneer at the noisy, materialistic idiots down in Who-ville, he also abuses his sad, disheartened dog, Max.[1]

Like many other "Christmas classics," the arc of the story concerns transformation and redemption.[2] At the beginning of the story the Grinch takes pleasure at the misfortune of others (schadenfreude), even proactively engaging in sadistic acts to harm others.[3] Specifically, to stop Christmas from coming the Grinch steals all physical manifestations of the holiday (such as ornaments, rich food, and presents) from the village. Despite having all their material goods plundered, the Who villagers unite in song. Witnessing their resilience as a community, the Grinch has an epiphany: "Maybe Christmas . . . perhaps . . . means a little bit more!" In atonement for his earlier hostile actions, he then returns all the stolen items. At the very end of the story, the Grinch has the honor of carving the roast beef at the celebratory Christmas feast.

Most readers interpret the *Grinch* as a moral lesson about the spirit of Christmas, namely that the holiday should unite family and community rather than encourage materialism. While such a reading is not wrong, *How the Grinch Stole Christmas* also tells the story of Ted Geisel's personal wartime trauma, his sources of internal resilience, and his eventual healing process.[4] We begin with the Grinch as an allegorical representation of the author. Because Ted Geisel viewed the Grinch as an embodiment of himself, we can reasonably assume that his portrayal of the Grinch reflected (to some degree) his own experiences, mental state, and emotions. Next, we argue that Ted Geisel experienced wartime trauma while trapped behind enemy lines in December 1944 during the Battle of the Bulge. He only mentioned his experience in the Ardennes forest to a single journalist during an interview in 1964, and then never discussed it again.[5] In fact, unlike many contemporary celebrities,[6] he almost never discussed his political views, sex life, or mental health in public. But statistical probabilities indicate that Ted Geisel probably experienced some type of trauma during his World War II military service and subsequently also experienced post-traumatic stress disorder (PTSD) after the war.[7] Then we evaluate the Grinch's symptoms of PTSD in light of the criteria defined by contemporary medical literature and the *Diagnostic and Statistical Manual* (DSM) (which is the standard medical reference book for the diagnosis and treatment of mental disorders). These criteria include exposure to traumatic events, progressive worsening of the disorder, physical alterations, emotional dysregulation, avoidance of triggers, social withdrawal, and reenactment of trauma.[8] Next we examine how individual resilience promotes post-traumatic healing. In the case of the Grinch (who appears to be completely cured of his PTSD by the end of the book), social acceptance by the citizens of Who-ville contributed to the Grinch's personal resilience, which eventually allowed him to heal. But what about the trauma suffered by the citizens of Who-ville at the hands of the Grinch? Finally, we examine the specific attributes of Who society that enabled their resilience and recovery (otherwise known as social resilience). Although not generally recognized as a classic manuscript in the scientific literature of trauma, *How the Grinch Stole Christmas* nevertheless offers an astonishingly accurate portrayal of wartime trauma and recovery.

THE GRINCH AS ALLEGORY

The character of the Grinch, as Theodor Geisel revealed, was based on himself. "I was brushing my teeth on the morning of the 26th of last December," as Geisel observed in December 1957, "when I noticed a very Grinch-ish countenance in the mirror. It was Seuss! So I wrote about my sour friend, the Grinch, to see if I could rediscover something about Christmas that obviously I'd lost."[9] So deep was this personification of himself in the Grinch that Ted Geisel had a vanity license plate on his car that stated "GRINCH."[10]

Indeed, the biographical details of the Grinch's life track with those of Ted Geisel. With the application of some basic math (and a little imagination), we can determine the Grinch's eligibility to serve in World War II. Early in the book, the Grinch unequivocally identifies his age. "Why, for fifty-three years I've put up with it now!" growls the Grinch. "I MUST stop this Christmas from coming!" Since the book was published in 1957, at which point the Grinch was fifty-three years old, he must have been born around 1904. (Ted Geisel was also born in 1904.) From these facts, we can extrapolate that the Grinch was in his late thirties or early forties during World War II. Therefore it seems logical to assume that the Grinch would have served in the military during the war.

Given Ted Geisel's admission regarding the origin of the character and the overlapping time frames of their life events, the Grinch does appear to be an allegorical representation of the author. But what is an allegory? In the field of literary theory, an allegory is a rhetorical device that conveys a deeper truth not explicitly contained within the boundaries of the narrative.[11] Allegories typically have three elements: metaphor (one thing is substituted for another), personification (abstract qualities are represented as though they were real persons), and moral conflict (the personified metaphors engage in inner battles).[12] In the *Grinch*, all three elements of allegory are met: Whoville substitutes for US civil society (metaphor), the character of the Grinch personifies the grouchiness of Ted Geisel (personification), and a moral conflict occurs between the forces of misanthropy and redemption (inner battle). In the *Grinch*, Ted Geisel develops a complex, dual, autobiographical allegory that demonstrates a real mastery of rhetorical forms,[13] which is almost certainly a legacy from his brief educational experience at Oxford studying Jonathan Swift.[14]

WARTIME TRAUMA

Dr. Seuss never reveals the Grinch's backstory, leaving readers to guess the etiology of his misery. Nobody starts out living in an ice cave; something caused the Grinch's torment, most likely the experience of bloodshed and violence during World War II. Although we lack precise biographical details regarding the Grinch's exposure to traumatic events, we know that the Grinch symbolizes Geisel, and Geisel was exposed to the harsh and devastating conditions of war,[15] having been trapped behind German lines in the Ardennes forest during the Battle of the Bulge. Very few military personnel survive this type of experience alive and/or unscathed. During the Vietnam War, for example, the US Army Special Operations Group, deployed behind enemy lines, suffered a 100 percent casualty rate.[16] More recently, Spetsnaz forces suffered a 95 percent casualty rate in the Ukraine.[17]

That Ted Geisel had PTSD appears likely based on the statistics on other veteran populations. Although the diagnostic criteria were not determined until thirty-five years after World War II, "combat stress" and "battle fatigue" (as PTSD was previously known) were common among military personnel who served during World War II. One study of World War II veterans concluded that the lifetime prevalence of PTSD was 23 percent.[18] Another study found that 46 percent of World War II veterans suffered from PTSD.[19] The actual numbers are probably much higher, given that military personnel of the "greatest generation" underreported their personal mental health issues.[20] Moreover, the stigma associated with combat stress resulted in incorrect diagnoses as anxiety disorder, melancholia, and even schizophrenia.[21] Psychiatrists have observed conclusively that the more severe and traumatic the wartime experience (such as being trapped behind enemy lines), the more likely a service member is to develop PTSD.[22]

DIAGNOSTIC CRITERIA

Thus far, we have established that the Grinch is an allegorical representation of the author, and that the author almost certainly experienced trauma during his World War II service. But does the Grinch display any symptoms of PTSD according to the diagnostic criteria commonly employed by medical professionals? In fact, *How the Grinch Stole Christmas* clearly portrays the following criteria: progressive worsening of the disorder, physical alterations, emotional dysregulation, avoidance of triggers, social withdrawal, and reenactment of trauma.

Progressive Worsening of the Disorder

Certain diseases and disorders—mental or physical—require almost no medical intervention to return the body to a healthy, homeostatic state.[23] But other conditions classified as "progressive" worsen over time without medical treatment.[24] Recent scientific research indicates that PTSD should be classified as a progressive disease; it is neither self-healing nor does it gradually improve over time. Veterans with untreated PTSD rarely recover spontaneously on their own. In one study of war veterans from the former Yugoslavia who received no psychiatric or psychological treatment for PTSD, 83.7 percent still had symptoms a decade after the war's end.[25] Other studies have determined that symptoms of combat trauma may persist up to forty-five years after the event.[26] Further complicating matters, PTSD often has a "delayed onset," meaning that symptoms may not manifest for years after the event or injury.[27] Particularly in military and emergency service personnel, "a progressive escalation of distress or a later emergence of symptoms" often occurs,[28] sometimes more than thirty years after the event.[29] On the other hand, however, patients who are treated for PTSD (with psychotherapy or medications) have an improved long-term prognosis.[30]

In the Grinch's case, documentary evidence indicates that his PTSD symptoms grew more severe over time. A comparison of the 1955 and 1957 versions of the Grinch clearly demonstrates a rapid degeneration of his mental and moral state. In 1955 the Grinch makes his first appearance in a thirty-three-line poem entitled "The Hoobub and the Grinch," in which he attempts to deceive a sunbathing Hoobub into purchasing a green string. In the 1955 "Hoobub" story, the Grinch commits deceptive business practices (or perhaps fraud if we really stretch the definition). But this type of manipulative marketing, which is typically associated with consumer capitalism, rarely results in prosecution. As the narrator of the 1955 story reminds us, "I'm sorry to say / that Grinches sell Hoobubs such things every day."

In 1957, with the publication of *How the Grinch Stole Christmas*, on the other hand, the Grinch seeks to destroy the *entire system* of consumer capitalism. Living in a remote, frozen wilderness, the Grinch has fully divested himself of normative social constraints. Like any fervent revolutionary, the Grinch has no scruples about employing criminal methods to attain his objective. To subvert the political system of Who-ville, the Grinch perpetrates multiple acts of animal cruelty (e.g., forces his poor dog Max to pull an overloaded sled), commits theft of Who property (e.g., steals the Christmas gifts), violates zoning laws (e.g., dwells in a cave), commits burglary (e.g., enters a house through the chimney with intent to steal), violates municipal anti-dumping ordinances (e.g., disposes of the Whos' property over the cliff), drives recklessly (e.g., drives a nonmotorized conveyance [sled] at excessive speed), commits an act of child abuse (has unauthorized contact with a minor child when he pats Cindy-Lou's head), and so on.

As this inventory of criminal allegations demonstrates, the Grinch's PTSD symptoms, left untreated, degenerated significantly between 1955 and 1957. Clearly, the Grinch received little therapeutic help from the psychologists down in Who-ville. In addition, certain lifestyle factors probably exacerbated the Grinch's symptoms and accelerated his decline, such as his older age (e.g., 53 years); adverse living conditions (e.g., dwelling in an unheated, unfurnished ice cave); unemployment (e.g., having no visible means of financial support); and loneliness (e.g., having no companionship except Max the dog).[31]

The progressive worsening of the Grinch's mental state between his first appearance in the 1955 "Hoobub" and his 1957 appearance in *How the Grinch Stole Christmas* is characteristic of PTSD. Moreover, the Grinch's environment includes all the psychosocial elements leading to an adverse prognosis. On this basis, it seems reasonable to conclude that the Grinch suffered from PTSD. By extension, Ted Geisel probably suffered PTSD too.

Physical Alterations

Among the diagnostic criteria for PTSD, the DSM includes a category that pertains primarily to physical changes in the body (rather than mood, intrusion, or avoidance symptoms). This category involves "marked alterations in arousal and reactivity" as evidenced by exaggerated startle response, sleep disturbance, and so on. Indeed, recent medical literature on PTSD indicates that poor physical health correlates strongly with PTSD symptoms.[32] Experiencing traumatic events is believed to alter a "variety of physiological and behavioral pathways, all of which contribute to allostatic load, which in turn increases susceptibility to illness."[33]

What physical symptoms of the Grinch might be indicative of PTSD?[34] When reading the text of the story, the first illustration of the Grinch shows him standing at the mouth of his ice cave. Physically, he appears to be debilitated and unwell. Not only is he balding and overweight, his eyes are completely red and his skin is a lurid green.[35] The Grinch also has a very serious cardiovascular defect (e.g., his heart is two sizes too small).

Oddly enough, green skin may actually be associated with PTSD. As early as the sixteenth century, medical texts described a condition known as "chlorosis," in which an individual's skin assumed a markedly green hue. Typically diagnosed in girls experiencing puberty, chlorosis also occurred more rarely in males.[36] Before advances in blood testing proved that chlorosis resulted from low hemoglobin levels, "green skin disease" was attributed to a variety of improbable causes, such as unrequited love, masturbation, sweet coffee, constipation, virginity, constricted blood vessels, and intestinal blockages. More importantly, doctors believed that chlorosis (like PTSD) resulted from shock and could be caused by "an 'injurious cause' which provokes a moral shock that 'discharges the disorder into the organism.'"[37] In other words, the medical world accepted that a traumatic event could be so shocking that it would cause physical changes in human beings.

Although it may seem odd to contemporary readers, until the mid-1930s, chlorosis was an exceptionally common diagnosis. As we know, Ted Geisel was born in 1904 and therefore grew up in a society where green skin disease would have been associated with trauma in the popular imagination. Therefore, within the logic of the allegory at the core of the book, the Grinch's green skin signifies his exposure to traumatic events and further confirms Ted Geisel's own traumatic experiences during the war. (On a positive note, green skin disease seemed to "disappear spontaneously," in the late 1930s at the same time that the diet of women in Western Europe and the United States improved with an increase in iron and other nutrients.)[38]

The Grinch's green skin also evokes the mythological figure of the Green Man, a decorative motif found commonly in European churches from the late eleventh to the sixteenth centuries. The Green Man typically has a human face surrounded by (or composed of) leaves, with vines or snakes emerging from his forehead or mouth. Although the exact origin of the Green Man remains obscure, scholars believe that this medieval imagery has pre-Christian, pagan origins. The Green Man has been characterized as a symbol of transformation, representing "opposing metamorphic trajectories held in tension: leaves transforming into animal, and at the same time animal transforming into leaves. Both and neither animal and plant, . . . this is a creature ever in a state of disjunctive becoming."[39]

During the medieval era the Green Man became deeply and directly associated with Christmas,[40] precisely as the Grinch is associated with Christmas in the present era. In the late fourteenth-century poem "Sir Gawain and the Green Knight," a green knight rides a green horse into King Arthur's hall at Camelot. Like the Grinch, the Green Knight interrupts the Christmas festivities, introducing pagan chaos and disorder on the ordered spectacle of a Christian holiday. Although the Green Knight does not transform physically, he does exist betwixt and between states of being. "The Green Knight belongs simultaneously to the vegetable world (with his green skin and beard like a bush), the non-human animal world (his skin the same verdant shade as his horse's), and the world of things (his green clothes and armour matching his epidermis)."[41]

Like the Green Man and the Green Knight, the Grinch exists perpetually in a liminal state, transforming from evil to good, from exterior to interior, from sickness to health, and so on. Like these other green men of pagan origin, the Grinch is "one of the old gods,"[42] "evocative of processes unknown and unknowable, located at the outermost edges of experience."[43]

Emotional Dysregulation

At the beginning of *How the Grinch Stole Christmas*, the narrator theorizes about the origins of the Grinch's aversion to Christmas, admitting that "no one quite knows the reason." Then, the narrator offers two possible origins: "It *could* be his head wasn't screwed on just right. / It *could* be, perhaps, that his shoes were too tight." The Grinch's hatred of Christmas is thus initially attributed to either his individual psychology ("his head") or his individual physical being ("his shoes"). In the next verse, the narrator provides a third possibility: "But I think that the most likely reason of all / May have been that his heart was two sizes too small." The last possibility is the correct answer: the Grinch's dislike of Christmas results from neither mental nor physical trauma, but rather from emotional suffering ("his heart").

The association of PTSD with the heart is not as farfetched as one might imagine. Before PTSD was formally identified as a diagnosable condition in 1980, a cluster of symptoms now associated with PTSD was commonly named "soldier's heart" (also known at various times as DaCosta's syndrome, effort syndrome, and disordered action of the heart, among other terms). During the Crimean War, British Army doctors noted that some soldiers experienced palpitations, fatigue, chest pain, and other irregularities of the heart. Later during the Civil War, American doctors referred to "soldier's heart" in more formal terms as "cardiac muscular exhaustion."[44] As one might expect before the advent of psychology, doctors and scientists believed that physical agents (such as excessive glandular secretions, pack straps, or toxic bacteria) caused the disorder. In 1941, writing in the *British Medical Journal*, universally respected pulmonary physician and scientist Paul Wood entirely shifted the paradigmatic understanding of the disease,[45] observing that among doctors and patients "the symptoms were misinterpreted and the heart was blamed." But in fact the syndrome was "unrelated" to any physical factor and could instead "be traced to emotional causes."[46] In Paul Wood's opinion, the origin of what would now be called PTSD was *emotional*.

As a result of Dr. Paul Wood's very influential 1941 article, the medical establishment came to realize that *physical ailments* might not have *physical causes*. Following this epiphany, doctors ceased diagnosing soldier's heart in their patients and by the mid-1940s, no well-informed medical professional viewed soldier's heart as a disease.[47] The terminology, however, continued to be commonly used by lawyers, poets, high school students, Shakespeare scholars, and art critics until the 1950s.[48] Anyone who served in World War II (such as Ted Geisel) would have certainly been familiar with the term since military medicine lagged behind civilian medical practice in this domain. In fact, new US Army recruits were given a "mark time test" for soldier's heart, which essentially meant marching in place until fatigued.[49] In military hospitals, soldiers were "cured" of the disease by an operation that involved cutting into the nerves.[50] Nurses continued to be trained to treat patients diagnosed with soldier's heart.[51] A diagnosis of soldier's heart continued to be used as a reason by the Army for medical discharge.[52]

In sum, by attributing the origin of the Grinch's PTSD to his emotions (rather than to physical constraints or brain physiology), Ted Geisel was fundamentally correct. The Grinch suffered from an "emotional disturbance and not from any disease or alteration of visceral function."[53] The Grinch's physical symptoms were symptomatic of his emotional state.

Avoidance of Triggers

As defined by the DSM, the third criteria for PTSD is "persistent avoidance of stimuli associated with the traumatic event(s)," including external reminders (people, places, conversations, activities, objects, situations) that arouse distressing memories, thoughts, or feelings associated with the traumatic event. The "avoidance cluster" (as it is referred to in medical literature) contains a range of responses beyond simple avoidance, such as "failure to recall important aspects of the trauma, loss of interest in significant activities, subjective detachment or estrangement from others, restricted range of affect, and sense of foreshortened future."[54] More than the intrusive thoughts or alteration of mood, avoidance is the classic, characteristic symptom of PTSD and is often dispositive to the diagnosis of PTSD.[55]

Although *How the Grinch Stole Christmas* was written long before PTSD had been identified as a disorder, the Grinch's avoidance of certain stimuli can nevertheless be recognized as one of the most common PTSD symptoms. Specifically, the celebratory holiday noise of the Whos functions as the primary stimulus (also known as a "trauma trigger") for the Grinch, causing an involuntary negative response, "Oh, the noise! Oh, the Noise!," complains the Grinch. "Noise! Noise! Noise! / That's *one* thing he hated! The NOISE! NOISE! NOISE! NOISE!" Interestingly, when our green-skinned protagonist agonizes about the "NOISE" of the Whos' Christmas celebration, the party has not even started. The Grinch's horror at the noise of Christmas celebrations does not result from any actual noise, but from the *anticipation of the triggering stimulus*. Like the Grinch, patients with PTSD tend to be "overly sensitive to unpredictable threats," experiencing normal "fear" but enhanced "anxiety."[56] Interestingly, for PTSD patients "anticipated anxiety of these aversive events" is often more painful than the actual triggers.[57]

Like veterans who avoid Fourth of July fireworks, the Grinch seeks to avoid the triggering stimulus of loud noise. But instead of retreating to his ice cave to avoid the Whos' holiday ruckus, the Grinch commits a violent act to *destroy the source* of the triggering stimulus. As he declares, "I MUST find some way to stop Christmas from coming!" Exposure to trauma-related triggers (such as loud noise) may indeed provoke sudden, intense, and automatic emotional responses.[58] Among military veterans, moreover, PTSD correlates strongly with increased anger, aggressive behavior, and actual violence.[59] In one study, 33 percent of the veteran sample reported committing at least one act of violence or aggression in the community, leading to the conclusion that a probable diagnosis of PTSD yielded higher odds of severe violence or physical aggression.[60]

In sum, the Grinch's PTSD symptoms track perfectly with the specific manifestation of the disorder in military veterans, including enhanced anxiety, strong emotional responses, and potential for violence against the source of the trigger.

Withdrawal and Isolation

Having apparently fled the triggering stimulus of Who-ville noise, the Grinch took up residence in an isolated ice cave on top of a mountain. Completely estranged from the community of Whos, the Grinch has no other companionship than Max, the abused dog.

Clinical research indicates that for individuals with PTSD, avoidance symptoms like these may be the most detrimental symptoms to psychosocial functioning and quality of life.[61] Unlike physical symptoms of PTSD (such as green skin), which have a limited scope of effect, avoidance symptoms extend to every aspect of an individual's life. The Grinch withdrew from family and community in order to avoid triggering his PTSD, but found himself alone. Although lonely exile may not have been the Grinch's actual objective, it was nevertheless the result of his avoidant behavior.

As the example of the Grinch makes clear, the avoidance cluster of PTSD symptoms often has a cascading negative effect on an individual's well-being. An individual's efforts to avoid triggers may result in restricted social networks and social isolation.[62] Reclusive sequestration in turn may cause veterans to view themselves as outsiders with no community. (The Grinch, for example, perceived the community from the *outside*: "Staring down from his cave with a sour, Grinchy frown / At the warm lighted windows below in their town.") In turn, self-perception as an outsider may reduce the likelihood that an individual will seek help from the health-care system.[63] Lack of health care may exacerbate PTSD symptoms, and so on. For this reason, early avoidance symptoms may predict subsequent development of PTSD.[64]

Reenactment of Trauma

Individuals who have survived traumatic events sometimes *reenact* the conditions, situations, or actions of the original experience in their present lives.⁶⁵ Psychologists who specialize in trauma do not fully understand the cause (or the effect) of traumatic reenactment.⁶⁶ In some cases, the repetition of the traumatic experience may help an individual "achieve mastery" over the event, allowing psychological integration and emotional healing to take place.⁶⁷ In other cases, victims of trauma become perpetrators in an endless cycle of pain. Indeed, several studies have linked childhood physical abuse to violent and aggressive behavior as an adult.⁶⁸ In particular, sexual offenders against children are more likely to have been sexually abused as children.⁶⁹

In real life, reenactment of traumatic events rarely results in resolution or mastery.⁷⁰ In *How the Grinch Stole Christmas,* however, the Grinch's recapitulation of his wartime trauma has a positive result, leading to psychological healing and community reintegration. In what way does the *Grinch* involve reenactment of war trauma? To put it simply, the protagonist plans and executes a brutal attack on a civilian community, with the intention of wounding them physically and emotionally. More specifically, to stop Christmas from coming, the Grinch mounts a covert nighttime incursion into hostile territory and sabotages the enemy's supplies and logistics in order to disrupt enemy activity.

The Grinch's preparation for Operation Xmas begins with deception. His first step is to design a costume that will enable him to impersonate Santa Claus. "What a great Grinchy trick!" the Grinch says to himself as he admires his red-and-white costume. "With this coat and this hat, I look just like Saint Nick!" In addition to an appropriate style of dress, effective deception requires suitable props. "All I need is a reindeer . . .' / The Grinch looked around. / But, since reindeer are scarce, there was none to be found." Military operational planning requires significant adaptability.⁷¹ Here, the Grinch employs existing environmental resources to solve his problem.⁷² "If I can't *find* a reindeer, I'll *make* one instead!" In the illustration that accompanies the verse, the Grinch—with an evil smile—uses thread to affix a fake horn to the top of Max's head.

Ted Geisel clearly possessed significant skills and knowledge regarding military deception and military transport. Having perfected his deception, the Grinch hooks his dog Max up to the sleigh and proceeds swiftly down the mountain into enemy territory. While the use of hoofed animals to transport personnel and equipment (such as the donkeys during World Wars I and II) is well known,⁷³ dogs have also been extensively used in warfare.⁷⁴ Max, as a working military dog, is part of a long tradition of canine service. During World War I, for example, the Belgian Army used dogs to tow Maxim guns on wheeled carriages, and the Soviet Army used dogs during World War II to drag wounded men to medical stations.⁷⁵ Max has clearly undergone extensive training for combat operations,⁷⁶ which is made apparent by Max's patience and silence while the Grinch conducts the raid.

To achieve the objective of Operation Xmas, the Grinch must remain undetected during his nighttime infiltration into enemy territory. In order to remain undetected, most military planners recommend using unexpected entry points to occupied structures.[77] The Grinch chooses to make his initial incursion through the chimney, which is "a rather tight pinch. / But, if Santa could do it, then so could the Grinch."

The aim of the Grinch's raid is not to capture and kill enemy forces, but rather to disrupt the enemy's holiday logistics. Attacking the logistics infrastructure of enemy forces has historically been a successful technique for disruption or destruction of enemy combat capacity.[78] To that end, the Grinch "slithered and slunk, with a smile most unpleasant, / Around the whole room, and he took every present!" In addition to the pop guns and bicycles, roller skates and drums, and so on, the Grinch also sabotages the Christmas celebration by stealing the comestible supplies. As the narrator informs us, "Then he slunk to the icebox. He took the Whos' feast! / He took the Who-pudding! He took the roast beast!"

No plan survives contact with reality (as they say in the military), and despite the Grinch's excellent operational planning, he is discovered by a civilian. "He turned around fast, and he saw a small Who! / Little Cindy-Lou Who, who was not more than two." Civilians in the battlespace pose a continual challenge to armed forces, in some cases impeding an objective and in other cases causing total mission failure.[79] A well-known, highly effective strategy for dealing with local civilians who ask too many questions is denial. When caught red-handed while destroying the holiday materiel of the Whos (an offense punishable under the Uniform Code of Military Justice and also prohibited by the Geneva Convention), the Grinch denies the accusation with an alternative explanation of the facts. In answer to Cindy-Lou Who's question ("Santy Claus, why, / *Why* are you taking our Christmas tree? WHY?), the Grinch tells her that he will fix the malfunctioning light and return the tree. She appears to accept this lie as plausible. Clearly, Ted Geisel understood the basic principle of black operations behind enemy lines, namely that effective denial must be believable to the victim.[80]

In sum, Operation Xmas reenacted the Grinch's wartime trauma, which tracks perfectly with the criteria established in the DSM and the medical literature.

RESILIENCE AND HEALING OF THE GRINCH

Despite the gravity of his apparent psychological and emotional wounds, the Grinch heals from his PTSD in just sixty-four pages—a rather remarkable achievement! In the field of psychology, the capacity to recover from trauma and resurrect the self is called resilience. According to the American Psychological Association, "resilience" is the "process of adapting well in the face of adversity, trauma, tragedy, threats or even significant sources of stress."[81] The inadequacies of this definition have been noted by a number of psychologists and social scientists, in particular the American medical establishment's view of resilience as an individual personality trait rather than as a social process involving both individual and community.[82]

Since World War I, PTSD and its precursor disorders, such as shell shock, have been viewed by clinicians as a psychological malady caused by a specific event in an individual soldier.[83] In the United States, the postdeployment mental health framework of the Veterans Administration has also focused on the individual as an "isolated node,"[84] rather than an embedded member of various communities. Thus, a substantial body of scientific research has sought to identify the particular personality traits and psychological orientation that influence the level of individual resilience, such as optimism, cognitive flexibility, active problem solving, moral code, and altruism.[85]

Because the scientific establishment tends to view PTSD as an individual problem, resilience also tends to be viewed as a trait of individuals. Of course, veterans do not exist in isolation but live in a social context with myriad connections to family, community, professional groups, and so on.[86] Indeed, "the social context or situation of veterans matters greatly in their resilience."[87] Recent scientific research on military veterans conclusively demonstrates the link between social support, psychological resilience, and mental health. For example, postdeployment social support and community integration increased the resilience of veterans (i.e., high number of lifetime traumas, low current psychological distress).[88] Conversely, poor social support strongly correlates with the development of PTSD.[89]

Although psychologists who study veterans' mental health took more than sixty years after the classification of the disorder to recognize just how critical community can be to healing from trauma, Dr. Seuss recognized how social factors influence resilience in 1957. In *How the Grinch Stole Christmas*, two social factors figure prominently in determining the Grinch's resilience: social connectedness and social acceptance.

Social Connectedness

During his attempted burglary of a pine tree, the Grinch has a surprise encounter with a junior member of the local civil society. "And the Grinch grabbed the tree, and he started to shove / When he heard a small sound like the coo of a dove. / He turned around fast, and he saw a small Who!/ Little Cindy-Lou Who, who was not more than two." Although discovering an intruder in the home would probably frighten most people, Cindy-Lou lacks the life experience that would cause her to mistrust the Grinch. Rather than assuming he's a thief, Cindy-Lou questions the Grinch innocently, "Why are you taking our Christmas tree? WHY?"

What significance does this short conversation have, especially since it is the only encounter between the Grinch and another person in the book? The narrator subtly reveals its importance, stating that the "Grinch had been *caught* by this tiny Who daughter."[90] While the verb "catch" might refer to witnessing the perpetration of a crime (e.g., "caught in the act"), on a deeper symbolic level "catching" might also refer to rescuing someone (e.g., "catch me, I'm falling"). In the *Grinch*, Cindy-Lou catches the protagonist in both senses of the term. During their brief conversation, Cindy-Lou catches the Grinch in the act of theft but also rescues the lonely, isolated Grinch by creating a social connection that eventually becomes a bridge to the entire Who community.

In *How the Grinch Stole Christmas*, neither hatha yoga, mindfulness training, meditation, primal scream therapy, nor any other individual practice heals the Grinch's trauma. Rather, the protagonist's ability to overcome trauma results from establishment of a new social connection with the local community (in the form of Cindy-Lou Who). Indeed, scientific research shows that social connectedness, which refers to an individual's feeling of belonging to the surrounding social environment,[91] improves psychological functioning, reduces suicide risk, increases self-esteem, boosts personal happiness, and lowers substance abuse rates.[92] Social connectedness has a direct effect on PTSD symptoms, resulting in improved healing and fewer PTSD symptoms.[93]

Acceptance

In addition to social connectedness, *How the Grinch Stole Christmas* identifies social acceptance as a key factor in veterans' resilience after traumatic experiences. When the book was published in 1957, veterans returning from World War II were valorized with parades and patriotic speeches. The effect of public scorn on veterans' mental health only became a topic of interest for psychologists following the Vietnam War (making Dr. Seuss's identification of social acceptance all the more prescient). For Vietnam veterans, the "social rejection, abandonment, and even betrayal" they experienced after coming home "contributed to feelings of alienation and a growing sense of isolation in their social communities."[94] The stress associated with homecoming after the Vietnam War predicted the development of PTSD more than the level of combat exposure, stressful life events, or childhood and civilian traumas.[95]

Guided by his personal experience and powerful intuition, Ted Geisel understood that a lack of social acceptance inhibits resilience, and inversely that social acceptance promotes resilience. How are these ideas expressed in the *Grinch*? Most veterans with PTSD struggle with low self-esteem, lack of trust, and sensitivity to rejection.[96] They tend to avoid situations that exacerbate these feelings. But the Grinch cannot escape from Cindy-Lou when she catches him red-handed during the robbery. As a veteran with full-blown PTSD, the Grinch probably expected scorn and rejection. But instead, Cindy-Lou believes his lies. His "fib fooled the child. Then he patted her head / And he got her a drink and he sent her to bed." As she toddles off, she waves and smiles at him, accepting the Grinch completely as he is, without condemnation or disgust.

Given the allegorical structure of the book, we can reasonably infer that Cindy-Lou symbolizes the Who community. Her nonjudgmental acceptance of him (and by extension, civil society's acceptance) established the preconditions that enable the Grinch to heal from trauma. In the final illustration, the Grinch has completely recovered from his various maladies and looks emotionally stable and physically robust. He wears a satisfied smile as he carves the roast beast. The entire Who community has accepted this wounded, isolated veteran into their society, as signified by the fact that Who-ville parents trust the Grinch with a large knife around their small children. When the Grinch carves the roast beast at the Christmas table, instead of avoiding social contact, he assumes a place of honor in the Who community.

TRAUMA AND RESILIENCE OF WHO-VILLE

Individuals can be traumatized and/or resilient, but what about whole societies? Ted Geisel's portrayal of Who society indicates that he conceptualized trauma and resilience not just as an individual phenomenon but also as a social one. It is worth noting that social scientists have only recently begun to conduct research on social resilience,[97] a topic that Ted Geisel theorized about seventy years ago in the *Grinch*.

As a society, the Whos have experienced collective social trauma. Published in 1954, *Horton Hears a Who* chronicled the extreme, traumatic events that befell this community. After being dehumanized in a hostile propaganda campaign conducted by a Sour Kangaroo, their community was almost obliterated when the kangaroos plunged into the pool in the Jungle of Nool. Even more devastating, the indigenous Who population was faced with complete annihilation by the Wickersham Brothers, who threatened to boil them in Beezlenut oil. In addition to violence committed against them by simian armed forces, the Whos also experienced an earthquake of great magnitude "when that black-bottomed birdie let go and we dropped."

Although these traumatic events took place only three years prior to publication of the *Grinch* in 1957, the Whos appear to be almost entirely recovered from their collective tragedies. The community repaired all the physical damage documented in *Horton*; they prospered economically, as evidenced by their furnishings and foodstuffs; they repaired their electric infrastructure ("the warm lighted windows below in their town"); they resumed their normal holiday practices ("hanging a mistletoe wreath"); and they reestablished trade connections with the global market to accommodate their surplus production ("Then the Whos, young and old, would sit down to a feast. / And they'd feast! *And they'd feast!* / And they'd FEAST! / FEAST! / FEAST! / FEAST! They would feast on Who-pudding, and rare Who-roast-beast.").

What enabled the Whos to recover after multiple devastating events that would have crushed many other communities? Among social scientists, the current answer would be social resilience. While no standard definition of "social resilience" exists, most social scientists seem to agree that a definition should include (1) the ability to maintain institutions, functions, and structure during a major national crisis; (2) the ability to recover from adverse conditions or events; and (3) the ability to adapt in response to events, or in other words to make innovative changes to address the crisis.[98]

Because social scientists discovered this area of study only recently, the literature on the social attributes of resilient societies is slim. Most studies have focused on the response to natural and man-made disasters, such as floods, hurricanes, famine, and fire.[99] Unfortunately for our purposes here, few studies have focused on social resilience in the context of war.[100]

Horton, however, provides some insight into social attributes that promote social resilience: namely, citizen engagement in governance, community solidarity, and unification in response to external threats. Indeed, social scientists have confirmed that citizen engagement in governance contributes to the overall resilience of a society.[101] In particular, societies that have collaborative decision-making processes tend to be the most resilient.[102] The Whos have a relatively advanced political system, with democratic representation (given that they have a "mayor," which is usually an elected office) and a guarantee of minority rights ("Please don't harm all my little folks who / Have as much right to live as us bigger folks do!"). In addition, the Whos apparently practice consensus decision-making in which the entire community participates ("the scared little Mayor / Quick called a big meeting in Who-ville Town Square.").

Regarding the second attribute, resilient societies generally possess community solidarity, meaning networks, social processes, and activities that provide support, create opportunities, and unite diverse groups.[103] In *How the Grinch Stole Christmas*, the Who community builds and maintains unity through communal music. As the narrator tells us, "Every Who down in Who-ville, the tall and the small, / Was singing! Without any presents at all!" In the accompanying illustration, the arch of singing Whos includes everyone: a mother holding an infant, an elderly man wearing a waistcoat, and a young girl with a bow in her hair. Indeed, scientific studies have shown that community singing promotes resilience, prevents chronic disease, promotes self-esteem, and reduces loneliness.[104]

The third factor supporting Who resilience is unity of the social collective when faced with an existential threat. Throughout history, societies have collapsed as the result of natural disasters (ninth-century Maya civilization), ecological catastrophe (eighth-century Greenland), or external threats (fourth-century Roman Empire).[105] The oldest surviving societies are neither states nor nations but segmentary lineage systems such as tribes, clans, and moieties.[106] Their survival for thousands of years—despite disease, colonization, natural disasters, and war—demonstrates their incredible social resilience. What sets tribes apart from every other type of social structure is their tendency to unify and mobilize in response to an external threat.[107] Even the most well-equipped, well-trained, and well-armed military organization have had their strategic ambitions quashed by the collective mobilization of tribes.[108] While the Whos do not appear to have had a segmentary lineage system per se, they united as a political community to defend their existence in the face of mass annihilation as described in *Horton Hears a Who*. Despite their fear when confronted by opposing military forces ("some poor little person who's shaking with fear"), the indigenous inhabitants did not capitulate to threats of incineration by Beezlenut oil. Instead, even the youngest child among them (e.g., Jo-Jo with the yo-yo) fought back, making enough noise to preserve their existence. Indeed, the continued existence of any political community requires some (and perhaps all) members to fight for its existence, as the Whos did in *Horton*.

In short, long before social scientists coined the term "social resilience," Ted Geisel accurately and succinctly identified the social attributes that contribute to resilience: citizen engagement in governance, community solidarity, and unification in response to external threats.

IN CONCLUSION: REDEMPTION IN "WAR STORIES"

If the actual plot of *How the Grinch Stole Christmas* is segregated from the innocuous banality of children's literature as a genre, the *Grinch* can be recognized as more than just a parable for children about capitalist consumer culture. Examined closely, the actual structure of the narrative becomes readily apparent; the planning, execution and aftermath of Operation Xmas indicate that the *Grinch* is quintessentially a war story.

As a war story the Grinch falls into the subgenre of tales of redemption, which can include novels, films, and memoirs. In this particular subgenre, the protagonist seeks atonement for a personality fault (such as cowardice in A. E. W. Mason's 1902 *Four Feathers*), perpetration of harm (such as causing the death of comrades in Steven Spielberg's 1998 film *Saving Private Ryan*), or loss of faith (such as doubting God in Elie Wiesel's 1958/1960 *Night*). In wartime tales of redemption, the protagonist's path to redemption may involve committing symbolic acts to reverse one's shame, rituals of remembrance to honor the dead, and/or honoring God by resisting despair. Unlike tales of redemption from other genres (such as prison films, family melodrama, or horror), wartime redemption stories emphasize the restoration of personal honor.

Remarkably, the Grinch contains all three types of error commonly portrayed in wartime tales of redemption. In the Grinch, the wounded, lonely protagonist must atone for a personality fault (failure to control his sadistic impulses), for causing harm (plundering the Who village), and for loss of faith (capitulating to despair). The Grinch also contains all three types of redemption commonly portrayed in the wartime subgenre. Having lost his personal honor by committing acts of violence against innocent civilians in Who-ville, the Grinch must redeem himself and restore that honor by committing a symbolic act to reverse his shame (e.g., returning the Whos' stolen property), participating in a ritual that honors the community (e.g., the Christmas feast), and affirming communal joy instead of wallowing in lonely despair (e.g., participating in the communal feast).

As discussed at the beginning of this chapter, the *Grinch* has an allegorical structure in which the protagonist personifies Ted Geisel and a narrative arc of transformation and redemption. Given the literary construction, presumably the Grinch's journey from the lonely agony of his ice cave (where his emotions remained frozen) to the festive holiday celebrations (at which he is the guest of honor) maps to Ted Geisel's own life. Although the specific details of his own personal journey remain unknown, the depth of knowledge regarding the need for atonement (whether because of personal failures, harm done to others, or loss of faith) shown in the *Grinch* tend to indicate that Ted Geisel had direct personal experience of transformation and redemption. Recovery from PTSD demands the restoration of honor through acts of redemption (whether a symbolic reversal, a communal ritual, or the affirmation of joy). Perhaps by writing a children's book called *How the Grinch Stole Christmas*, Ted Geisel attained his own personal redemption?

NOTES

1. Animal abuse correlates with antisocial and aggressive behaviors. James Serpell, "Factors Influencing Human Attitudes to Animals and Their Welfare," *Animal Welfare* 13, no. 1 (2004): S145–52.

2. For example, see Martin H. Sable, "The Day of Atonement in Charles Dickens' 'A Christmas Carol,'" *Tradition: A Journal of Orthodox Jewish Thought* 22, no. 3 (1986): 66–76; and Lorraine Mortimer, "The Grim Enchantment of 'It's a Wonderful Life,'" *Massachusetts Review* 36, no. 4 (1995): 656–86.

3. A. Ben-Ze'ev, "The Personal Comparative Concern in Schadenfreude," in *Schadenfreude: Understanding Pleasure at the Misfortune of Others*, ed. W. Van Dijk and J. Ouwerkerk (Cambridge: Cambridge University Press, 2014).

4. Thanks to Erika "Ann" Jeschke for input on the early draft of this chapter.

5. E. J. Kahn, "Children's Friend," *New Yorker*, December 17, 1960.

6. Peter Pels, "The Confessional Ethic and the Spirits of the Screen. Reflections on the Modern Fear of Alienation," *Etnofoor* 15, nos. 1/2 (2002): 91–119.

7. PTSD estimates for military service members range from 8.1 to 23 percent. J. J. Fulton et al., "The Prevalence of Posttraumatic Stress Disorder in Operation Enduring Freedom/Operation Iraqi Freedom (OEF/OIF) Veterans: A Meta-Analysis," *Journal of Anxiety Disorders* 31 (2014): 98–107; J. Muller, S. Ganeshamoorthy, and J. Myers, "Risk Factors Associated with Posttraumatic Stress Disorder in US Veterans: A Cohort Study," *PLoS ONE* 12 (2017): e0181647; and L. K. Richardson, B. C. Frueh, and R. Acierno, "Prevalence Estimates of Combat—Related Post-Traumatic Stress Disorder: Critical Review," *Australia and New Zealand Journal of Psychiatry* 44 (2010): 4–19.

8. Sarah A. Mustillo and Ashleigh Kysar-Moon. "Race, Gender, and Post-Traumatic Stress Disorder in the U.S. Military: Differential Vulnerability?" *Armed Forces & Society* 43, no. 2 (2017): 322–45; and Benjamin R. Karney et al., "Predicting the Immediate and Long-Term Consequences of Post-Traumatic Stress Disorder, Depression, and Traumatic Brain Injury in Veterans of Operation Enduring Freedom and Operation Iraqi Freedom," in *Invisible Wounds of War: Psychological and Cognitive Injuries, Their Consequences, and Services to Assist Recovery*, ed. Terri Tanielian and Lisa H. Jaycox (Santa Monica, CA: RAND, 2008).

9. William B. Hart, "Between the Lines," *Redbook*, December 1957, 4.

10. Caroline M. Smith, *Dr. Seuss: the Cat Behind the Hat* (New York: Andrews McMeel Publishing, 2012), 218.

11. Cleanth Brooks and Robert Penn Warren, *Understanding Poetry*, 3rd ed. (New York: Holt, Rinehart and Winston, 1960).

12. B. B. Stern, "Medieval Allegory: Roots of Advertising Strategy for the Mass Market," *Journal of Marketing* 52, no. 3 (1988): 84–94.

13. Other scholars who have written about Dr. Seuss have noted the consistent use of metaphor and allegory. Peter Nicolas, "The Sneetches as an Allegory for the Gay Rights Struggle: Three Prisms," *New York Law School Law Review* 58, no. 525 (2013); and Jennifer Zicht, "In Pursuit of the Lorax," *EPA Journal* 17, no. 4 (September 1991): 27–30.

14. Edward Connery Lathem, *The Beginnings of Dr. Seuss: An Informal Reminiscence* (Hanover, NH: Dartmouth College, 1976), 17.

15. Thomas Fensch, *The Man Who Was Dr. Seuss: The Life and Work of Theodor Geisel* (Woodlands, TX: New Century Books, 2001).

16. John L. Plaster, *SOG: The Secret Wars of America's Commandos In Vietnam* (New York: Simon & Schuster, 1997).

17. Alex Horton, "Russia's Commando Units Gutted by Ukraine War, U.S. Leak Shows," *Washington Post*, April 14, 2023.

18. Jonathan Herrmann and Goran Eryavec, "Posttraumatic Stress Disorder in Institutionalized World War II Veterans," *American Journal of Geriatric Psychiatry* 2, no. 4 (1994): 324–31.

19. Paul R. J. Falger et al., "Current Posttraumatic Stress Disorder and Cardiovascular Disease Risk Factors in Dutch Resistance Veterans from World War II," *Psychotherapy and Psychosomatics* 57, no. 4 (1992): 164–71.

20. Committee on the Assessment of Ongoing Effects in the Treatment of Posttraumatic Stress Disorder, *Treatment for Posttraumatic Stress Disorder in Military and Veteran Populations: Initial Assessment* (Washington, DC: National Academies Press, 2012), ch. 2.

21. Ron Langer, "Combat Trauma, Memory, and the World War II Veteran," *War, Literature & the Arts: An International Journal of the Humanities* 23, no.1 (2011): 50–58.

22. Herrmann and Eryavec, "Posttraumatic Stress Disorder."

23. M. D. Caldwell, "Wound Surgery," *Surgical Clinics of North America* 90, no. 6 (2010): 1125–32.

24. Gjumrakch Aliev et al., "Neurophysiology and Psychopathology Underlying PTSD and Recent Insights into the PTSD Therapies—A Comprehensive Review." *Journal of Clinical Medicine* 9, no. 9 (2020), https://doi.org/10.3390/jcm9092951.

25. Stefan Priebe et al., "Consequences of Untreated Posttraumatic Stress Disorder Following War in Former Yugoslavia: Morbidity, Subjective Quality of Life, and Care Costs," *Croatian Medical Journal* 50, no. 5 (2009): 465–75.

26. J. M. Cook, "Post-Traumatic Stress Disorder In Older Adults," *PTSD Research Quarterly* 12 (2001): 1–7.

27. Z. Solomon and M. Mikulincer, "Trajectories of PTSD: A 20-Year Longitudinal Study," *American Journal of Psychiatry* 163, no. 4 (2006): 659–66.

28. A. C. McFarlane, "The Long-Term Costs of Traumatic Stress: Intertwined Physical and Psychological Consequences," *World Psychiatry* 9, no. 1 (2010): 3–10.

29. C. Van Dyke, N. J. Zilberg, and J. A. McKinnon, "Posttraumatic Stress Disorder: A Thirty-Year Delay in a World War II Veteran," *American Journal of Psychiatry 142, no. 9* (1985): 1070–73.

30. Isaac R. Galatzer-Levy et al., "Early PTSD Symptom Trajectories: Persistence, Recovery, and Response to Treatment: Results from the Jerusalem Trauma Outreach and Prevention Study (J-TOPS)," *PLoS One* 8, no. 8 (2013), https://doi.org/10.1371/journal.pone.0070084; and J. D. Markowitz, "Post-Traumatic Stress Disorder in an Elderly Combat Veteran: A Case Report," *Military Medicine* 172, no. 6 (2007): 659–62.

31. S. Priebe et al., "Consequences of Untreated Posttraumatic Stress Disorder."

32. Tracy Xavia Karner, "Post-Traumatic Stress Disorder and Older Men: If Only Time Healed All Wounds," *Generations: Journal of the American Society on Aging* 32, no. 1 (2008): 82–87; and K. M. Magruder et al., "PTSD Symptoms, Demographic Characteristics, and Functional Status among Veterans Treated in VA Primary Care," *Journal of Social Psychology* 137 (2004): 618–28.

33. S. B. Norman et al., "Associations between Psychological Trauma and Physical Illness in Primary Care," *Journal of Traumatic Stress* 19 (2006): 461–67.

34. Neil Prior, "PTSD: Eyes Can Reveal Previous Trauma, Study Reveals," *BBC News*, July 26, 2020.

35. Although the book has only black-and-white renditions of the Grinch, in Geisel's original drawings the Grinch is green. See, for example, the sold items archive at https://dndgalleries.com.

36. L. J. Witts, "Chlorosis in Males," *Guy's Hospital Report* 10, no. 4 (1930): 417.

37. Jean Starobinski, "Chlorosis—The 'Green Sickness,'" *Psychological Medicine* 11 (1981): 459–68.

38. Robert P. Hudson, "The Biography of Disease: Lessons from Chlorosis," *Bulletin of the History of Medicine* 51, no. 3 (1977): 448–63.

39. Carolyn Dinshaw, "Black Skin, Green Masks: Medieval Foliate Heads, Racial Trauma, and Queer World-making," in *Middle Ages in the Modern World*, ed. Bettina Bildhauer and Chris Jones (Oxford: Oxford University Press, 2017).

40. Jean Louise Carrière, "Sir Gawain and the Green Knight as a Christmas Poem," *Comitatus: A Journal of Medieval and Renaissance Studies* 1, no. 1 (1970): 25–42.

41. Carrière, "Sir Gawain and the Green Knight."

42. Dinshaw, "Black Skin, Green Masks."

43. Dinshaw, "Black Skin, Green Masks."

44. H. Hartshorne (1864), quoted in Edgar Jones, "Historical Approaches to Post-Combat Disorders," *Philosophical Transactions of the Royal Society B: Biological Sciences* 361, no. 1468 (2006): 533–42.

45. J. Somerville, "The Master's Legacy: The First Paul Wood Lecture," *Heart* 80, no. 6 (1998): 612–18.

46. Paul Wood, "Aetiology of Da Costa's Syndrome," *British Medical Journal* 1, no. 4196 (1941): 845–51.

47. Royal Society of Medicine, "Discussion on the Nature and Treatment of the Effort Syndrome," *Proceedings of the Royal Society of Medicine* 34 (1941): 37–50.

48. Hubert Winston Smith, "Relation of Emotions to Injury and Disease: Legal Liability for Psychic Stimuli," *Virginia Law Review* 30, no. 2 (1944): 193–317; H. A. Wilkins, "The Soldier of Auvergne," in *Selections from Canadian Poets: With Occasional Critical and Biographical Notes and an Introductory Essay on Canadian Poetry,* ed. Edward Hartley Dewart, and Douglas Lochhead (Toronto: University of Toronto Press, 1973); Lincoln High School, *Lincoln High School Annual 1968,* 6, https://www.e-yearbook.com/yearbooks/Lincoln_High_School_Yearbook/1968/Page_1.html; Richard D. Altick, "Hamlet and the Odor of Mortality," *Shakespeare Quarterly* 5, no. 2 (1954): 167–76; and Delphine Fitz Darby, "Review of *Ribera* by Elizabeth du Gué Trapier," *Art Bulletin* 35, no. 1 (1953): 68–74.

49. Science Service, "In Science Fields," *Science News-Letter* 37, no. 24 (1940): 376–77.

50. Science Service, "'Soldier's Heart' Cured by Cutting Nerves to Gland," *Science News-Letter* 38, no. 18 (1940): 286.

51. Louis Faugeres Bishop, "Soldier's Heart," *American Journal of Nursing* 42, no. 4 (1942): 377–80.

52. Science Service, "Man with Soldier's Heart Discharged for Second Time," *Science News-Letter* 43, no. 23 (1943): 367.

53. Wood, "Aetiology of Da Costa's Syndrome."

54. American Psychiatric Association, *Diagnostic and Statistical Manual of Mental Disorders*, 5th ed. (Washington, DC: American Psychiatric Press, 2000).

55. Carol S. North et al., "Psychiatric Disorders Among Survivors of the Oklahoma City Bombing," *Journal of the American Medical Association* 282 (1999): 755–62.

56. C. Grillon et al., "Increased Anxiety during Anticipation of Unpredictable Aversive Stimuli in Posttraumatic Stress Disorder But Not in Generalized Anxiety Disorder," *Biological Psychiatry* 66, no. 1 (2009): 47–53.

57. N. Pole et al., "Fear-Potentiated Startle and Posttraumatic Stress Symptoms in Urban Police Officers," *Journal of Trauma Stress* 16 (2003): 471–79.

58. Anke Ehlers and David Clark, "A Cognitive Model of Posttraumatic Stress Disorder," *Behaviour Research and Therapy* 15 (2000): 249–75.

59. Anthony J. Rosellini et al., "Predicting Non-Familial Major Physical Violent Crime Perpetration in the US Army from Administrative Data," *Psychological Medicine* 46 (2016): 303–16.

60. Joshua Camins, "Military PTSD and Post-Service Violence: A Review of the Evidence," *Trauma Psychology News*, November 9, 2018.

61. Z. Solomon and M. Mikulincer, "Posttraumatic Intrusion, Avoidance, and Social Functioning: A 20-Year Longitudinal Study," *Journal of Consulting and Clinical Psychology* 75 (2007): 316–24; and C. A. Lunney and P. P. Schnurr, "Domains of Quality of Life and Symptoms in Male Veterans Treated for Posttraumatic Stress Disorder," *Journal of Traumatic Stress* 20, no. 6 (2007): 955–64.

62. L. Lu, "Social Support, Reciprocity, and Well-Being," *Journal of Social Psychology* 137 (1997): 618–28.

63. Lu, "Social Support, Reciprocity."

64. R. A. Bryant et al., "A Prospective Study of Psychophysiological Arousal, Acute Stress Disorder, and Posttraumatic Stress Disorder," *Journal of Abnormal Psychology* 2 (2000): 341–44.

65. Judith Lewis Herman, *Trauma and Recovery* (New York: Basic Books, 1992).

66. M. S. Levy, "A Helpful Way to Conceptualize and Understand Reenactments," *Journal of Psychotherapy Practice and Research* 7, no. 3 (1998): 227–35.

67. Herman, *Trauma and Recovery*.

68. Robert T. Muller and Terry Diamond, "Father and Mother Physical Abuse and Child Aggressive Behaviour in Two Generations," *Canadian Journal of Behavioural Science* 31, no. 4 (1999): 221–28; and C. S. Widom and R. L. Shepard, "Accuracy of Adult Recollections of Childhood Victimization: Part 1, Childhood Physical Abuse," *Psychological Assessment* 8, no. 4 (1996): 412–21.

69. A. F. Jespersen, M. L. Lalumière, and M. C. Seto, "Sexual Abuse History among Adult Sex Offenders and Non-Sex Offenders: A Meta-Analysis," *Child Abuse & Neglect* 33 (2009): 179–92.

70. B. A. van der Kolk and M. S. Greenberg, "The Psychobiology of the Trauma Response: Hyperarousal, Constriction, and Addiction to Traumatic Reexposure," in *Psychological Trauma*, ed. B. A. van der Kolk (Washington, DC: American Psychiatric Press, 1987).

71. Robert M. Klein, "Adaptive Planning," *Joint Forces Quarterly* 45 (2007): 84–88.

72. Clem Maginniss, *A Great Feat of Improvisation: Logistics and the British Expeditionary Force in France 1939–1940* (London: Helion Company, 2021); and Center of Military History, *Military Improvisations during the Russian Campaign* (Washington, DC: United States Army, 2004).

73. Dhananjay Khadilkar, "How Donkeys Changed the Course of Human History," *BBC*, January 16, 2023; and Cate Folsom, *Smoke the Donkey A Marine's Unlikely Friend* (Washington, DC: Potomac Books, 2016).

74. E. S. Forster, "Dogs In Ancient Warfare," *Greece & Rome* 10 (1941): 114–17; John Varner, *Dogs of the Conquest* (Norman: University of Oklahoma Press, 1983); and Rebecca Frankel, *War Dogs* (New York: Palgrave Macmillan, 2014).

75. H. P. Willmott, *World War I* (New York: Dorling Kindersley, 2003).

76. Maria Goodavage, *Soldier Dogs* (New York: Dutton, 2012).

77. Rick Baillergeon and John Sutherland, "Tactics 101: Infiltration in History and Practice," *Armchair General*, March 12, 2013, http://armchairgeneral.com/tactics-101-082-infiltraton-in-history-and-practice.htm.

78. Martin Van Creveld, *Supplying War: Logistics from Wallenstein to Patton* (Cambridge, MA: Cambridge University Press, 1977); and Wicky W. K. Tse, "Cutting the Enemy's Line of Supply: The Rise of the Tactic and Its Use in Early Chinese Warfare," *Journal of Chinese Military History* 6 (2017): 131–56.

79. Andrew Bell, "Civilians, Urban Warfare, and US Doctrine," *Parameters* 50, no. 4 (Winter 2020): 33–55.

80. Rory Cormac and Richard J. Aldrich, "Grey Is the New Black: Covert Action and Implausible Deniability," *International Affairs* 94, no. 3 (May 2018): 477–94.

81. American Psychological Association, *The Road to Resilience* (Washington, DC: American Psychological Association, 2014).

82. Michael Ungar, ed., *The Social Ecology of Resilience: A Handbook of Theory and Practice* (New York: Springer, 2012).

83. Orla T. Muldoon and Robert D. Lowe, "Social Identity, Groups, and Post-Traumatic Stress Disorder," *Political Psychology* 33, no. 2 (April 2012): 259–73.

84. S. Kintzle et al., "PTSD in U.S. Veterans: The Role of Social Connectedness, Combat Experience and Discharge," *Healthcare* 6, no. 3 (2018): 102.

85. Dennis S. Charney, "Psychobiological Mechanisms of Resilience and Vulnerability: Implications for Successful Adaptation to Extreme Stress," *American Journal of Psychiatry* 161 (2004): 368–91.

86. Muldoon and Lowe, "Social Identity."

87. R. M. Lee, M. Draper, and S. Lee, "Social Connectedness, Dysfunctional Interpersonal Behaviors, and Psychological Distress: Testing a Mediator Model," *Journal of Counseling Psychology* 48 (2002): 310–18.

88. Robert H. Pietrzak and Joan M. Cook, "Psychological Resilience in Older U.S. Veterans: Results from the National Health and Resilience in Veterans Study," *Depression and Anxiety* 30, no. 5 (2013): 432–43.

89. Anthony Charuvastra and Marylene Cloitre, "Social Bonds and Posttraumatic Stress Disorder," *Annual Review of Psychology* 59 (2008): 301–28.

90. Emphasis added.

91. Lee, Draper, and Lee, "Social Connectedness."

92. K. L. Williams and R. V. Galliher, "Predicting Depression and Self-Esteem from Social Connectedness, Support and Competence," *Journal of Clinical Social Psychology* 25 (2006): 855–74.

93. Kintzle, "PTSD in U.S. Veterans."

94. Miraj U. Desai et al., "'I Want to Come Home': Vietnam-Era Veterans' Presenting for Mental Health Care, Roughly 40 Years after Vietnam," *Psychiatric Quarterly* 87 (2015): 229–39; and Lynne McCormack and Joseph Stephen, "Psychological Growth in Aging Vietnam Veterans: Redefining Shame and Betrayal," *Journal of Humanistic Psychology* 54, no. 3 (2014): 336–55.

95. David Read Johnson et al., "The Impact of the Homecoming Reception on the Development of Posttraumatic Stress Disorder: The West Haven Homecoming Stress Scale (WHHSS)," *Journal of Traumatic Stress* 10, no. 2 (1997): 259–77.

96. T. B. Kashdan et al., "Fragile Self-Esteem and Affective Instability in Posttraumatic Stress Disorder," *Behavioral Research and Therapy Journal* 44 (2006): 1609–19; E. L. Zurbriggen, R. L. Gobin, and L. A. Kaehler, "Trauma, Attachment, and Intimate Relationships," *Journal of Trauma Dissociation* 13 (2012):127–33; and G. Nietlisbach and A. Maercker, "Effects of Social Exclusion in Trauma Survivors with Posttraumatic Stress Disorder," *Psychological Trauma* 1 (2009): 323–31.

97. Philippe Bourbeau, "Resilience and International Politics: Premises, Debates, Agenda," *International Studies Review* 17, no. 3 (2015): 374–95.

98. Brad Allenby and Jonathan Fink, "Toward Inherently Secure and Resilient Societies," *Science* 309, no. 5737 (2005): 1034–36; and Reuven Gal, "Social Resilience in Times of Protracted Crises," *Armed Forces & Society* 40, no. 3 (2014): 452–75.

99. C. Aradau, "Security That Matters: Critical Infrastructure and Objects of Protection," *Security Dialogue* 41, no. 5 (2010): 491–514.

100. Bruce E Goldstein, ed., *Collaborative Resilience: Moving through Crisis to Opportunity* (Cambridge, MA: MIT Press, 2011); Paul Williams, "Protection, Resilience and Empowerment: United Nations Peacekeeping and Violence Against Civilians in Contemporary War Zones," *Politics* 33 (2013): 287–98; Erika Cudworth, "Armed Conflict, Insecurity and Gender: The Resilience of Patriarchy?," *Resilience: International Policies, Practices and Discourses* 1, no. 1 (2013): 69–75; and Ami C. Carpenter, "Havens in a Firestorm: Perspectives from Baghdad on Resilience to Sectarian Violence," *Civil Wars Journal* 14, no. 2 (2012): 182–204.

101. Regina Bateson, "Crime Victimization and Political Participation," *American Political Science Review* 106, no. 3 (2012): 570–87; Christopher Blattman, *Civil War*

(Berkeley, CA: UC Berkeley Center for International and Development Economics Research, 2009); and John Bellows and Edward Miguel, "War and Local Collective Action in Sierra Leone," *Journal of Public Economics* 93, nos. 11–12 (2009): 1144–57.

102. Kirsten Maclean, Michael Cuthill, and Helen Ross, "Six Attributes of Social Resilience," *Journal of Environmental Planning and Management* 57, no. 1 (2014): 144–56.

103. Lauren M. Sippel et al., "How Does Social Support Enhance Resilience in the Trauma-Exposed Individual?," *Ecology and Society* 20, no. 4 (2015), http://www.jstor.org/stable/26270277; and Braden Leap and Diego Thompson, "Social Solidarity, Collective Identity, Resilient Communities: Two Case Studies from the Rural U.S. and Uruguay," *Social Sciences* 7, no. 12 (2018), https://doi.org/10.3390/socsci7120250.

104. Jing Sun and Nicholas Jan Buys, "Improving Aboriginal and Torres Strait Islander Australians' Well-Being Using Participatory Community Singing Approach," *International Journal on Disability and Human Development* 12, no. 3 (2013), https://doi.org/10.1515/ijdhd-2012-0108.

105. Robert Edgerton, *Sick Societies: Challenging the Myth of Primitive Harmony* (New York: Free Press, 1992).

106. Scott Cane, *First Footprints: The Epic Story of the First Australians* (New York: Allen and Unwin, 2013).

107. Philip Carl Salzman, *Culture and Conflict in the Middle East* (New York: Humanity Books, 2008); and Marshall D Sahlins, "The Segmentary Lineage: An Organization of Predatory Expansion," *American Anthropologist* 63, no. 2 (1961): 322–45.

108. Michael G. Findley, and Scott Edwards. "Accounting for the Unaccounted: Weak-Actor Social Structure in Asymmetric Wars." *International Studies Quarterly* 51, no. 3 (2007): 583–606; and Kristian Berg Harpviken "Transcending Traditionalism: The Emergence of Non-State Military Formations in Afghanistan," *Journal of Peace Research* 34, no. 3 (1997): 271–87.

Bibliography

Ajdukovic, Dean. "Social Contexts of Trauma and Healing." *Medicine, Conflict and Survival* 20, no. 2 (2004): 120–35.

Akerlof, George A., and Robert J. Shiller. *Phishing for Phools: The Economics of Manipulation and Deception*. Princeton, NJ: Princeton University Press, 2015.

Al Rasheed, Madawi. *A History of Saudi Arabia*. 2nd ed. New York: Cambridge University Press, 2010.

Albrecht, Holger. "Does Coup-Proofing Work? Political–Military Relations in Authoritarian Regimes amid the Arab Uprisings." *Mediterranean Politics* 20, no. 1 (2015): 36–54.

Albright, Madeline, and William Woodward. *Fascism: A Warning*. New York: HarperCollins, 2018.

Alfoneh, Ali. "The Basij Resistance Force: A Weak Link in the Iranian Regime?" Washington Institute for Near East Policy Studies, February 5, 2010. https://www.washingtoninstitute.org/policy-analysis/view/the-basij-resistance-force-a-weak-link-in-the-iranian-regime.

Al-Hathloul, Alia, and Lina Al-Hathloul. "Loujain Al-Hathloul Is Not a Terrorist." *Marie Claire*, January 11, 2021. https://www.marieclaire.com/politics/a35122269/loujain-alhathloul-prison-sentence-essay/.

Aliev, Gjumrakch, Narasimha M. Beeraka, Vladimir N. Nikolenko, Andrey A. Svistunov, Tatyana Rozhnova, Svetlana Kostyuk, Igor Cherkesov, et al. "Neurophysiology and Psychopathology Underlying PTSD and Recent Insights into the PTSD Therapies—A Comprehensive Review." *Journal of Clinical Medicine* 9, no. 9 (2020). https://doi.org/10.3390/jcm9092951.

Allenby, Brad, and Jonathan Fink. "Toward Inherently Secure and Resilient Societies." *Science* 309, no. 5737 (2005):1034–36.

Alperovitch, Dmitri, and Ian Ward. "How Should the U.S. Respond to the Solarwinds and Microsoft Exchange Hacks?" *Lawfare Blog*, March 12, 2021. https://

www.lawfaremedia.org/article/how-should-us-respond-solarwinds-and-microsoft-exchange-hacks.

Alpers, Benjamin L. "This Is the Army: Imagining a Democratic Military in World War II." *Journal of American History* 85, no. 1 (June 1998): 148.

Alter, Alexandra, and Elizabeth A. Harris. "Dr. Seuss Books Are Pulled, and a 'Cancel Culture' Controversy Erupts." *New York Times*, March 4, 2021.

Altick, Richard D. "Hamlet and the Odor of Mortality." *Shakespeare Quarterly* 5, no. 2 (1954): 167–76.

American Psychiatric Association. *Diagnostic and Statistical Manual of Mental Disorders.* 5th ed. Washington, DC: American Psychiatric Press, 2000.

American Psychological Association. *The Road to Resilience.* Washington, DC: American Psychological Association, 2014.

Anderson, Lisa. "Muammar al-Qaddafi: The 'King' of Libya." *Journal of International Affairs* 54, no. 2 (Spring 2001): 515–17.

Anderson, Mary B. *Do No Harm: How Aid Can Support Peace—Or War.* Boulder, CO: Lynne Rienner Publishers, 1999.

Aradau, C. "Security That Matters: Critical Infrastructure and Objects of Protection." *Security Dialogue* 41, no. 5 (2010): 491–514.

Art, Robert J. *A Grand Strategy for America.* Ithaca, NY: Cornell University Press, 2003.

"Artist Profile: Galal Al-Behairy." Artists at Risk Connection, November 17, 2020. https://artistsatriskconnection.org/story/galal-el-behairy.

Arts, Karin, and Atabongawung Tamo. "The Right to Development in International Law: New Momentum Thirty Years Down the Line?" *Netherlands International Law Review* 63, no. 3 (2016): 221–49.

Associated Press. "A Look at Estonia's Cyber Attack in 2007." *NBC News*, July 8, 2009. https://www.nbcnews.com/id/wbna31801246.

Atkinson, Rick. *Guns at Last Light.* New York: Henry Holt, 2013.

Axelrod, Robert. *Evolution of Cooperation.* New York: Basic Books, 1984.

Baillergeon, Rick, and John Sutherland. "Tactics 101 082: Infiltration in History and Practice." *Armchair General*, March 12, 2013. http://armchairgeneral.com/tactics-101-082-infiltraton-in-history-and-practice.htm.

Baiocchi, David, and William Wesler IV. "Democratization of Space: New Actors Need New Rules." *Foreign Affairs*, May/June 2015.

Balzacq, Thierry, Peter Dombrowski, and Simon Reich. "Introduction: Comparing Grand Strategies in the Modern World." In *Comparative Grand Strategy: A Framework and Cases,* edited by Thierry Balzacq, Peter Dombrowski, and Simon Reich, 1–22. New York: Oxford University Press, 2019.

Banville, John. "The Power and the Glory of Pablo Picasso." *New Republic*, November 16, 2021. https://newrepublic.com/article/164381/picasso-biography-richardson-review-power-glory.

Barajas, Julia. "How Teachers in L.A. and Beyond Turned Away from Dr. Seuss." *Los Angeles Times*, March 5, 2021.

Barany, Zoltan. "Comparing the Arab Revolts: The Role of the Military." *Journal of Democracy* 22, no. 4 (October 2011): 24–35.

Bargu, Banu. "Why Did Bouazizi Burn Himself? The Politics of Fate and Fatal Politics." *Constellations* 23, no. 1 (March 2016): 27–36.
Barkin, Samuel. "Realism, Prediction, and Foreign Policy." *Foreign Policy Analysis* 5, no. 3 (2009): 233–46. http://www.jstor.org/stable/24909777.
Barkin, Steve M. "Fighting the Cartoon War: Information Strategies in World War II." *Journal of American Culture* 7, nos. 1–2 (Spring/Summer 1984): 113–17.
Barnhill, John H. "Watts Riots (1965)." In *Revolts, Protests, Demonstrations, and Rebellions in American History*, edited by Steven L. Danver. Santa Barbara, CA: ABC-CLIO, 2011.
Bastone, Nick. "Chinese Electronics Giant Huawei Allegedly Offered Bonuses to Any Employee Who Stole Trade Secrets." *Business Insider*, January 28, 2019. https://www.businessinsider.com/huawei-indictment-trade-secrets-2019-1.
Bateson, Gregory. *Steps to an Ecology of Mind: Collected Essays in Anthropology, Psychiatry, Evolution and Epistemology*. Northvale, NJ: Jason Aronson, 1987.
Bateson, Regina. "Crime Victimization and Political Participation." *American Political Science Review* 106, no. 3 (2012): 570–87.
"Battle of the Bulge: The Greatest American Battle of the War." National Veteran's Memorial and Museum, December 16, 2020. https://nationalvmm.org/battle-of-the-bulge-the-greatest-american-battle-of-the-war/.
Bayat, Asef. "Un-civil Society: The Politics of the Informal People." *Third World Quarterly* 18, no. 1 (March 1997): 53–72.
Baylis, John, James Wirtz, Eliot Cohen, and Colin S. Gray, eds. *Strategy in the Contemporary World*. Oxford: Oxford University Press, 2002.
Beblawi, Hazem. "The Rentier State in the Arab World." *Arab Studies Quarterly* 9 no. 4 (Fall 1987): 383–98.
Beck, Jerry, and Will Friedwald. *Looney Tunes and Merrie Melodies: A Complete Guide to the Warner Bros. Cartoons*. New York: Holt, 1989.
Bell, Andrew. "Civilians, Urban Warfare, and US Doctrine." *Parameters* 50, no. 4 (Winter 2020): 33–45.
Bellin, Eva. "Reconsidering the Robustness of Authoritarianism in the Middle East: Lessons from the Arab Spring." *Comparative Politics* 44, no. 2 (January 2012): 127–49.
———. "The Robustness of Authoritarianism in the Middle East: Exceptionalism in Comparative Perspective." *Comparative Politics* 36, no. 2 (January 2004): 139–57.
Bellows, John, and Edward Miguel. "War and Local Collective Action in Sierra Leone." *Journal of Public Economics* 93, nos. 11–12 (2009): 1144–57.
Bender, Bryan. "More U.S. Troops Dying Outside of Combat." *Boston Globe*, May 3, 2009.
Benford, Robert D., and David A. Snow. "Framing Processes and Social Movements: An Overview and Assessment." *Annual Review of Sociology* 26 (2000): 611–39.
Benton, Michael J., and Richard J. Twitchett, "How to Kill (Almost) All Life: The End-Permian Extinction Event." *Trends in Ecology and Evolution* 18, no. 7 (July 2003): 358–65.

Ben-Ze'ev, A. "Personal Comparative Concern in Schadenfreude." In *Schadenfreude: Understanding Pleasure at the Misfortune of Others*, edited by W. Van Dijk and J. Ouwerkerk, 77–90. Cambridge: Cambridge University Press, 2014.

Bergen, Peter. *Holy War, Inc.: Inside the Secret World of Bin Laden.* New York: Free Press, 2001.

Berger, Eric. "In a Consequential Decision, Air Force Picks Its Rockets for Mid-2020s Launches." *ArsTechnica*, August 7, 2020. https://arstechnica.com/science/2020/08/the-air-force-selects-ula-and-spacex-for-mid-2020s-launches/.

Bergman, Ronen, and Mark Mazzetti. "The Battle for the World's Most Powerful Cyberweapon," *New York Times*, January 28, 2022. https://www.nytimes.com/2022/01/28/magazine/nso-group-israel-spyware.html.

Best, Richard A. *Intelligence Reform after Five Years: The Role of the Director of National Intelligence.* Washington, DC: Congressional Research Service, 2010.

Beyer, Rick. "Seuss on the Loose." *Dartmouth Alumni Magazine*, November–December 2004. https://dartmouthalumnimagazine.com/articles/seuss-loose.

Bhattacharya, Debapriya, and Andrea Ordóñez Llanos. *Southern Perspectives on the Post-2015 International Development Agenda.* New York: Routledge, 2017.

Bibler Coutin, Susan. "Citizenship and Clandestine among Salvadoran Immigrants." *Political and Legal Anthropology Review* 22, no. 2 (November 1999): 53–63.

Birdwell, Michael. "Technical Fairy First Class? Is This Any Way to Run An Army? Private Snafu and WWII." *Historical Journal of Film, Radio, and Television* 25, no. 2 (June 2005): 203–12.

Bishop, Louis Faugeres. "Soldier's Heart." *American Journal of Nursing* 42, no. 4 (1942): 377–80.

Bland, Lucie M., David A. Keith, Rebecca M. Miller, Nicholas J. Murray, and Jon Paul Rodriguez, eds. *Guidelines for the Application of IUCN Red List of Ecosystems Categories and Criteria.* International Union for the Conservation of Nature, 2016. https://portals.iucn.org/library/node/45794.

Blattman, Christopher. *Civil War.* Berkeley, CA: UC Berkeley Center for International and Development Economics Research, 2009.

Bloomfield, Gary. *George S. Patton: On Guts, Glory, and Winning.* Guilford, CT: Lyons Press, 2017.

Booth, Ken. *Strategy and Ethnocentrism.* New York: Holmes & Meier, 1979.

Bourbeau, Philippe. "Resilience and International Politics: Premises, Debates, Agenda." *International Studies Review* 17, no. 3 (2015): 374–95.

Bowman, Tom. "On the Hunt for Roadside Bombs in Afghanistan." *NPR*, October 27, 2009.

Boyd, Brian. "The Origin of Stories: Horton Hears a Who." *Philosophy and Literature* 29, no. 2 (October 2001): 197–214.

Bozeman, Adda Bruemmer. *The Future of Law in a Multicultural World.* Princeton, NJ: Princeton University Press, 2015. First published 1971.

———. *Strategic Intelligence and Statecraft.* New York: Brassey's, 1992.

Brands, Hal W. *American Dreams: The United States since 1945.* New York: Penguin, 2010.

———. *Promise and Pitfalls of Grand Strategy*. Carlisle Barracks, PA: Strategic Studies Institute, US Army War College, 2012.

———. *What Good Is Grand Strategy? Power and Purpose in American Statecraft from Harry S. Truman to George W. Bush*. Ithaca, NY: Cornell University Press, 2014.

Branfill-Cook, Roger. *Torpedo: The Complete History of the World's Most Revolutionary Naval Weapon*. Annapolis, MD: Naval Institute Press, 2014.

Brennan, Martha. "Subversive as Hell: Political Satire in the Works of Dr. Seuss." *Waterloo Historical Review* 9 (April 2017). https://doi.org/10.15353/whr.v9.153.

British Broadcasting Corporation. "Ukraine Cyber-Attack: Software Firm MeDoc's Servers Seized." *BBC News*, July 4, 2017. https://www.bbc.com/news/technology-40497026.

Brooks, Cleanth, and Robert Penn Warren. *Understanding Poetry*. 3rd ed. New York: Holt, Rinehart and Winston, 1960.

Brown, L. Carl. *Religion and State: The Muslim Approach to Politics*. New York: Columbia University Press, 2001.

Bruce, James B., and Roger Z. George. "Intelligence Analysis—The Emergence of a Discipline." In *Analyzing Intelligence: Origins, Obstacles, and Innovations*, edited by James Bruce and Roger George. Washington DC: Georgetown University Press, 2008.

Bryant, R. A., A. G. Harvey, R. M. Guthrie, and M. L. Moulds. "A Prospective Study of Psychophysiological Arousal, Acute Stress Disorder, and Posttraumatic Stress Disorder." *Journal of Abnormal Psychology* 2 (2000): 341–44.

Brzezinski, Zbigniew. *The Grand Chessboard: American Primacy and Its Geostrategic Imperatives*. New York: Basic Books, 1997.

———. *Second Chance: Three Presidents and the Crisis of American Superpower*. New York: Basic Books, 2007.

———. *Strategic Vision: America and the Crisis of Global Power*. New York: Basic Books, 2012.

Buchanan, Ben. *The Hacker and the State: Cyber Attacks and the New Normal of Geopolitics*. Cambridge, MA: Harvard University Press, 2020.

Caddell, Joseph W. *Deception 101—Primer on Deception*. Carlisle, PA: Strategic Studies Institute, US Army War College, 2004. http://www.jstor.com/stable/resrep11327.

Caddick-Adams, Peter. *Snow & Steel: The Battle of the Bulge, 1944–45*. Oxford: Oxford University Press 2015.

Caldwell, M. D. "Wound Surgery." *Surgical Clinics of North America* 90, no. 6 (2010): 1125–32.

Camins, Joshua. "Military PTSD and Post-service Violence: A Review of the Evidence." *Trauma Psychology News*, November 9, 2018.

Cane, Scott. *First Footprints: The Epic Story of the First Australians*. New York: Allen and Unwin, 2013.

Caputo, Philip. *A Rumor of War*. New York: Ballantine, 1978.

Caraccilo, Dominic J. *Beyond Guns and Steel: A War Termination Strategy*. Santa Barbara, CA: Praeger, 2011.

Carnegie Middle East Center. "Syria in Crisis: The Damascus Spring." *Diwan*, April 1, 2012. https://carnegie-mec.org/diwan/48516?lang=en.

Carpenter, Ami C. "Havens in a Firestorm: Perspectives from Baghdad on Resilience to Sectarian Violence." *Civil Wars Journal* 14, no. 2 (2012): 182–204.

Carrière, Jean Louise. "Sir Gawain and the Green Knight as a Christmas Poem," *Comitatus: A Journal of Medieval and Renaissance Studies* 1, no. 1 (1970): 25–42.

Casas, Gustavo de la. "Is Nationalism Good for You?" *Foreign Policy*, October 8, 2009. https://foreignpolicy.com/2009/10/08/is-nationalism-good-for-you/.

Casey, Constance. "Moose: The Final Frontier." *Landscape Architecture* 104, no. 11 (2014): 66–70. http://www.jstor.org/stable/44796105.

Cashman, Greg. *What Causes War?* Lanham, MD: Lexington Books, 1993.

Casti, J. L. *Complexification: Explaining a Paradoxical World through the Science of Surprise*. New York: Abacus, 1994.

Ceballos, Gerardo, Paul R. Ehrlich, and Peter H. Raven. "Vertebrates on the Brink as Indicators of Biological Annihilation and the Sixth Mass Extinction." *PNAS* 117, no. 24 (2020): 13596–602. https://doi.org/10.1073/pnas.1922686117.

Center of Military History. *Military Improvisations during the Russian Campaign*. Washington, DC: US Army, 2004.

Chappell, Bill. "Dr. Seuss Enterprises Will Shelve 6 Books, Citing 'Hurtful' Portrayals." *NPR*, March 2, 2021.

Charney, Dennis S. "Psychobiological Mechanisms of Resilience and Vulnerability: Implications for Successful Adaptation to Extreme Stress." *American Journal of Psychiatry* 161 (2004): 368–91.

Charuvastra, Anthony, and Marylene Cloitre. "Social Bonds and Posttraumatic Stress Disorder." *Annual Review of Psychology* 59 (2008): 301–28.

Chesney, Robert, and Max Smeets, eds. *Deter, Disrupt, or Deceive: Assessing Cyber Conflict as an Intelligence Contest*. Washington, DC: Georgetown University Press, 2023.

Chisholm, Roderick M., and Thomas D Feehan. "Intent to Deceive." *Journal of Philosophy* 74, no. 3 (1977): 143–59.

Churchill, Winston. *River War: An Account of the Reconquest of the Sudan*. Mineola, NY: Dover, 2006.

Cislaghi, Beniamino, Diane Gillespie, and Gerry Mackie. *Values Deliberation and Collective Action: Community Empowerment in Rural Senegal*. London: Palgrave Macmillan, 2016.

Claeys, Priscilla. "Right to Land and Territory: New Human Right and Collective Action Frame." *Interdisciplinary Journal of Legal Studies* 75, no. 2 (2015): 115–37.

Clark, Robert M., and William L. Mitchell. *Deception: Counterdeception and Counterintelligence*. Washington, DC: CQ Press, 2018.

Clark, Roger S. "Is *The Butter Battle Book*'s Bitsy Big-Boy Boomeroo Banned? What Has International Law to Say about Weapons of Mass Destruction." *New York Law School Law Review* 58 (2013–14).

Clausewitz, Carl von. *On War*. Edited and translated by Michael Howard and Peter Paret. Princeton, NJ: Princeton University Press, 1989. First published 1976.

Clemens, Elisabeth S. "Of Asteroids and Dinosaurs: The Role of the Press in the Shaping of Scientific Debate." *Social Studies of Science* 16, no. 3 (1986): 421–56. http://www.jstor.org/stable/285026.
Cohen, Li. "'I Still Like Dr. Seuss': Kevin McCarthy Releases Video of Himself Reading 'Green Eggs and Ham.'" *CBS News*, March 6, 2021.
Committee on the Assessment of Ongoing Effects in the Treatment of Posttraumatic Stress Disorder. *Treatment for Posttraumatic Stress Disorder in Military and Veteran Populations: Initial Assessment*. Washington, DC: National Academies Press, 2012.
Cook, J. M. "Post-Traumatic Stress Disorder in Older Adults." *PTSD Research Quarterly* 12 (2001): 1–7.
Cook, Steven A. *Ruling but Not Governing: The Military and Political Development in Egypt, Algeria, and Turkey*. Baltimore, MD: Johns Hopkins University Press, 2007.
Cook, Timothy. "Another Perspective on Political Authority in Children's Literature: The Fallible Leader in L. Frank Baum and Dr. Seuss." *Western Political Quarterly* 36, no. 2 (June 1983): 326–36.
Cooper, Anthony J., Chris N. Bayer, and Mark S. Winters. *It Takes a Community: Collective Action Initiatives Confronting Corruption and Forced Labour*. N.p.: Development International and Konung International, 2019.
Cordesman, Anthony H. "The 'End State' Fallacy: Setting the Wrong Goals for Warfighting." Center for Strategic and International Studies, September 26, 2016. https://www.csis.org/analysis/end-state-fallacy-setting-wrong-goals-war-fighting.
Cordesman, Anthony H., and Nicholas Harrington. *The Arab Gulf States and Iran: Military Spending, Modernization, and the Shifting Military Balance*. Washington, DC: Center for Strategic and International Studies, December 12, 2018. https://www.csis.org/analysis/arab-gulf-states-and-iran-military-spending-modernization-and-shifting-military-balance.
Corera, Gordon. "Huawei: MPs Claim 'Clear Evidence of Collusion' with Chinese Communist Party—the Inquiry Focused on Huawei's Telecoms Kit Rather Than Its Consumer Handset Business." *BBC*, October 7, 2020. https://www.bbc.com/news/technology-54455112.
Cormac, Rory, and Richard J. Aldrich. "Grey Is the New Black: Covert Action and Implausible Deniability," *International Affairs* 94, no. 3 (May 2018): 477–94.
Corradi, Giselle, Eva Brems, and Mark Goodale, eds. *Human Rights Encounter Legal Pluralism: Normative and Empirical Approaches*. London: Hart Publishing, 2017.
Cott, Jonathan. "The Good Dr. Seuss." In *Of Sneetches and Whos and the Good Dr. Seuss: Essays on the Writings and Life of Theodor Geisel*, edited by Thomas Fensch, 99–124. Chesterfield, VA: New Century Books, 1997.
———. *Pipers at the Gates of Dawn: The Wisdom of Children's Literature*. New York: Random House, 1983.
———. "'Somebody's Got to Win' in Kids' Books: An Interview with Dr. Seuss on His Books for Children, Young and Old." In *Of Sneetches and Whos and the Good Dr. Seuss: Essays on the Writings and Life of Theodor Geisel*, edited by Thomas Fensch, 125–28. Chesterfield, VA: New Century Books, 1997.

Council on Foreign Relations. *Agent.btz* (report). November 2008. https://www.cfr.org/cyber-operations/agentbtz#:~:text=of%20USB%20drives.-,Agent.,operations%20for%20the%20U.S.%20military.

Crane, Conrad. *Cassandra in Oz: Counterinsurgency and Future War*. Annapolis, MD: Naval Institute Press, 2016.

Crane, Keith W., Evan Linck, Bhavya Lal, and Rachel Y. Wei. *Measuring the Space Economy: Estimating the Value of Economic Activities in and for Space*. Alexandria, VA: Institute for Defense Analyses, March 2020.

Croft, M. "Hidden Meanings in Classic Dr. Seuss Books." Best of New Jersey, February 25, 2016. https://bestofnj.com/features/entertainment/hidden-meanings-dr-seusss-classic-books/.

Crumm, Robin K. *Information Warfare: An Air Force Policy for the Role of Public Affairs*. Maxwell AFB, AL: Air University Press, 1996.

Cudworth, Erika. "Armed Conflict, Insecurity and Gender: The Resilience of Patriarchy?" *Resilience: International Policies, Practices and Discourses* 1, no. 1 (2013): 69–75.

Culbert, David. "'Why We Fight': Social Engineering for a Democratic Society at War." In *Readings in Propaganda and Persuasion: New and Classic Essays*, edited by Garth S. Jowett and Victoria O'Donnell, 169–88. Thousand Oaks, CA: Sage Publications, 2006.

Cybersecurity and Infrastructure Security Agency. "Defining Insider Threats." Department of Homeland Security. Accessed September 20, 2023. https://www.cisa.gov/topics/physical-security/insider-threat-mitigation/defining-insider-threats.

Darby, Delphine Fitz. "Review of *Ribera* by Elizabeth du Gué Trapier." *Art Bulletin* 35, no. 1 (1953): 68–74.

Davenport, Christian. *Space Barons: Elon Musk, Jeff Bezos, and the Quest to Colonize the Cosmos*. New York: Public Affairs, 2018.

Dedijer, Vladimir. *The Road to Sarajevo*. New York: Simon & Schuster, 1966.

Demchak, Chris C. "Achieving Systemic Resilience in a Great Systems Conflict Era." *Cyber Defense Review* 6, no. 2 (Spring 2021). https://cyberdefensereview.army.mil/Portals/6/Documents/2021_spring_cdr/COVID_CDR_V6N2_Spring_2021_r4.pdf.

———. "Defending Democracies in a Cybered World." *Brown Journal of World Affairs* 24, no. 1 (2017): 139–58. https://www.jstor.org/stable/27119084.

———. "Resilience, Disruption, and a 'Cyber Westphalia': Options for National Security in a Cybered Conflict World." In *Securing Cyberspace: A New Domain for National Security*, edited by Nicholas Burns and Jonathon Price, 59–94. Washington, DC: Aspen Institute, 2012.

Demchak, Chris C., and Peter J. Dombrowski. "Rise of a Cybered Westphalian Age." *Strategic Studies Quarterly* 5, no. 1 (March 1, 2011): 31–62.

Derickson, Alan. "'No Such Thing as a Night's Sleep': The Embattled Sleep of American Fighting Men from World War II to the Present." *Journal of Social History* 47, no. 1 (2013): 1–26. http://www.jstor.org/stable/43306043.

Desai, Miraj U., Anthony J Pavlo, Larry Davidson, Ilan Harpaz-Rotem, and Robert Rosenheck. "'I Want to Come Home': Vietnam-Era Veterans' Presenting for Mental Health Care, Roughly 40 Years after Vietnam." *Psychiatric Quarterly* 87 (2015): 229–39.

Dinshaw, Carolyn. "Black Skin, Green Masks: Medieval Foliate Heads, Racial Trauma, and Queer World-Making." In *Middle Ages in the Modern World*, edited by Bettina Bildhauer and Chris Jones, 276–304. Oxford: Oxford University Press, 2017.

Dominy, Nathaniel J., Sandra Winters, Donald E. Pease, and James P. Higham. "Dr. Seuss and the Real Lorax." *Nature Ecology & Evolution* 2 (2018): 1196–98.

Dow, Christopher. "Private Snafu's Hidden War: Historical Survey and Analytical Perspective." *Bright Lights Film Journal*, May 24, 2014. https://brightlightsfilm.com/private-snafus-hidden-war-historical-survey-analytical-perspective/.

Downing Assessment Task Force. *Force Protection Assessment of USCENTCOM AOR and Khobar Towers*. August 30, 1996.

Dreier, Peter. "Dr. Seuss's Progressive Politics." *Tikkun* 26, no. 4 (Fall 2011): 28–47.

Drengson, Alan, and Yuichi Inoue. *Deep Ecology Movement*. Berkeley, CA: North Atlantic Books, 1995.

du Picq, Ardant. *Battle Studies: Ancient and Modern Battle*. New York: Macmillan, 1921.

Duffy, Michael. "So Much for the WMD." *Time*, February 9, 2004. https://content.time.com/time/subscriber/article/0,33009,993272,00.html.

Durac, Vincent. "A Flawed Nexus? Civil Society and Democratization in the Middle East and North Africa." Middle East Institute, October 15, 2015. https://www.mei.edu/publications/flawed-nexus-civil-society-and-democratization-middle-east-and-north-africa.

Durbin, Jeffrey L. "Expressions of Mass Grief and Mourning: The Material Culture of Makeshift Memorials." *Material Culture* 35, no. 2 (2003): 22–47. http://www.jstor.org/stable/29764188.

Dyer, Hugh. "Environmental Security and International Relations: The Case for Enclosure." *Review of International Studies* 27, no. 3 (2001): 441–50.

Earle, Edward Mead. "Introduction to *Makers of Modern Strategy: Military Thought from Machiavelli to Hitler*, vii–xi. Edited by Edward Mead Earle. Princeton, NJ: Princeton University Press, 1943.

Echevarria, Antulio J., II. *Military Strategy: A Very Short Introduction*. New York: Oxford University Press, 2017.

Edgerton, Robert. *Sick Societies: Challenging the Myth of Primitive Harmony*. New York: Free Press, 1992.

Egloff, Florian J. *Semi-State Actors in Cybersecurity*. New York: Oxford University Press, 2021.

Ehlers, Anke, and David Clark. "A Cognitive Model of Posttraumatic Stress Disorder." *Behaviour Research and Therapy* 15 (2000): 249–75.

Ehrenreich, Barbara. *Blood Rites: Origins and History of the Passions of War*. New York: Metropolitan Books, 1997.

Eisenhower, David. *Eisenhower: At War, 1943–1945*. New York: Random House, 1986.

Eldredge, H. Wentworth. "Biggest Hoax of the War: Operation FORTITUDE; the Allied Deception Plan That Fooled the Germans about Normandy." *Air Power History* 37, no. 3 (1990): 15–22. http://www.jstor.org/stable/26271165.

Eldridge, Thomas R., Susan Ginsburg, Walter T. Hempel II, Janice L. Kephart, Kelly Moore, and Joanne M. Accolla. *9/11 and Terrorist Travel: Staff Report of* the *National Commission on Terrorist Attacks Upon the United States*. Edited by Alice Falk. National Commission on Terrorist Attacks Upon the United States. August 21, 2004. http://govinfo.library.unt.edu/911/staff_statements/911_TerrTrav_Monograph.pdf.

Ellul, Jacques. *Propaganda: The Formation of Men's Attitudes*. New York: Alfred Knopf, 1971.

Elvin, Mark. "Mandarins and Millenarians: Reflections on the Boxer Uprising of 1899–1900." *Journal of the Anthropological Society of Oxford* 10, no. 3 (1979): 115–38.

Emery, Frank. *The Red Soldier*. Johannesburg: Jonathan Ball, 1977.

Emmott, Robin, and Ben Blanchard. "Wary of Trump, China Launches EU Charm Offensive: Diplomats." *Reuters*, March 28, 2017. http://www.reuters.com/article/us-eu-china-idUSKBN16Z22S.

Encyclopaedia of Islam Vol. 10. Leiden: E. J. Brill, 2000.

Engels, Jens Ivo. "Modern Environmentalism." In *Turning Points of Environmental History*, edited by Frank Uekoetter, 119–31. Pittsburgh: University of Pittsburgh Press, 2010. https://doi.org/10.2307/j.ctt5hjsg1.11

Erwin, Sandra. "Air Force Laying Groundwork for Future Military Use of Commercial Megaconstellations." *SpaceNews*, February 28, 2019. https://spacenews.com/air-force-laying-groundwork-for-future-military-use-of-commercial-megaconstellations/.

———. "New Studies Provide Fresh Insights into the Escalating Space Arms Race." *SpaceNews*, April 4, 2019. https://spacenews.com/new-studies-provide-fresh-insights-into-the-escalating-space-arms-race/.

———. "SpaceX President Gwynne Shotwell: 'We Would Launch a Weapon to Defend the U.S.'" *SpaceNews*, September 17, 2018. https://spacenews.com/spacex-president-gwynne-shotwell-we-would-launch-a-weapon-to-defend-the-u-s/.

———. "U.S. Military Eyes a Role in the Great Power Competition for Lunar Resources." *SpaceNews*, August 20, 2020. https://spacenews.com/u-s-military-eyes-a-role-in-the-great-power-competition-for-lunar-resources/.

Esposito, John, and John Voll. *Islam and Democracy*. Oxford: Oxford University Press, 1995.

Esposito, John L., ed. *The Oxford Dictionary of Islam*. Oxford: Oxford University Press, 2003.

Falger, Paul R. J., Wybrand Op den Velde, Johannes E. Hovens, Erik G. W. Schonten, Johannes H. M. De Groen, and Hans Van Duijn. "Current Posttraumatic Stress Disorder and Cardiovascular Disease Risk Factors in Dutch Resistance Veterans from World War II." *Psychotherapy and Psychosomatics* 57, no. 4 (1992): 164–71.

Falliere, Nicolas, Liam O. Murchu, and Eric Chien. "W32.Stuxnet Dossier, Version 1.4." Symantec, February 4, 2011. https://www.wired.com/images_blogs/threatlevel/2011/02/Symantec-Stuxnet-Update-Feb-2011.pdf.

Fallis, Don. "The Varieties of Disinformation." In *Philosophy of Information Quality*, edited by Luciano Floridi and Phyllis Illari, 135–61. Cham: Springer International Publishing, 2014. https://doi.org/10.1007/978-3-319-07121-3_8.

Feigin, Svetlana, Glynn Owens, and Felicity Goodyear-Smith. "Theories of Human Altruism: A Systematic Review." *Annals of Neuroscience and Psychology* 1, no. 1 (2014): 1–8. http://dx.doi.org/10.7243/2055-3447-1-5.

Fensch, Thomas. *The Man Who Was Dr. Seuss: The Life and Work of Theodor Geisel*. Woodlands, TX: New Century Books, 2001.

Findley, Michael G., and Scott Edwards. "Accounting for the Unaccounted: Weak-Actor Social Structure in Asymmetric Wars." *International Studies Quarterly* 51, no. 3 (2007): 583–606.

Fingar, Thomas. "Analysis in the US Intelligence Community: Missions, Masters, and Methods." In *Intelligence Analysis: Behavioral and Social Scientific Foundations*, National Research Council. Washington, DC: National Academies Press, 2011.

Fischer, Meredith. "Capturing the Animated Soldier: Private Snafu and the Docile Body Assemblage." *Studies in Popular Culture* 41, no. 1 (2018): 94–127.

Flanagan, Stephen J. "Managing the Intelligence Community." *International Security* 10, no. 1 (1985): 58–95. https://doi.org/10.2307/2538790.

Flood, Alison. "Dr Seuss's Thank-You Letter to Man Who Saved His First Book." *Guardian UK*, January 30, 2019.

Flynn, George Q. *The Draft, 1940–1973*. Lawrence: University Press of Kansas, 1993.

Folsom, Cate. *Smoke the Donkey A Marine's Unlikely Friend*. Washington, DC: Potomac Books, 2016.

Force Protection Assessment of USCENTCOM AOR and Khobar Towers. Report of the Downing Assessment Task Force. Washington, DC: US Government Printing Office, August 30, 1996.

Ford, Ronnie L. *Tet 1968: Understanding the Surprise*. London: Frank Cass, 1995.

Forster, E.S. "Dogs in Ancient Warfare." *Greece & Rome* 10 (1941): 114–17.

Forsythe, David P. "The United Nations, Human Rights, and Development." *Human Rights Quarterly* 19, no. 2 (May 1997): 334–49.

Fouda, Yosri, and Nick Fielding. *Masterminds of Terror: The Truth behind the Most Devastating Terrorist Attack the World Has Ever Seen*. London: Mainstream Publishing, 2003.

Foust, Jeff. "Commerce Department to Develop New Estimate of the Size of the Space Economy." *Spacenews*, January 2, 2020. https://spacenews.com/commerce-department-to-develop-new-estimate-of-the-size-of-the-space-economy/.

———. "With Pentagon Award, SpaceX Joins the Establishment." *Spacenews*, August 7, 2020. https://spacenews.com/news-analysis-with-pentagon-award-spacex-joins-the-establishment/.

Frankel, Rebecca. *War Dogs*. New York: Palgrave Macmillan, 2014.

Freedman, Lawrence. *Deterrence*. Malden, MA: Polity Press, 2004.

———. *Evolution of Nuclear Strategy*. 3rd ed. New York: Palgrave Macmillan, 2003.

———. *Strategy: A History*. Oxford: Oxford University Press, 2013.

Freedom House. "Freedom in the World 2019: Democracy in Retreat." Accessed September 18, 2023. https://freedomhouse.org/report/freedom-world/2019/democracy-retreat.

Friedrich, Carl J. "Education and Propaganda." *Atlantic Monthly*, June 1937, 693–701. https://www.theatlantic.com/magazine/archive/1937/06/education-and-propaganda/652484/.

Fritz, Stephen. *Frontsoldaten: The German Soldier in World War II*. Lexington: University Press of Kentucky, 1995.

Froehlich, Thomas J. "A Not-So-Brief Account of Current Information Ethics: The Ethics of Ignorance, Missing Information, Misinformation, Disinformation and Other Forms of Deception or Incompetence." *BiD*, no. 39 (December 2017). https://bid.ub.edu/en/39/froehlich.htm.

Froese, Katrine. "Beyond Liberalism: The Moral Community of Rousseau's Social Contract." *Canadian Journal of Political Science* 34, no. 3 (September 2001): 579–600.

Fulton, J. J., P. S. Calhoun, H. R. Wagner, A. R. Schry, L. P. Hair, N. Feeling, and J. C. Beckham. "Prevalence of Posttraumatic Stress Disorder in Operation Enduring Freedom/Operation Iraqi Freedom (OEF/OIF) Veterans: A Meta-Analysis." *Journal of Anxiety Disorders* 31 (2014): 98–107.

Gaddis, John Lewis. *On Grand Strategy*. New York: Penguin Press, 2018.

———. *Strategies of Containment: A Critical Appraisal of American National Security Policy during the Cold War*. New York: Oxford University Press, 1981.

———. *Surprise, Security, and the American Experience*. Cambridge, MA: Harvard University Press, 2005.

Gal, Reuven. "Social Resilience in Times of Protracted Crises." *Armed Forces & Society* 40, no. 3 (2014): 452–75.

Galatzer-Levy, Isaac R., Yael Ankri, Sara Freedman, Yossi Israeli-Shalev, Pablo Roitman, Moran Gilad, and Arieh Y. Shalev. "Early PTSD Symptom Trajectories: Persistence, Recovery, and Response to Treatment: Results from the Jerusalem Trauma Outreach and Prevention Study (J-TOPS)." *PLoS One* 8, no. 8 (2013). http://doi.org/10.1371/journal.pone.0070084.

Garfinkle, Adam. *Telltale Hearts: The Origins and Impact of the Vietnam Antiwar Movement*. New York: St. Martin's, 1995.

Garner, Bryan A., ed. *Black's Law Dictionary*. 10th ed. St. Paul, MN: Thompson Reuters, 2014.

Garnett, John. "The Causes of War and Conditions of Peace." In *Strategy in the Contemporary World*, edited by John Baylis, James Wirtz, Eliot Cohen, and Colin S. Gray, 75–76. Oxford: Oxford University Press, 2002.

Gartzke, Erik, and Jon R. Lindsay. "Weaving Tangled Webs: Offense, Defense, and Deception in Cyberspace." *Security Studies* 24, no. 2 (2015): 316–48.

Gause, F. Gregory, III. "The Persistence of Monarchy in the Arabian Peninsula: A Comparative Analysis." In *Middle East Monarchies: The Challenge of Modernity*, edited by Joseph Kostiner, 167–87. Boulder, CO: Lynne Rienner Publishers, 2000.

———. "Why Middle East Studies Missed the Arab Spring: The Myth of Authoritarian Stability." *Foreign Affairs* 90, no. 4 (July/August 2011): 81–90.

Geisel, Audrey. "Living with the Cat." In *Your Favorite Seuss: Thirteen Best Loved Stories*, compiled by Janet Schulman and Cathy Goldsmith, 338. New York: Random House, 2004.

Geisel, Theodor. "Japan's Young Dreams." *Life*, March 29, 1954.

George, Alexander L., and Richard Smoke. *Deterrence in American Foreign Policy: Theory and Practice*. New York: Columbia University Press, 1974.

Gewirtzman, Doni. "The Seussian Dead Hand: Concluding Remarks to Exploring Civil Society through the Writings of Dr. Seuss." *New York Law School Law*

Review 58 (2013–14). https://digitalcommons.nyls.edu/nyls_law_review/vol58/iss3/11/.

Ghosh, Bishnupriya. "Animating Uncommon Life: Malaria Films (1942–1945) and the Pacific Theater." In *Animating Film Theory*, edited by Karen Beckman, 264–86. Durham, NC: Duke University Press, 2014.

Gibney, Mark, Linda Cornett, Reed Wood, Peter Haschke, Daniel Arnon, Attilio Pisanò, Gray Barrett, and Baekkwan Park. "Political Terror Scale," accessed September 21, 2023, http://www.politicalterrorscale.org

Gilbert, Sophie. "The Complicated Relevance of Dr. Seuss's Political Cartoons." *Atlantic*, January 31, 2017. https://www.theatlantic.com/entertainment/archive/2017/01/dr-seuss-protest-icon/515031/.

Gilje, Paul. *Rioting in America*. Bloomington: Indiana University Press, 1996.

Gill, Paul. *Lone-Actor Terrorists: A Behavioural Analysis*. London: Routledge, 2015.

Gilpin, Robert. *War and Change in World Politics*. Cambridge: Cambridge University Press, 1981.

Gilpin, Robert, and Jean M. Gilpin. *Global Political Economy: Understanding the International Economic Order*. Princeton, NJ: Princeton University Press, 2001.

Girard, Rene. *Violence and the Sacred*. Baltimore, MD: Johns Hopkins University Press, 1972.

Gitlin, Todd. *The Whole World Is Watching: Mass Media in the Making and Unmaking of the New Left*. Berkeley: University of California Press, 1980.

Giunta, Andrea. "The Power of Interpretation (or How MoMA Explained Guernica to Its Audience)." Translated by Jane Brodie. *Artelogie*, October 2017. https://doi.org/10.4000/artelogie.953.

Godson, Roy. *Dirty Tricks or Trump Cards*. Washington, DC: Brassey's, 1995.

Godson, Roy, and James J. Wirtz. "Strategic Denial and Deception." *International Journal of Intelligence and Counterintelligence* 13 (2000): 424–37.

Goffman, Erving. *Frame Analysis*. New York: Harper & Row, 1974.

Goldberg, Jeffrey. "The Unknown: The C.I.A. and the Pentagon Take Another Look at Al Qaeda and Iraq." *New Yorker*, February 10, 2003, 40–47.

Golding, William. *Lord of the Flies*. London: Faber & Faber, 1954.

Goldman, Emily O. "The Cyber Paradigm Shift." In *Ten Years In: Implementing New Strategic Approaches to Cyberspace*, edited by Emily Goldman, Jacqueline Schneider, and Michael Warner, 31–46. Newport, RI: US Naval War College Press, 2020.

Goldstein, Bruce E., ed. *Collaborative Resilience: Moving through Crisis to Opportunity*. Cambridge, MA: MIT Press, 2011.

Goldsworthy, Adrian. *The Punic Wars*. London: Cassel & Co., 2001.

Goodavage, Maria. *Soldier Dogs*. New York: Dutton, 2012.

"Good Luck Belts." *New York Times Magazine*, January 11, 1942.

Goodin, Dan. "There's a Vexing Mystery Surrounding the 0-Day Attacks on Exchange Servers—a Half-Dozen Groups [Including Hafnium] Exploiting the Same 0-Days Is Unusual, If Not Unprecedented." *Ars Technica*, March 11, 2021. https://arstechnica.com/gadgets/2021/03/security-unicorn-exchange-server-0-days-were-exploited-by-6-apts/.

Gottlieb, Carla. "The Meaning of Bull and Horse in Guernica." *Art Journal* 24, no. 2 (1964): 106–12. https://doi.org/10.2307/774777.

Government Accountability Office. *Information Environment: Opportunities and Threats to DOD's National Security Mission*. September 2022. https://www.gao.gov/assets/gao-22-104714.pdf.

Graham, Thomas. "Let Russia Be Russia." *Foreign Affairs* 98 (2019): 134.

Grange, Judith K., J. Battilega, J. Bennett, and S. Summers. *Soviet Measurements of Strategic Balance and Arms Control*. Washington, DC: Defense Nuclear Agency, April 21, 1986. https://apps.dtic.mil/sti/pdfs/ADA221221.pdf.

Gray, Colin S. *Modern Strategy*. Oxford: Oxford University Press, 1999.

———. *National Security Dilemmas: Challenges & Opportunities*. Washington, DC: Potomac Books, 2009.

———. *The Sheriff: America's Defense of the New World Order*. Lexington: University Press of Kentucky, 2009.

———. "Strategic Culture as Context: The First Generation of Theory Strikes Back." *Review of International Studies* 25, no 1 (January 1999): 1–34. https://doi.org/10.1017/S0260210599000492.

———. "Strategic Stability Reconsidered." *Daedalus* 109, no. 4 (Fall 1980): 135–54.

———. *The Strategy Bridge: Theory for Practice*. Oxford: Oxford University Press, 2010.

———. *Theory of Strategy*. Oxford: Oxford University Press, 2018.

———. *War, Peace and International Relations: An Introduction to Strategic History*. New York: Routledge, 2011.

Greenberg, Andy. "The Untold Story of NotPetya, the Most Devastating Cyberattack in History." *Wired Magazine*, August 22, 2018. https://www.wired.com/story/notpetya-cyberattack-ukraine-russia-code-crashed-the-world/.

———. "The Untold Story of the 2018 Olympics Cyberattack, the Most Deceptive Hack in History." *Wired Magazine*, October 17, 2019. https://www.wired.com/story/untold-story-2018-olympics-destroyer-cyberattack/.

Griggs, Mary Beth. "Trump's Space Force Aims to Create 'American Dominance in Space' by 2020." *Popular Science*, August 10, 2018. https://www.popsci.com/space-force-2020/.

Grillon, C., D. S. Pine, S. Lissek, S. Rabin, O. Bonne, and M. Vythilingam. "Increased Anxiety during Anticipation of Unpredictable Aversive Stimuli in Posttraumatic Stress Disorder but Not in Generalized Anxiety Disorder." *Biological Psychiatry* 66, no. 1 (2009): 47–53.

Gruber, Frank. *Pulp Jungle*. Los Angeles: Sherbourne Press, 1967.

"Guernica." *Treasures of the World*. Public Broadcasting System, 1999. https://www.pbs.org/treasuresoftheworld/guernica/glevel_1/5_meaning.html.

Guerrero-Saade, Juan Andrés. "Draw Me Like One of Your French APTs—Expanding Our Descriptive Palette for Cyber Threat Actors." Paper presented at *Virus Bulletin* conference, Montreal, October 3–5, 2018. https://www.virusbulletin.com/uploads/pdf/magazine/2018/VB2018-Guerrero-Saade.pdf.

———. "King of the Hill: Nation-State Counterintelligence for Victim Deconfliction." Paper presented at *Virus Bulletin* conference, London, October 2–4, 2019.

https://www.virusbulletin.com/virusbulletin/2020/01/vb2019-paper-king-hill-nation-state-counterintelligence-victim-deconfliction/.

Guha, Ramachandra. *Environmentalism: A Global History*. New York: Longman, 2000.

Guy, Jack. "Everything You Thought about 'The Scream' Is Wrong." *CNN*, March 21, 2019. https://www.cnn.com/style/article/munch-scream-british-museum-gbr-scli-intl/index.html.

Haass, Richard. *Foreign Policy Begins at Home: The Case for Putting America's House in Order*. New York: Basic Books, 2013.

———. *War of Necessity, War of Choice: A Memoir of Two Iraq Wars*. New York: Simon & Schuster, 2010.

Hachemaoui, Mohammed, and Michael O'Mahony. "Does Rent Really Hinder Democracy? A Critical Review of the 'Rentier State' and 'Resource Curse' Theories." *Revue Française De Science Politique* (English ed.) 62, no. 2 (2012): 207–30.

Hajdu, David. *Ten-Cent Plague: The Great Comic-Book Scare and How It Changed America*. New York: Farrar, Straus, and Giroux, 2008.

Halper, Stefan, and Jonathan Clarke. *America Alone: The Neo-Conservatives and the Global Order*. New York: Cambridge University Press, 2004.

Halsey, William. *Admiral Halsey's Story*. New York: Wittlesley House, 1947.

Hardin, Garrett. "Tragedy of the Commons." *Science* 162, no. 3859 (December 13, 1968): 1243–48.

Harjo, Suzan Shown, ed. *Nation to Nation: Treaties between the United States and American Indian Nations*. Washington, DC: Smithsonian Institution, 2014.

Harpviken, Kristian Berg "Transcending Traditionalism: The Emergence of Non-State Military Formations in Afghanistan," *Journal of Peace Research* 34, no. 3 (1997): 271–87.

Harris, Lee. "*Al Qaeda's* Fantasy Ideology." *Policy Review* 114 (August–September 2002): 19–36.

Harris, Nick. "Bad Omen for Italy as Their Unlucky Number Comes Up." *UK Independent*, November 15, 2007.

Harrison, Simon. "Skull Trophies of the Pacific War: Transgressive Objects of Remembrance." *Journal of the Royal Anthropological Institute* 12, no. 4 (2006): 817–36. http://www.jstor.org/stable/4092567.

Hart, B. H. Liddell. *Strategy*. 2nd ed. New York: Meridian, 1954.

Hart, William B. "Between the Lines." *Redbook*, December 1957.

Harvard Kennedy School of Government. "The Labor Market in Saudi Arabia: Background, Areas of Progress, & Insights for the Future." 2019. https://epod.cid.harvard.edu/sites/default/files/2019-08/EPD_Report_Digital.pdf.

Heatherly, Chris. "Dr. Seuss and the Operational Art of War." *Armchair General*, April 27, 2010. http://armchairgeneral.com/dr-seuss-and-the-operational-art-of-war.htm.

Heinl, Robert. *Dictionary of Military and Naval Quotations*. Annapolis, MD: Naval Institute Press, 1966.

Held, Jacob M., and Eric N. Wilson. "What Would You Do if Your Mother Asked You? A Brief Introduction to Ethics." In *Dr. Seuss and Philosophy: Oh, the Things*

You Can Think!, edited by Jacob M. Held, 103–17. Lanham, MD: Rowman & Littlefield, 2011.

Held, Jutta, and Alex Potts. "How Do the Political Effects of Pictures Come About? The Case of Picasso's *Guernica*." *Oxford Art Journal* 11, no. 1 (1988): 33–39. http://www.jstor.org/stable/1360321.

Hemingway, Ernest, ed. *Men at War*. New York: Brammel House, 1979.

Henry, Caleb. "SpaceX Raises $1.9 Billion in Equity." *SpaceNews*, August 18, 2020. https://spacenews.com/spacex-raises-1-9-billion-in-equity/.

Herb, Michael. "No Representation without Taxation? Rents, Development, and Democracy." *Comparative Politics* 37, no. 3 (April 2005): 297–316.

Herman, Judith Lewis. *Trauma and Recovery*. New York: Basic Books, 1992.

Herrmann, Jonathan, and Goran Eryavec. "Posttraumatic Stress Disorder in Institutionalized World War II Veterans." *American Journal of Geriatric Psychiatry* 2, no. 4 (1994): 324–31.

Hesford, Wendy S. "Human Rights Rhetoric of Recognition." *Rhetoric Society Quarterly* 41, no. 3 (2011) 282–89.

Heuer, Richard J., Jr. *Psychology of Intelligence Analysis*. Washington, DC: Center for the Study of Intelligence, 1999.

Hibou, Béatrice. *The Force of Obedience: The Political Economy of Repression in Tunisia*. Translated by Andrew Brown. Cambridge, UK: Polity Press, 2011.

Hicks, Louis. "Normal Accidents in Military Operations." *Sociological Perspectives* 36, no. 4 (1993): 377–91.

Hitchens, Theresa. "Russia Builds New Co-Orbital Satellite: SWF, CSIS Say." *Breaking Defense*, April 4, 2019. https://breakingdefense.com/2019/04/russia-builds-new-co-orbital-satellite-swf-csis-say/.

———. "Space Lasers for Satellite Defense Top New French Space Strategy." *Breaking Defense*, July 26, 2019. https://breakingdefense.com/2019/07/france-envisions-on-orbit-lasers-for-satellite-defense/.

———. "US, Allies Agree on Threats in Space but Struggle with Messaging." *Breaking Defense*, September 11, 2020. https://breakingdefense.com/2020/09/us-allies-agree-on-threats-in-space-but-struggle-with-messaging/

Hobbes, Thomas. *Leviathan*. Edited by Edwin Curley. Indianapolis, IN: Hackett, 1994.

Holmes, Richard. *Acts of War: The Behavior of Men in Battle*. New York: Free Press, 1989.

———. *Sahib: The British Soldier in India 1750–1914*. London: HarperCollins, 2005.

Horton, Alex. "Russia's Commando Units Gutted by Ukraine War, U.S. Leak Shows." *Washington Post*, April 14, 2023.

Howard, Marc Morjé, and Meir R. Walters. "Explaining the Unexpected: Political Science and the Surprises of 1989 and 2011." *Perspectives on Politics* 12, no. 2 (June 2014): 394–408.

Howard, Philip, and Muzammil M. Hussain. "The Role of Digital Media." *Journal of Democracy* 22, no. 3 (July 2011): 35–48.

Hoyos, Dexter. *Mastering the West: Rome and Carthage at War*. Oxford, UK: Oxford University Press, 2015.

Hubbell, John G., Andrew Jones, and Kenneth Y. Tomlinson. *P.O.W.: A Definitive History of the American Prisoner-of-War Experience in Vietnam, 1964–1973*. New York: Thomas Y. Crowell Company, 1976.

Hudson, Robert P. "The Biography of Disease: Lessons from Chlorosis." *Bulletin of the History of Medicine* 51, no. 3 (1977): 448–63.

Hugo, Victor. *Battle of Waterloo*. East Aurora, NY: The Roycrofters, 1907.

Hunter, R. F. "Hill Sheep and Their Pasture: A Study of Sheep-Grazing in South-East Scotland." *Journal of Ecology* 50, no. 3 (1962): 651–80. https://doi.org/10.2307/2257476.

Huntington, Samuel. "Religion and the Third Wave." *National Interest*, no. 24 (Summer 1991): 29–42.

Huntington, Samuel P. *The Third Wave: Democratization in the Late Twentieth Century*. Norman: University of Oklahoma Press, 1991.

Hursh, John. "International Law, Armed Conflict, and the Construction of Otherness: A Critical Reading of Dr. Seuss's *The Butter Battle Book* and a Renewed Call for Global Citizenship." *New York Law School Law Review* 58, no. 617 (2014): 618–52.

Hutchinson, William. "Information Warfare and Deception." *Informing Science* 9, no. 20 (2006). https://doi.org/10.28945/480.

Huxley, Aldous. *Island*. New York: Harper, 1962.

Ikle, Fred C. *Every War Must End*. New York: Columbia University Press, 1991.

"Instructions for the Last Night." *Frontline*, PBS, n.d. Accessed April 29, 2023. https://www.pbs.org/wgbh/pages/frontline/shows/network/personal/instructions.html.

International Global Change Program, "The Anthropocene Era." *IGBP Science* 4 (2001): 11–15.

International Rescue Committee. *Policy Brief: Overview of Right to Work for Refugees, Syria Crisis Response in Lebanon and Jordan*. IRC, January 1, 2016. https://www.rescue.org/report/policy-brief-overview-right-work-refugees-syria-crisis-response-lebanon-and-jordan.

Irwin, H. J. "The Measurement of Superstitiousness as a Component of Paranormal Belief—Some Critical Reflections." *European Journal of Parapsychology* 22, no. 2 (2007): 95–120.

"Is the News Media Making Too Much of Karen Twins, Johnny and Luther?" *Straits Times Interactive*, August 11, 2000. www.hartford-hwp.com/archives/54/170.html.

Ishizuka, Katie, and Ramón Stephens. "The Cat Is Out of the Bag: Orientalism, Anti-Blackness, and White Supremacy in Dr. Seuss's Children's Books." *Research on Diversity in Youth Literature* 1, no. 2 (2019). https://sophia.stkate.edu/rdyl/vol1/iss2/4.

Jamal, Amaney A. *Barriers to Democracy: The Other Side of Social Capital in Palestine and the Arab World*. Princeton, NJ: Princeton University Press, 2007.

Jasper, James M. *The Art of Moral Protest: Culture, Biography, and Creativity in Social Movements*. Chicago: University of Chicago Press, 1997.

———. *Protest: A Cultural Introduction to Social Movements*. New York: Polity Press, 2014.

Javers, Eamon. "Cyberattacks: Why Companies Keep Quiet." *CNBC*, February 25, 2013. https://www.cnbc.com/id/100491610.

Jeffcoat, Tanya. "From There to Here, from Here to There, Diversity Is Everywhere." In *Dr. Seuss and Philosophy: Oh, the Things You Can Think!*, edited by Jacob M. Held, 93–102. Lanham, MD: Rowman & Littlefield, 2011.

Jenkins, Henry. "'No Matter How Small': The Democratic Imagination of Dr. Seuss." In *Hop on Pop: The Politics and Pleasures of Popular Culture*, edited by Henry Jenkins, Tara McPherson, and Jane Shattuc, 187–208. Durham, NC: Duke University Press, 2002.

Jervis, Robert. *Illogic of American Nuclear Strategy*. Ithaca, NY: Cornell University Press, 1984.

———. "War and Misperception." *The Journal of Interdisciplinary History* 18, no. 4 (1988): 675–700. https://doi.org/10.2307/204820.

Jespersen, A. F., M. L. Lalumière, and M. C. Seto. "Sexual Abuse History among Adult Sex Offenders and Non-Sex Offenders: A Meta-Analysis." *Child Abuse & Neglect* 33 (2009): 179–92.

Johnson, David Read, Hadar Lubin, Robert Rosenheck, Alan Fontana, Steven Sonthwick, and Dennis Charney. "The Impact of the Homecoming Reception on the Development of Posttraumatic Stress Disorder: The West Haven Homecoming Stress Scale (WHHSS)." *Journal of Traumatic Stress* 10, no. 2 (1997): 259–77.

Johnson, Loch. "The CIA's Weakest Link: What Our Intelligence Agencies Need Are More Professors." *Washington Monthly*, July/August 2001.

Johnson, Luke. "Muddling through Is a Strategy That Works." *Financial Times*, May 8, 2012.

Johnson, Robert H. "Periods of Peril: The Window of Vulnerability and Other Myths." *Foreign Affairs* 61, no. 4 (Spring 1983). https://doi.org/10.2307/20041562.

Johnson-Freese, Joan. *Space as a Strategic Asset*. New York: Columbia University Press, 2007.

———. *Space Warfare in the 21st Century: Arming the Heavens*. New York: Routledge, 2017.

Joint Chiefs of Staff. *Information Operations.* Joint Publication 3.13. Washington, DC: Government Printing Office, 2012.

Joint Chiefs of Staff. *Joint Campaigns and Operations*. Joint Publication 3.0. Washington, DC: Government Printing Office, 2022.

Jones, Andrew. "China Carries Out Secretive Launch of 'Reusable Experimental Spacecraft.'" *SpaceNews*, September 4, 2020. https://spacenews.com/china-carries-out-secretive-launch-of-reusable-experimental-spacecraft/.

———. "Chinese Space Launch Firm iSpace Raises $173 Million in Series B Funding." *SpaceNews*, August 25, 2020. https://spacenews.com/chinese-space-launch-firm-ispace-raises-173-million-in-series-b-funding/.

Jones, Brian Jay. *Becoming Dr. Seuss: Theodor Geisel and the Making of an American Imagination*. New York: Dutton, 2019.

Jones, Edgar. "Historical Approaches to Post-Combat Disorders." *Philosophical Transactions of the Royal Society B: Biological Sciences* 361, no. 1468 (2006): 533–42.

Jones, Michael Owen. "What's Disgusting, Why, and What Does It Matter?" *Journal of Folklore Research* 37, no. 1 (2000): 53–71. http://www.jstor.org/stable/3814665.

Kahn, E. J. "Children's Friend." *New Yorker*, December 17, 1960, 47–93.

Kamrava, Mehran. "Military Professionalization and Civil-Military Relations in the Middle East." *Political Science Quarterly* 115, no. 1 (Spring 2000): 67–92.

Karner, Tracy Xavia. "Post-Traumatic Stress Disorder and Older Men: If Only Time Healed All Wounds." *Generations: Journal of the American Society on Aging* 32, no. 1 (2008): 82–87.

Karney, Benjamin R., Rajeev Ramchand, Karen Chan Osilla, Leah Barnes Caldarone, and Rachel M. Burns. "Predicting the Immediate and Long-Term Consequences of Post-Traumatic Stress Disorder, Depression, and Traumatic Brain Injury in Veterans of Operation Enduring Freedom and Operation Iraqi Freedom." In *Invisible Wounds of War: Psychological and Cognitive Injuries, Their Consequences, and Services to Assist Recovery*, edited by Terri Tanielian and Lisa H. Jaycox, 119–66. Santa Monica, CA: RAND, 2008.

Kashdan, T. B., G. Uswatte, M. F. Steger, and T. Julian, "Fragile Self-Esteem and Affective Instability In Posttraumatic Stress Disorder." *Behavioral Research and Therapy Journal* 44 (2006): 1609–19.

Kaspersky Labs. "Equation: The Death Star of Malware Galaxy." Securelist, February 16, 2015. https://securelist.com/equation-the-death-star-of-malware-galaxy/68750/.

Katzenstein, Peter, ed. *Culture of National Security: Norms and Identity in World Politics*. New York: Columbia University Press, 1996.

Kaufmann, Mark David. "Ignorant Armies: Private Snafu Goes to War." *Public Domain Review*, March 15, 2015. https://publicdomainreview.org/essay/ignorant-armies-private-snafu-goes-to-war.

Kean, Thomas H., and Lee H. Hamilton, eds. *The 9/11 Commission Report: Final Report of the National Commission on Terrorist Attacks Upon the United States*. New York: W. W. Norton, 2004.

Keddie, Nikki. "Can Revolutions Be Predicted? Can Their Causes be Understood?" In *Debating Revolutions*, edited by Nikki Keddie, 3–26. New York: New York University Press, 1995.

Kedourie, Elie. *Democracy and Arab Political Culture*. London: Frank Cass, 1994.

Keegan, John. *Intelligence in War: Knowledge of the Enemy from Napoleon to Al-Qaeda*. New York: Knopf, 2003.

Keenan, Salla. "Private Snafu: What Can a Cartoon Tell U.S. about the U.S. Military in World War II?" *Primary Source* 3, no. 2 (2014): 1–4.

Keeter, Hunter. "Cartwright: Threat Location, Prediction Capability Should Be Priorities." *Defense Daily*, April 30, 2003.

Kelly, Ray. "Dr. Seuss' Great-Nephew Calls Museum Mural Removal 'Extreme,' Criticism 'a Lot of Hot Air Over Nothing.'" *Mass Live*, October 11, 2017. www.masslive.com/news/index.ssf/2017/10/dr_seuss_great_nephew_calls_mu.html.

Kennedy, Paul. "American Grand Strategy, Today and Tomorrow: Learning from the European Experience." In *Grand Strategies in War and Peace*, edited by Paul Kennedy, 167–84. New Haven, CT: Yale University Press, 1991.

———. "Grand Strategy in War and Peace: Toward a Broader Definition." In *Grand Strategies in War and Peace*, edited by Paul Kennedy, 1–7. New Haven, CT: Yale University Press, 1991.

———. *Rise and Fall of the Great Powers: Economic Change and Military Conflict from 1500 to 2000*. New York: Vintage, 1987.

Kent, Sherman. *Strategic Intelligence for American World Policy*. Princeton, NJ: Princeton University Press, 1949.

Keo, Kounila. "Is the News Media Making Too Much of Karen Twins, Johnny and Luther?" *Straits Times Interactive*, August 11, 2000.

———. "Modernity Counteracts Magic Tattoos in Cambodia." *Daily Telegraph UK*, November 18, 2009.

Khadilkar, Dhananjay. "How Donkeys Changed the Course of Human History." *BBC*, January 16, 2023. https://www.bbc.com/future/article/20230116-how-donkeys-changed-the-course-of-human-history.

King, Gary, Robert O. Keohane, and Sidney Verba. *Designing Social Inquiry: Scientific Inference in Qualitative Research*. Princeton, NJ: Princeton University Press, 1994.

Kintzle, S., N. Barr, G. Corletto, and C. A. Castro. "PTSD in U.S. Veterans: The Role of Social Connectedness, Combat Experience and Discharge." *Healthcare* 6, no. 3 (2018): 102.

Kippenberg, Hans G. "'Consider That It Is a Raid on the Path of God': The Spiritual Manual of the Attackers of 9/11." *Numen* 52, no. 1 (2005): 29–58.

Kissinger, Henry. *Does America Need a Foreign Policy? Toward a Diplomacy for the 21st Century*. New York: Simon & Schuster, 2002.

———. *Nuclear Weapons and Foreign Policy*. New York: W. W. Norton, 1957.

———. *The World Restored*. London: V. Gollancz, 1974.

Klein, Norman M. *Seven Minutes: The Life and Death of the American Animated Cartoon*. New York: Verso, 1998.

Klein, Robert M. "Adaptive Planning." *Joint Forces Quarterly* 45 (2007): 84–88.

Kleinberg, Mindy. *Statement to the First Public Hearing of the National Commission on Terrorist Attacks Upon the United States*, March 31, 2003. https://govinfo.library.unt.edu/911/hearings/hearing1/witness_kleinberg.htm.

Knox, MacGregor, and Alvin Bernstein, eds. *Making of Strategy: Rulers, States, and War*. New York: Cambridge University Press, 1994.

Koehler, Kevin. "Political Militaries in Popular Uprisings: A Comparative Perspective on the Arab Spring." *International Political Science Review* 38, no. 3 (June 2017): 363–77.

Kolirin, Lianne. "Venus Is a Russian Planet—Say the Russians." *CNN*, September 18, 2020. https://www.cnn.com/2020/09/18/world/venus-russian-planet-scn-scli-intl/index.html.

Kollars, Nina A. "Taking Nonstate Actors Seriously (No, Seriously)." In *Deter, Disrupt, Or Deceive: Assessing Cyber Conflict as an Intelligence Contest*, edited by Robert Chesney and Max Smeets, 261–72. Washington, DC: Georgetown University Press, 2023.

Kovach, Karen. "Genocide and the Moral Agency of Ethnic Groups." *Metaphilosophy* 37, nos. 3/4 (July 2006): 618–38.

Kowalski, Dean A. "Horton Hears You, Too! Seuss and Kant on Respecting Persons." In *Dr. Seuss and Philosophy: Oh, the Things You Can Think!*, edited by Jacob M. Held, 119–131. Lanham, MD: Rowman & Littlefield, 2011.

Kramer, Andrew E. "Russia Acknowledges Antisatellite Missile Test That Created a Mess in Space." *New York Times*, November 16, 2021. https://www.nytimes.com/2021/11/16/world/europe/russia-antisatellite-missile-test.html.

Kramer, Martin. "Islam vs. Democracy." *Commentary* 95 (January 1993): 35–42.

Kurzman, Charles. *Unthinkable Revolution in Iran*. Cambridge, MA: Harvard University Press, 2004.

Lahoud, Nelly. *Bin Laden Papers: How the Abbottabad Raid Revealed the Truth about al-Qaeda, Its Leader, and His Family*. New Haven, CT: Yale University Press, 2022.

Lamoreaux, Naomi R. "The Mystery of Property Rights: A U.S. Perspective." *Journal of Economic History* 71, no. 2 (2011): 275–306. http://www.jstor.org/stable/23018300.

Lange, Kendall N. "Oh, The Things You Can Find (If Only You Analyze): A Close Textual Analysis of Dr. Seuss' Rhetoric for Children." MA thesis, Kansas State University, 2007.

Langer, Ron. "Combat Trauma, Memory, and the World War II Veteran." *War, Literature & the Arts: An International Journal of the Humanities* 23, no.1 (2011): 50–58.

Larsen, Paul. "Outer Space: How Shall the World's Governments Establish Order among Competing Interests?" *Washington International Law Journal* 29, no. 1 (2019): 1–60. https://digitalcommons.law.uw.edu/wilj/vol29/iss1/3

Lathem, Edward Connery. *The Beginnings of Dr. Seuss: An Informal Reminiscence*. Hanover, NH: Dartmouth College, 1976.

Latimer, Jon. *Deception in War: The Art of the Bluff, the Value of Deceit, and the Most Thrilling Episodes of Cunning in Military History from the Trojan Horse to the Gulf War*. New York: Abrams Press, 2003.

Leap, Braden, and Diego Thompson. "Social Solidarity, Collective Identity, Resilient Communities: Two Case Studies from the Rural U.S. and Uruguay." *Social Sciences* 7, no. 12 (2018). https://doi.org/10.3390/socsci7120250.

Lee, Brian Angelo. "Just Undercompensation: The Idiosyncratic Premium in Eminent Domain." *Columbia Law Review* 113, no. 3 (2013): 593–655. http://www.jstor.org/stable/23479386.

Lee, R. M., M. Draper, and S. Lee. "Social Connectedness, Dysfunctional Interpersonal Behaviors, and Psychological Distress: Testing A Mediator Model." *Journal of Counseling Psychology* 48 (2002): 310–18.

Lepgold, Joseph. "NATO's Post-Cold War Collective Action Problem." *International Security* 23, no. 1 (1998): 78–106. https://doi.org/10.2307/2539264.

Levi, Primo. *Survival in Auschwitz: The Nazi Assault on Humanity*. Translated by Stuart Woolf. New York: Simon & Schuster, 1996.

Levite, Ariel E., and June Lee. "Attribution and Characterization of Cyber Attacks." In *Managing U.S.-China Tensions over Public Cyber Attribution*, edited by Ariel E. Levite, Lu Chuanying, George Perkovich, and Fan Yang, 33–42. Washington, DC: Carnegie Endowment for International Peace, 2022.

Levy, M. S. "A Helpful Way to Conceptualize and Understand Reenactments." *Journal of Psychotherapy Practice and Research* 7, no. 3 (1998): 227–35.

Lewis, Bernard. *The Assassins*. New York: Basic Books, 1968.

———. *Islam and the West*. New York: Oxford University Press, 1993.
Lin, Herbert. "Doctrinal Confusion and Cultural Dysfunction in the Do." *Cyber Defense Review* 4, no. 2 (Summer 2020): 89–106.
Lindner, Robert. *Rebel without a Cause: The Story of a Criminal Psychopath*. New York: Grune & Stratton, 1944.
Lindsay, Jon R. "Restrained by Design: The Political Economy of Cybersecurity." *Digital Policy, Regulation and Governance* 19, no. 6 (2017): 493–514.
———. "Stuxnet and the Limits of Cyber Warfare." *Security Studies* 22, no. 3 (2013): 365–404.
Logan, David C., "Making Sense of China's Missile Forces." In *Chairman Xi Remakes the PLA*, edited by Phillip C. Saunders, Arthur S. Ding, Andrew Scobell, Andrew N. D. Yang, and Joel Wuthnow, 393–435. Washington, DC: NDU Press, 2019. https://ndupress.ndu.edu/Portals/68/Documents/Books/Chairman-Xi/Chairman-Xi_Chapter-11.pdf?ver=2019-02-08-112005-803.
Lonergan, Erica D., and Shawn W. Lonergan. *Escalation Dynamics in Cyberspace*. New York: Oxford University Press, 2023.
Louise, Mary. "Years of Military Service Helped Inform '2034: A Novel of the Next World War.'" *NPR*, March 18, 2021.
Lovelock, James. *Gaia: A New Look at Life on Earth*. Oxford: Oxford University Press, 1979.
Loveman, Mara. "High-Risk Collective Action: Defending Human Rights in Chile, Uruguay, and Argentina." *American Journal of Sociology* 104, no. 2 (September 1998): 477–525.
Lu, L. "Social Support, Reciprocity, and Well-Being." *Journal of Social Psychology* 137 (1997): 618–28.
Lucas, Laddie. *Out of the Blue: The Role of Luck in Air Warfare, 1917–1966*. London: Hutchinson, 1985.
Lunney, C. A., and P. P. Schnurr. "Domains of Quality of Life and Symptoms in Male Veterans Treated for Posttraumatic Stress Disorder." *Journal of Traumatic Stress* 20, no. 6 (2007): 955–64.
Lurie, Alison. "The Cabinet of Dr. Seuss." *New York Review of Books*, December 20, 1990. https://www.nybooks.com/articles/1990/12/20/the-cabinet-of-dr-seuss/
Lynch, Marc. *Arab Uprising: The Unfinished Revolutions of the New Middle East*. New York: PublicAffairs, 2012.
MacDonald, Alexander. *Long Space Age: The Economic Origins of Space Exploration*. New Haven, CT: Yale University Press, 2017.
Macdonald, Fiona. "The Surprisingly Radical Politics of Dr. Seuss." *BBC*, March 2, 2019. https://www.bbc.com/culture/article/20190301-the-surprisingly-radical-politics-of-dr-seuss.
MacDonald, J. Fred. "Propaganda and Order in Modern Society." In *Propaganda: A Pluralistic Perspective*, edited by Ted J. Smith III, 23–35. New York: Praeger, 1989.
Machemaoui, Mohammed, and Michael O'Mahony. "Does Rent Really Hinder Democracy? A Critical Review of the 'Rentier State' and 'Resource Curse' Theories." *Revue Française de Science Politique* (English Edition) 62, no. 2 (2012): 1–24.

Maciag, Mike. "Law Enforcement Officers Per Capita for Cities, Local Departments." Governing, August 31, 2012. https://www.governing.com/gov-data/safety-justice/law-enforcement-police-department-employee-totals-for-cities.html.

Maclean, Kirsten, Michael Cuthill, and Helen Ross, "Six Attributes of Social Resilience." *Journal of Environmental Planning and Management* 57, no. 1 (2014): 144–56.

Maginniss, Clem. *A Great Feat of Improvisation: Logistics and the British Expeditionary Force in France 1939–1940*. London: Helion Company, 2021.

Magruder, K. M., B. Christopher Frueh, Rebecca G. Knapp, Michael R. Johnson, James A. Vaughan 3rd, Toni Coleman Carson, Donald A. Powell, and Renée Hebert. "PTSD Symptoms, Demographic Characteristics, and Functional Status among Veterans Treated in VA Primary Care." *Journal of Social Psychology* 137 (2004): 618–28.

Mahshie, Abraham. "As China's Military Might Rises, the US Must Master Change to Prepare for Next War, Milley Says." *AirForceMag.com*, August 2, 2021. https://www.airandspaceforces.com/as-chinas-military-might-rises-the-us-must-master-change-to-prepare-for-next-war-milley-says/.

Malik, Adeel. "Rethinking the Rentier Curse." In *Combining Economic and Political Development: The Experience of MENA*, edited by Giacamo Luciani, 41–57. Leiden: Brill, 2017. https://journals.openedition.org/poldev/2266.

Maloney, Suzanne. "1979: Iran and America." Brookings Institution, January 24, 2019. https://www.brookings.edu/articles/1979-iran-and-america/.

Mandel, David R., and Daniel Irwin. "Uncertainty, Intelligence and National Security Decisionmaking." *International Journal of Intelligence and CounterIntelligence* 34, no. 3 (2021): 558–82.

Mann, Itamar. *Humanity at Sea: Maritime Migration and the Foundations of International Law*. New York: Cambridge University Press, 2016.

Mansoor, Peter R., and Williamson Murray. "Introduction: Grand Strategy and Alliances." In *Grand Strategy and Military Alliances*, edited by Peter R. Mansoor and Williamson Murray, 1–16. New York: Cambridge University Press, 2016.

Margasek, Larry. "Hollywood Went to War in 1941—and It Wasn't Easy." National Museum of American History, May 3, 2016. https://americanhistory.si.edu/blog/hollywood-went-war-1941.

Marino, Ben. "China and the US: The Arms Race in Space." *Financial Times*, August 27, 2020. https://www.ft.com/video/80b1eb31-6cbc-422d-865b-29686dc9b235.

Markowitz, J. D. "Post-Traumatic Stress Disorder in an Elderly Combat Veteran: A Case Report." *Military Medicine* 172, no. 6 (2007): 659–62.

Marks, Stephen. "The Human Right to Development: Between Rhetoric and Reality." *Harvard Human Rights Journal* 17 (2004): 137–68.

Marr, Phebe. *The Modern History of Iraq*. 3rd ed. Boulder, CO: Westview Press, 2012.

Marrin, Stephen and Jonathan D. Clemente. "Improving Intelligence Analysis by Looking to the Medical Profession." *International Journal of Intelligence and CounterIntelligence* 18, no. 4 (2005): 707–29.

Martel, William C. *Grand Strategy in Theory and Practice: The Need for an Effective American Foreign Policy*. Cambridge: Cambridge University Press, 2015.

Marzouki, Nadia. "Tunisia's Wall Has Fallen." *MERIP*, January 20, 2011. https://merip.org/2011/01/tunisias-wall-has-fallen/.

Masson-Zwaan, Tanja. 2019. "New States in Space," *American Journal of International Law Unbound* 113 (2019): 98–102. https://doi.org/10.1017/aju.2019.13.

Maurer, Tim. *Cyber Mercenaries: The State, Hackers, and Power*. Cambridge: Cambridge University Press, 2018.

Mayer, Arthur L. "Fact into Film." *Public Opinion Quarterly* 8, no. 2 (Summer 1944): 206–25.

McAdam, Douglas. *Political Process and the Development of Black Insurgency: 1930–1970*. Chicago: Chicago University Press, 1999.

McCay, Vernon and Marjie L. Baughman. "Art, Madness, and Human Interaction." *Art Journal* 31, no. 4 (1972): 413–20. https://doi.org/10.2307/775545.

McCormack, Lynne, and Joseph Stephen. "Psychological Growth in Aging Vietnam Veterans: Redefining Shame and Betrayal." *Journal of Humanistic Psychology* 54, no. 3 (2014): 336–55.

McCurdy, Howard E. *Space and the American Imagination*. Baltimore, MD: Johns Hopkins University Press, 2011.

McDougall, Walter A. 2010. "Can the United States Do Grand Strategy?" *Orbis* 54, no. 2 (2010): 165–84. https://doi.org/10.1016/j.orbis.2010.01.008.

McFarlane, A. C. "The Long-Term Costs of Traumatic Stress: Intertwined Physical and Psychological Consequences." *World Psychiatry* 9, no. 1 (2010): 3–10.

McFate, Montgomery. "The 'Memory of War': Tribes and the Legitimate Use of Force in Iraq." In *Armed Groups*, edited by Jeffery Norwich, 291–310. Newport, RI: Naval Institute Press, 2008.

———. *Military Anthropology: Soldiers, Scholars and Subjects at the Margins of Empire*. New York: Oxford University Press, 2018.

McGillivray, Anne. "Horton Hears a Twerp: Myth, Law, and Children's Rights in Horton Hears a Who!" *New York Law School Law Review* 58, no. 509 (2013): 569–600.

McLynn, Frank. *Napoleon: A Biography*. New York: Arcade Publishing, 2002.

McManus, John. *Deadly Sky: The American Combat Airman in World War II*. New York: Penguin, 2016.

McNamara, John. *Millville's Mac—The Life Story of a World War II Combat Marine*. N.p.: Changing Attitudes LLC, 2014.

Mead, Walter Russell. *Power, Terror, Peace, and War: America's Grand Strategy in a World at Risk*. New York: Vintage Books, 2005.

———. *Special Providence: American Foreign Policy and How It Changed the World*. London: Routledge, 2002.

Mearsheimer, John J. *The Tragedy of Great Power Politics*. New York: Norton, 2001.

Menoret, Pascal. *Graveyard of Clerics: Everyday Activism in Saudi Arabia*. Stanford, CA: Stanford University Press, 2020.

Microsoft. "Defending Ukraine: Early Lessons from the Cyber War." June 22, 2022. https://query.prod.cms.rt.microsoft.com/cms/api/am/binary/RE50KOK.

Miellet, Sara. "Human Rights Encounters in Small Places: The Contestation of Human Rights Responsibilities in Three Dutch Municipalities." *Journal of Unofficial Law and Legal Pluralism* 51, no. 2 (2019) 213–32.

Mikaya, Kanan. *Republic of Fear: The Politics of Modern Iraq*. Berkeley: University of California Press, 1998.

Milevski, Lukas. *Evolution of Modern Grand Strategic Thought*. New York: Oxford University Press, 2016.

Miller, Joshua Rhett. "Chicago Public Library to Yank Six Dr. Seuss Books from Shelves." *New York Post*, March 9, 2021.

Miller, William Ian. *Mystery of Courage*. Cambridge, MA: Harvard University Press, 2000.

Minear, Richard H. *Dr. Seuss Goes to War: The World War II Editorial Cartoons of Theodor Seuss Geisel*. New York: New Press, 2001.

Minear, Richard, and Sopan Deb. "The Dr. Seuss Museum and His Wartime Cartoons about Japan and Japanese Americans." *Asia-Pacific Journal* 15, no. 3 (August 15, 2017): 16. https://apjjf.org/2017/16/Minear.html.

Mizokami, Kyle. "Meet Russia's Imposing New Satellite-Destroying Missile." *Popular Mechanics*, April 16, 2020. https://www.popularmechanics.com/military/weapons/a32173824/nudol-missile-anti-satellite/.

Mochanloo, Mahtub. "Grief, Obligation and Connection in the Iranian Green Movement." BA honors thesis, Vanderbilt University, 2016.

Mollet, Tracy Louise. *Cartoons in Hard Times: The Animated Shorts of Disney and Warner Brothers in Depression and War 1932–1945*. New York: Bloomsbury, 2019.

Moltz, James Clay. *The Politics of Space Security: Strategic Restraint and the Pursuit of National Interests*. Stanford, CA: Stanford University Press, 2008.

Monroe, Kristen Renwick. *The Heart of Altruism: Perceptions of a Common Humanity*. Princeton, NJ: Princeton University Press, 1996.

Monte, Matthew. *Network Attacks and Exploitation: A Framework*. Indianapolis, IN: Wiley, 2015.

Morgan, Forrest E., Karl P. Mueller, Evan S. Medeiros, Kevin L. Pollpeter, and Roger Cliff. *Dangerous Thresholds: Managing Escalation in the 21st Century*. Santa Monica, CA: RAND, 2008. https://www.jstor.org/stable/pdf/10.7249/mg614af.9.pdf.

Morgan, Judith, and Neil Morgan. *Dr. Seuss & Mr. Geisel: A Biography*. New York: Random House, 1995.

Morgenthau, Hans. *Politics among Nations*. New York: McGraw-Hill, 1948.

Morriss, Roger. "Colonization, Conquest, and the Supply of Food and Transport: The Reorganization of Logistics Management, 1780–1795." *War in History* 14, no. 3 (2007): 310–24. http://www.jstor.org/stable/26070709.

Mortimer, Lorraine. "The Grim Enchantment of 'It's a Wonderful Life.'" *Massachusetts Review* 36, no. 4 (1995): 656–86.

Mu-Hyun, Cho. "Huawei Regularly Tried to Steal Apple Trade Secrets: Report." *ZDNet*, February 18, 2019. https://www.zdnet.com/article/huawei-regularly-tried-to-steal-apple-trade-secrets/.

Muldoon, Orla T., and Robert D. Lowe. "Social Identity, Groups, and Post-Traumatic Stress Disorder." *Political Psychology* 33, no. 2 (April 2012): 259–73.

Muller, J., S. Ganeshamoorthy, and J. Myers. "Risk Factors Associated with Post-traumatic Stress Disorder in US Veterans: A Cohort Study." *PLoS ONE* 12 (2017): e0181647.

Muller, Robert T., and Terry Diamond. "Father and Mother Physical Abuse and Child Aggressive Behaviour in Two Generations." *Canadian Journal of Behavioural Science* 31, no. 4 (1999): 221–28.

Murphy, Audie. *To Hell and Back*. New York: Henry Holt, 2002.

Murphy, Dennis M., and James F. White. "Propaganda: Can a Word Decide a War?" *Parameters* 37, no. 3 (August 2007): 15–27.

Murray, Williamson. "Thoughts on Grand Strategy." In *The Shaping of Grand Strategy: Policy, Diplomacy, and War*, edited by Williamson Murray, Richard Hart Heinrich, and James Lacey, 1–33. New York: Cambridge University Press, 2011.

———. *War, Strategy, and Military Effectiveness*. New York: Cambridge University, 2011.

Murray, Williamson, and Mark Grimsley. "Introduction: On Strategy." In *The Making of Strategy: Rulers, States, and War*, edited by Williamson Murray, MacGregor Knox, and Alvin Bernstein, 1–23. New York: Cambridge University Press, 1994.

Murray, Williamson, Richard Hart Heinrich, and James Lacey, eds. *The Shaping of Grand Strategy: Policy, Diplomacy, and War*. New York: Cambridge University Press, 2011.

Mustillo, Sarah A., and Ashleigh Kysar-Moon. "Race, Gender, and Post-Traumatic Stress Disorder in the U.S. Military: Differential Vulnerability?" *Armed Forces & Society* 43, no. 2 (2017): 322–45.

Naess, Arne. "Self-Realization: An Ecological Approach to Being in the World." In *Deep Ecology for the Twenty-First Century: Readings on the Philosophy and Practice of the New Environmentalism*, edited by George Sessions, 225–39. Berkeley, CA: Shambhala Press, 1995.

Nanuam, Wassana. "Soldiers Keep Faith with Sacred Charms." *Bangkok Post*, November 28, 2003.

———. "Troops Get Amulet Protection." *Bangkok Post*, January 1, 2008.

Naticchia, Chris. "Hobbesian Realism in International Relations: A Reappraisal." In *Hobbes Today: Insights for the 21st Century*, 241–63. Cambridge: Cambridge University Press, 2013.

National Academies of Sciences, Engineering, and Medicine. *PTSD Compensation and Military Service*. Washington, DC: National Academies Press, 2007.

National Army Museum. "Luck and Superstition." n.d. Accessed April 29, 2024. www.nam.ac.uk/explore/luck-and-superstition.

National Cyber Security Centre. "How Cyber Attacks Work." July 2023. https://www.ncsc.gov.uk/information/how-cyber-attacks-work.

National Institute of Justice. *Five Things about Deterrence*. U.S. Department of Justice, May 2016. https://www.ojp.gov/pdffiles1/nij/247350.pdf.

Nel, Philip. "Children's Literature Goes to War: Dr. Seuss, P. D. Eastman, Munro Leaf, and the Private SNAFU Films (1943–46)." *Journal of Popular Culture* 40, no. 3 (June 2007): 468–87.

———. "Dada Knows Best: Growing Up 'Surreal' with Dr. Seuss." *Children's Literature* 27, no. 1 (1999): 150–84.

———. *Dr. Seuss: American Icon*. New York: Continuum, 2003.

———. "'Said a Bird in the Midst of a Blitz': How World War II Created Dr. Seuss." *Mosaic: An Interdisciplinary Critical Journal* 34, no. 2 (June 2001): 65–85.

———. "Was the Cat in the Hat Black? Exploring Dr. Seuss's Racial Imagination." *Children's Literature* 42, no. 1 (2014): 71–98.

———. *Was the Cat in the Hat Black? The Hidden Racism of Children's Literature and the Need for Diverse Books*. Oxford: Oxford University Press, 2017.

Nelson, Soryaya. "Taliban Flees Marjah, Threat Remains for Marines." *NPR*, March 1, 2010.

Neumann John von, and Oskar Morgenstern. *Theory of Games and Economic Behavior*. Princeton, NJ: Princeton University Press, 1944.

Neve, Alex. "Amnesty International: 58 Years On, Collective Action for Human Rights Matters More Than Ever." Amnesty International Canada, May 28, 2019. https://www.csjcanada.org/blog/2019/5/29/amnesty-international-58-years-on-collective-action-for-huma.html.

Nevin, Patrick (command sergeant major, French Foreign Legion), interview with author, October 24, 2012.

Newman, Bernard. *Soviet Atomic Spies*. London: Robert Hale, 1952.

Nicolas, Peter. "The Sneetches as an Allegory for the Gay Rights Struggle: Three Prisms." *New York Law School Law Review* 58, no. 525 (2013): 525–45.

Nietlisbach, G., and A. Maercker. "Effects of Social Exclusion in Trauma Survivors with Posttraumatic Stress Disorder." *Psychological Trauma* 1 (2009): 323–31.

Nimitz, Chester W. *Command Summary of Fleet Admiral Chester W. Nimitz, USN: Nimitz "Graybook" 7 December 1941–31 August 1945* (Newport, RI: US Naval War College, 2013).

Nobel Price Laureate Tawakkol Karman. "About." Accessed August 18, 2023. https://www.tawakkolkarman.net/enabout.

Norman, S. B., A. J. Means-Christensen, M. G. Craske, C. D. Sherbourne, P. P. Roy-Byrne, and M. B. Stein. "Associations between Psychological Trauma and Physical Illness in Primary Care." *Journal of Traumatic Stress* 19 (2006): 461–67.

North, Carol S., Sara Jo Nixon, Sheryll Shariat Sue Mallonee, J. Curtis McMillen, Edward L. Spitznagel, and Elizabeth M. Smith. "Psychiatric Disorders among Survivors of the Oklahoma City Bombing." *Journal of the American Medical Association* 282 (1999): 755–62.

Norton, Richard. *Civil Society in the Middle East*. Leiden: E. J. Brill, 1995.

Novy, Ron. "Rebellion in Sala-ma-Sond." In *Dr. Seuss and Philosophy: Oh the Thinks You Can Think!*, edited by Jacob M. Held, 167–78. Lanham, MD: Rowman & Littlefield, 2011.

Nowak, Manfred, and Elizabeth McArthur. *The United Nations Convention against Torture: A Commentary*. Oxford: Oxford University Press, 2008.

Nye, Joseph S. "Deterrence and Dissuasion in Cyberspace." *International Security* 41, no. 3 (2016): 44–71.

———. *The Future of Power*. New York: Public Affairs, 2011.

———. *Soft Power: The Means to Success in World Politics*. New York: Public Affairs, 2004.

Oberhaus, Daniel. "India's Anti-Satellite Test Wasn't Really About Satellites." *Wired*, March 27, 2019. https://www.wired.com/story/india-anti-satellite-test-space-debris/.

O'Brien, John J. "Coup d'Oeil: Military Geography and the Operational Level of War." MS thesis, School of Advanced Military Studies, U.S. Army Command and General Staff College, May 16, 1991.

O'Callaghan, Jonathan. "The Risky Rush for Mega Constellations." *Scientific American*, October 31, 2019. https://www.scientificamerican.com/article/the-risky-rush-for-mega-constellations/.

Okruhlik, Gwenn. "Rentier Wealth, Unruly Law, and the Rise of Opposition: The Political Economy of Oil States." *Comparative Politics* 31, no. 3 (April 1999): 295–315.

Organisation for Economic Co-operation and Development. *Do No Harm: International Support for Statebuilding*. Paris: OECD, 2009. https://www.oecd-ilibrary.org/development/do-no-harm_9789264046245-en.

Orwell, George. *Animal Farm*. New York: Signet, 1946.

O'Sullivan, Patrick, and Jesse Miller. *The Geography of Warfare*. London: Croom Helm, 1983.

Oxford Constitutional Law Online Encyclopedia. https://oxcon.ouplaw.com/.

Packer, George. *The Assassins' Gate: America in Iraq*. New York: Farrar, Straus and Giroux, 2005.

Parker, Charles F., and Eric K. Stern. "Bolt from the Blue or Avoidable Failure? Revisiting September 11 and the Origins of Strategic Surprise." *Foreign Policy Analysis* 1, no. 3 (November 2005): 301–31.

Patterson, Ian. *Guernica and Total War*. Cambridge, MA: Harvard University Press, 2007.

Pearlman, Wendy. "Emotions and the Microfoundations of the Arab Uprisings." *American Political Science Review* 11, no. 2 (June 2013): 387–409.

———. "Moral Identity and Protest Cascades in Syria." *British Journal of Political Science* 48, no. 4 (October 2018): 877–901.

Pease, Donald E. "Dr. Seuss in Ted Geisel's Never-Never Land." *PMLA* 126, no. 1 (January 2011): 198.

———. "Dr. Seuss's (Un)Civil Imaginaries," *New York Law School Law Review* 58, no. 509 (2013): 511.

———. *Theodor Geisel: A Portrait of the Man Who Became Dr. Seuss*. New York: Oxford University Press, 2010.

Pekkanen, Saadia M. "China, Japan, and the Governance of Space: Prospects for Competition and Cooperation." *International Relations of the Asia-Pacific* 12, no. 1 (2020): 1–28. https://doi.org/10.1093/irap/lcaa007.

———. "China Leads the Quantum Race While the West Plays Catch Up." *Forbes*, September 30, 2016. https://www.forbes.com/sites/saadiampekkanen/2016/09/30/china-leads-the-quantum-race-while-the-west-plays-catch-up/#1f98d8eb5928.

———. "Governing the New Space Race." *American Journal of International Law Unbound* 113 (2019): 92–97. https://doi.org/10.1017/aju.2019.16.

———. "Thank You for Your Service: The Security Implications of Japan's Counterspace Capabilities." *Texas National Security Review*, Roundtable, edited by Jonathan D. Caverley and Peter Dombrowski. October 1, 2020. https://tnsr.org/roundtable/policy-roundtable-the-future-of-japanese-security-and-defense/.

———. "Zooming in on the Promise and Peril of Satellite Imagery." *Seattle Times*, August 26, 2022. https://www.seattletimes.com/opinion/zooming-in-on-the-promise-and-peril-of-satellite-imagery/.

Pekkanen, Saadia M., Setsuko Aoki, and John Mittleman, "Small Satellites-Big Data: Uncovering the Invisible in Maritime Security." *International Security* 47, no. 2 (2022): 177–216. https://doi.org/10.1162/isec_a_00445.

Pels, Peter. "The Confessional Ethic and the Spirits of the Screen: Reflections on the Modern Fear of Alienation." *Etnofoor* 15, nos. 1/2 (2002): 91–119.

Percin, Alexandre. *Le Massacre de notre Infanterie 1914–1918*. Paris: Albin Michel, 1921.

Perrow, Charles. *Normal Accidents*. New York: Basic Books, 1984.

Pew Research Center. *Two Decades Later, the Enduring Legacy of 9/11*. September 2, 2021. https://www.pewresearch.org/politics/2021/09/02/two-decades-later-the-enduring-legacy-of-9-11/.

"Philadelphia Centennial, The." *American Advocate of Peace and Arbitration* 49, no. 5 (1887): 125. http://www.jstor.org/stable/45404791.

"Pied Piper of Hamelin." *The Aldine* 4, no. 6 (1871): 90–91. http://www.jstor.org/stable/20636049.

Pietrzak, Robert H., and Joan M. Cook. "Psychological Resilience in Older U.S. Veterans: Results from the National Health and Resilience in Veterans Study." *Depression and Anxiety* 30, no. 5 (2013): 432–43.

Pilat, Oliver. *The Atom Spies*. New York: Putnam, 1952.

Pilati, Katia, Giuseppe Acconcia, David Leone Suber, and Henda Chennaoui. "Between Organization and Spontaneity of Protests: The 2010–2011 Tunisian and Egyptian Uprisings." *Social Movement Studies* 18, no. 4 (January 2019): 463–81.

Pillar, Paul R. "Intelligence, Policy, and the War in Iraq." *Foreign Affairs* 85, no. 2 (2006): 15–27. https://doi.org/10.2307/20031908.

———. "Where Luck Comes In," *New York Times*, May 3, 2011.

Pinker, Steven. *The Better Angels of Our Nature: Why Violence Has Declined*. New York: Penguin Books, 2012.

Plaster, John L. *SOG: The Secret Wars of America's Commandos in Vietnam*. New York: Simon & Schuster, 1997.

Pole, N., T. C. Neylan, S. R. Best, S. P. Orr, and C. R. Marmar. "Fear-Potentiated Startle and Posttraumatic Stress Symptoms in Urban Police Officers." *Journal of Trauma Stress* 16 (2003):471–79.

Pollack, Kenneth M. "The Arab Militaries: The Double-Edged Swords." In *The Arab Awakening: America and the Transformation of the Middle East*, edited by Kenneth M. Pollack, Daniel Byman, and Akram Al-Turk, 58–65. Washington, DC: Brookings Institution, 2011.

Pollpeter, Kevin. "China's Space Program: Making China Strong, Rich, and Respected." *Asia Policy* 15, no. 2 (2020): 12–18. https://www.nbr.org/publication/asia-in-space-the-race-to-the-final-frontier/.

Poznansky, Michael. *In the Shadow of International Law: Covert Intervention in the Postwar World*. New York: Oxford University Press, 2020.

Pratt, Marion. "Useful Disasters: The Complexity of Response to Stress in a Tropical Lake Ecosystem." *Anthropologica* 38, no. 2 (1996): 125–48.

Priebe, Stefan, Aleksandra Matanov, Jelena Janković Gavrilović, Paul McCrone, D. Ljubotina, G. Knezević, A. Kucukalić, et al. "Consequences of Untreated Posttraumatic Stress Disorder Following War in Former Yugoslavia: Morbidity, Subjective Quality of Life, and Care Costs." *Croatian Medical Journal* 50, no. 5 (2009): 465–75.

Prior, Neil. "PTSD: Eyes Can Reveal Previous Trauma, Study Reveals." *BBC News*, July 26, 2020.

Pruncunn, Hank. *Counterintelligence Theory and Practice*. New York: Rowman & Littlefield, 2019.

Putnam, Robert. *Bowling Alone: The Collapse and Revival of American Community*. New York: Simon and Schuster, 2000.

Quadri, Sami. "The Scream May Not Be Screaming." *Daily Mail*, March 20, 2019. https://www.dailymail.co.uk/news/article-6832823/The-Scream-not-screaming-Edward-Munchs-iconic-artwork-shows-no-thing.html.

Quammen, David. "The Weeds Shall Inherit the Earth." *Independent*, November 22, 1998, 30–39.

Quinlivan, James T. "Coup-Proofing: Its Practice and Consequences in the Middle East." *International Security* 24, no. 2 (Fall 1999): 131–65.

Ramsberger, Peter, Peter Legree, and Lisa Mills. *Evaluation of the Buddy Team Assignment Program*. Arlington, VA: US Army Research Institute for the Behavioral and Social Sciences, Study Note 2003-1, October 2002.

Ratcliffe, Rebecca. "Woman Reportedly Shot Dead as Myanmar Police Escalate Crackdown." *Guardian*, February 27, 2021. https://www.theguardian.com/world/2021/feb/26/myanmar-envoy-urges-un-to-use-any-means-necessary-to-restore-democracy.

Reese, Thomas H. "The Oath of Allegiance." Proceedings 91/9/751, September 1965.

Reif, Kingston, and Shannon Bugos. "U.S., Russia Extend New START for Five Years," *Arms Control Today*, March 2021. https://www.armscontrol.org/act/2021-03/news/us-russia-extend-new-start-five-years.

Richardson, L. K., B. C. Frueh, and R. Acierno. "Prevalence Estimates of Combat-Related Post-Traumatic Stress Disorder: Critical Review." *Australia and New Zealand Journal of Psychiatry* 44 (2010): 4–19.

Ricks, Tom. *Fiasco: The American Military Adventure in Iraq*. New York: Penguin, 2007.

Rid, Thomas, and Ben Buchanan. "Attributing Cyber Attacks." *Journal of Strategic Studies* 38, nos. 1–2 (2015). https://doi.org/10.1080/01402390.2014.977382.

Robbins, William G. "The Social Context of Forestry: The Pacific Northwest in the Twentieth Century." *Western Historical Quarterly* 16, no. 4 (1985): 413–27. https://doi.org/10.2307/968606.

Robertson, James. *General A.P. Hill: The Story of a Confederate Warrior*. New York: Vintage, 1992.

Rochester, Stuart L., and Frederick Kiley. *Honor Bound: The History of American Prisoners of War in Southeast Asia, 1961–1973*. Annapolis, MD: Naval Institute Press, 1998.

Rose, Gideon. *How Wars End: Why We Always Fight the Last Battle*. New York: Simon & Schuster, 2011.

Rosellini, A. J., John Monahan, Amy E. Street, Steven G. Heeringa, Eric D. Hill, Maria Petukhova, Ben Y. Reis, et al. "Predicting Non-Familial Major Physical Violent Crime Perpetration in the US Army from Administrative Data." *Psychological Medicine* 46 (2016): 303–16.

Rosen, Lawrence. "Expecting the Unexpected: Cultural Components of Arab Governance." *Annals of the American Academy of Political and Social Science* 603 (January 2006): 163–78.

Rosen, Stephen Peter. *Winning the Next War: Innovation and the Modern Military*. Ithaca, NY: Cornell University Press, 1991.

Ross, E. Wayne. "Dr. Seuss and Dangerous Citizenship." Keynote address to the 6th Annual Equity and Social Justice Conference, SUNY New Paltz, March 2, 2013. https://blogs.ubc.ca/ross/files/2013/03/Dr-Seuss-and-Dangerous-Citizenship-Talk.pdf.

Ross, Michael. "Does Oil Hinder Democracy?" *World Politics* 53, no. 3 (April 2001): 325–61.

Rotberg, Robert I. "Failed States, Collapsed States, Weak States: Causes and Indicators." In *State Failure and State Weakness in a Time of Terror*, edited by Robert Rotberg, 1–25. Washington, DC: Brookings Institution Press, 2003.

Rousseau, Jean Jacques. *The Origin of Civil Society* (1762). In *A World of Ideas: Essential Readings for College Writers*, edited by Lee A. Jacobus, 58–74. Boston: Bedford/St. Martin's, 2010.

Rovner, Joshua. "Is Politicization Ever a Good Thing?" *Intelligence and National Security* 28, no. 1 (2013): 55.

Royal Society of Medicine. "Discussion on the Nature and Treatment of the Effort Syndrome." *Proceedings of the Royal Society of Medicine* 34 (1941): 37–50.

Sable, Martin H. "Day of Atonement in Charles Dickens' 'A Christmas Carol.'" *Tradition: A Journal of Orthodox Jewish Thought* 22, no. 3 (1986): 66–76.

Sadiki, Larbi. "Ben Ali's Tunisia: Democracy by Non-Democratic Means." *British Journal of Middle Eastern Studies* 29, no. 1 (May 2002): 57–78.

Sahlins, Marshall D. "The Segmentary Lineage: An Organization of Predatory Expansion." *American Anthropologist* 63, no. 2 (1961): 322–45.

Salmoni, Barak, and Paula Holmes-Eber. *Operational Culture for the Warfighter: Principles and Applications*. Quantico, VA: Marine Corps University Press, 2011.

Salzman, Philip Carl. *Culture and Conflict in the Middle East*. New York: Humanity Books, 2008.
Sanger, David E. *Confront and Conceal: Obama's Secret Wars and Surprising Use of American Power*. New York: Broadway Paperbacks, 2012.
Schelling, Thomas. *Arms and Influence*. New Haven, CT: Yale University Press, 1966.
Schlauch, Wolfgang. "American Policy towards Germany, 1945." *Journal of Contemporary History* 5, no. 4 (1970): 113–28. http://www.jstor.org/stable/259868
Schmitt, Michael N., and Liis Vihul. "Proxy Wars in Cyberspace: The Evolving International Law of Attribution." *Fletcher Security Review* 1, no. 2 (2014): 55–73.
Schneider, Jacquelyn. "Blue Hair in the Gray Zone." *War on the Rocks*, January 2018. https://warontherocks.com/2018/01/blue-hair-gray-zone/.
Science Service. "In Science Fields." *Science News-Letter* 37, no. 24 (1940): 376–77.
———. "Man with Soldier's Heart Discharged for Second Time." *Science News-Letter* 43, no. 23 (1943): 367.
———. "'Soldier's Heart' Cured by Cutting Nerves to Gland." *Science News-Letter* 38, no. 18 (1940): 286.
Scobell, Andrew, Edmund J. Burke, Cortez A. Cooper III, Sale Lilly, Chad J. R. Ohlandt, Eric Warner, and John Davis Williams. *China's Grand Strategy: Trends, Trajectories, and Long-Term Competition*. Washington, DC: RAND, 2020.
Scoles, Sarah. "U.S. Air Force Cadets Study Idea of Space Force Bases on the Moon." *Science*, July 15, 2020. https://www.sciencemag.org/news/2020/07/us-air-force-cadets-study-idea-space-force-bases-moon.
Scroxton, Alex. "Biden Sanctions Russia over Solarwinds Cyber Attacks—US President Imposes New Sanctions on Russia Following Malicious Cyber Attacks against the US and Allies." *Computer Weekly*, April 15, 2021. https://www.computerweekly.com/news/366551317/UK-and-US-slap-fresh-sanctions-on-Conti-ransomware-crew.
Scully, Megan. "'Social Intel' New Tool for U.S. Military." *Defense News*, April 26, 2004, 21.
Seagren, Chad W., and David R. Henderson. "Why We Fight: A Study of U.S. Government War-Making Propaganda." *Independent Review* 23, no. 1 (Summer 2018): 69–90.
Sen, Amartya. *Development as Freedom*. New York: Anchor Books, 2000.
Sengupta, Arjun. "Right to Development as a Human Right." *Economic and Political Weekly* 26, no. 27 (July 7–13, 2001): 2527–36.
Serpell, James. "Factors Influencing Human Attitudes to Animals and Their Welfare." *Animal Welfare* 13, no. 1 (2004): S145–51.
Set, Shounak. "India's Space Power: Revisiting the Anti-Satellite Test." Carnegie India, Carnegie Endowment for International Peace, September 6, 2019. https://carnegieindia.org/2019/09/06/india-s-space-power-revisiting-anti-satellite-test-pub-79797.
Seuss, Dr. *Bartholomew and the Oobleck*. New York: Random House, 1949.
———. *The Butter Battle Book*. New York: Random House, 1984.
———. *Cat in the Hat*. New York: Random House, 1957.
———. *The 500 Hats of Bartholomew Cubbins*. New York: Vanguard Press, 1965.

———. *Fox in Socks*. New York: Random House, 1965.
———. *Horton Hears a Who*. New York: Random House, 1954.
———. *Hunches in Bunches*. New York: Random House, 1982.
———. *I Had Trouble in Getting to Solla Sollew*. New York: Random House, 1965.
———. *The King's Stilts*. New York: Random House, 1939.
———. *McElligot's Pool*. New York: Random House, 1947.
———. *Oh, the Places You'll Go!* New York: Random House, 1990.
———. *On Beyond Zebra!* New York: Random House, 1955.
———. *One Fish, Two Fish, Red Fish, Blue Fish*. New York: Beginner Books/Random House, 1960.
———. *Sleep Book*. New York: Random House, 1962.
———. *Yertle the Turtle and Other Stories*. London: HarperCollins, 1986.
Shaw, Malcolm. *International Law*. 8th ed. Cambridge: Cambridge University Press, 2017.
Sheehan, Neil. *A Bright Shining Lie: John Paul Vann and America in Vietnam*. New York: Random House, 1988.
Sheetz, Michael. "How OneWeb's $1 Billion Bankruptcy Rescue Changes the Competitive Landscape for Elon Musk's Starlink." *CNBC*, July 10, 2020. https://www.cnbc.com/2020/07/10/onewebs-bankruptcy-rescue-changes-the-competition-for-elon-musks-spacex-starlink.html.
Sherman, Justin. "Untangling the Russian Web: Spies, Proxies, and Spectrums of Russian Cyber Behavior." *Atlantic Council Issue Brief*, September 19, 2022. https://www.atlanticcouncil.org/in-depth-research-reports/issue-brief/untangling-the-russian-web/.
Shirabi, Hisham. *Neopatriarchy: A Theory of Distorted Change in Arab Society*. New York: Oxford University Press, 1988.
Shulimson, Jack, Leonard A. Blasiol, Carles R. Smith, and David Dawson. *U.S. Marines in Vietnam: The Defining Year, 1968*. Washington, DC: US Marine Corps, History and Museums Division, 1997.
Shull, Michael S., and David E. Wilt. *Doing Their Bit: Wartime Animated Short Films, 1939–1945*. 2nd ed. Jefferson, NC: McFarland, 2004.
Siddiqui, Zeba, and Christopher Bing. "Chinese Hackers Spying on US Critical Infrastructure, Western Intelligence Says." *Reuters*, May 25, 2023. https://www.reuters.com/technology/microsoft-says-china-backed-hacker-targeted-critical-us-infrastructure-2023-05-24/.
Simkins, J. D. "'On War' Is Now a Children's Book Featuring 'Hare' Clausewitz Teaching Woodland Critters." *Military Times*, July 8, 2020.
Singer, P. W., and Ma Xiu. "China's Ambiguous Missile Strategy Is Risky." *Popular Science*, May 11, 2020. https://www.popsci.com/story/blog-network/eastern-arsenal/china-nuclear-conventional-missiles/.
Sippel, Lauren M., Robert H. Pietrzak, Dennis S. Charney, Linda C. Mayes, and Steven M. Southwick. "How Does Social Support Enhance Resilience in the Trauma-Exposed Individual?" *Ecology and Society* 20, no. 4 (2015). http://www.jstor.org/stable/26270277.
Skinner, B. F. *Walden Two*. New York: Macmillan, 1948.

Skoble, Aeon J. "Thidwick the Big-Hearted Bearer of Property Rights." In *Dr. Seuss and Philosophy: Oh, the Thinks You Can Think!*, edited by Jacob M. Held, 159–66. New York: Rowman & Littlefield, 2011.

Skocpol, Theda. *States and Social Revolutions: A Comparative Analysis of France, Russia, and China*. New York: Cambridge University Press, 1979.

Slaughter, Joseph. "A Question of Narration: The Voice in International Human Rights Law." *Human Rights Quarterly* 19, no. 2 (May 1997): 406–30.

Slayton, Rebecca. "What Is a Cyber Warrior? The Emergence of U.S. Military Cyber Expertise, 1967–2018." *Texas National Security Review* 4, no. 1 (January 11, 2021). http://tnsr.org/2021/01/what-is-a-cyber-warrior-the-emergence-of-u-s-military-cyber-expertise-1967-2018/.

———. "What Is the Cyber Offense-Defense Balance? Conceptions, Causes, and Assessment." *International Security* 41, no. 3 (January 1, 2017): 72–109.

Smeets, Max. "Cyber Arms Transfer: Meaning, Limits and Implications." *Security Studies* 31, no. 1 (2022): 65–91.

Smith, Caroline M. *Dr. Seuss: The Cat Behind the Hat*. New York: Andrews McMeel Publishing, 2012.

Smith, Hubert Winston. "Relation of Emotions to Injury and Disease: Legal Liability for Psychic Stimuli." *Virginia Law Review* 30, no. 2 (1944): 193–317.

Smith, W. Thomas. "The Roads Are Hell." *National Review*, April 16, 2007.

Snow, David A., and Robert D. Benford. "Ideology, Frame Resonance and Participant Mobilization." *International Social Movement Research* 1 (1988): 197–217.

Snow, David A., E. Burke Rochford Jr., Steven K. Worden, and Robert D. Benford. "Frame Alignment Processes, Micromobilization, and Movement Participation." *American Sociological Review* 51, no. 4 (August 1986): 464–81.

Snyder, Glenn. *Alliance Politics*. Ithaca, NY: Cornell University Press, 1997.

———. *Deterrence and Defense*. Princeton, NJ: Princeton University Press, 1961.

"Soldiers' Superstitions," *New York Times Magazine*, April 2, 1944.

Solomon, Z., and M. Mikulincer. "Posttraumatic Intrusion, Avoidance, and Social Functioning: A 20-Year Longitudinal Study." *Journal of Consulting and Clinical Psychology* 75 (2007): 316–24.

———. "Trajectories of PTSD: A 20-Year Longitudinal Study." *American Journal of Psychiatry* 163, no. 4 (2006): 659–66.

Somerville, J. "The Master's Legacy: The First Paul Wood Lecture." *Heart* 80, no. 6 (1998): 612–18.

Sorenson, David S. "Civil–Military Relations in North Africa." *Middle East Policy* 14, no. 4 (December 2007): 99–114.

Springborg, Robert. "Arab Militaries." In *The Arab Uprisings Explained: New Contentious Politics in the Middle East*, edited by Marc Lynch, 142–59. New York: Columbia University Press, 2014.

Sproull, L. S., S. B. Kiesler, and S. Kiesler. *Connections: New Ways of Working in the Networked Organization*. Cambridge, MA: MIT Press, 1992.

St. Rebor, Jack. "Butter Battle Book." *Seussblog*, December 30, 2013. https://seussblog.wordpress.com/2013/12/30/the-butter-battle-book/.

Stanford Encyclopedia of Philosophy. Edited by Edward N. Zalta and Uri Nodelman. Stanford, CA: Stanford University Philosophy Department, Metaphysics Research Lab. https://plato.stanford.edu.

Stanley, E., and J. Sawyer. "The Equifinality of War Termination." *Journal of Conflict Resolution* 53, no. 5 (October 2009): 651–76.

Starobinski, Jean. "Chlorosis—The 'Green Sickness.'" *Psychological Medicine* 11 (1981): 459–68.

"State of Civil Society, 2011." CIVICUS, April 2012. https://www.civicus.org/downloads/2011StateOfCivilSocietyReport/State_of_civil_society_2011-web.pdf.

Stavridis, James. "Years of Military Service Helped Inform '2034: A Novel of the Next World War.'" Interview by Mary Louise Kelly. *All Things Considered*. National Public Radio, March 18, 2021.

Stern, B. B. "Medieval Allegory: Roots of Advertising Strategy for the Mass Market." *Journal of Marketing* 52, no. 3 (1988): 84–94.

Stevenson, Angus, ed. *Shorter Oxford English Dictionary on Historical Principles*. 6th ed. Oxford: Oxford University Press, 2007.

Stewart, Patrick. "Civil Wars and Transnational Threats: Mapping the Terrain, Assessing the Links." *Daedalus* 146, no. 4 (2017): 45–58. https://www.jstor.org/stable/48563902.

Stewart, Rory, and Gerald Knaus. *Can Intervention Work?* New York: W. W. Norton, 2012.

Stillman, Richard Joseph. *U.S. Infantry: Queen of Battle*. New York: F. Watts, 1965.

Stofflet, Mary. *Dr. Seuss from Then to Now: A Catalogue of the Retrospective Exhibition*. San Diego, CA: San Diego Museum of Art, 1986.

Stolfi, Russel. "Chance in History: The Russian Winter of 1941–1942." *History* 65, no. 214 (1980): 214–28.

Stoneman, Paul, and Giuliana Battisti. "The Diffusion of New Technology." In *Handbook of the Economics of Innovation*, edited by Bronwyn H. Hall and Nathan Rosenberg, 2:733–60. Amsterdam: Elsevier, 2010.

Stouffer, Samuel A., Arthur A. Lumsdaine, Marion Harper Lumsdaine, Robin M. Williams Jr., M. Brewster Smith, Irving L. Janis, Shirley A. Star, et al. *American Soldier*. Princeton, NJ: Princeton University Press, 1949.

Strachan, Hew. "Clausewitz and the Dialectics of War." In *Clausewitz in the Twenty-First Century*, edited by Hew Strachan and Andreas Herberg-Rothe, 14–44. Oxford, UK: 2007.

Strassler, Robert B., ed. *The Landmark Thucydides: A Comprehensive Guide to the Peloponnesian War*. New York: Free Press, 1996.

Strong, Elizabeth. "Literary Rape in America: A Post-Coital Study of the Writings of Dr. Seuss," *Journal of Contemporary Satire* 4 (1977): 35.

Strout, Nathan. "The Space Force Doesn't Want to Send a Human to Do a Robot's Job." C4ISRNET. Accessed September 30, 2020. https://www.c4isrnet.com/battlefield-tech/space/2020/09/29/no-the-space-force-wont-be-sending-humans-into-space-anytime-soon/.

Sturmberg, J. P., and M. Carmel. *Handbook of Systems and Complexity in Health*. New York: Springer, 2013.

Sun, Jing, and Nicholas Jan Buys. "Improving Aboriginal and Torres Strait Islander Australians' Well-Being Using Participatory Community Singing Approach." *International Journal on Disability and Human Development* 12, no. 3 (2013). https://doi.org/10.1515/ijdhd-2012-0108.

Sunstein, Cass R. *Why Societies Need Dissent.* Cambridge, MA: Harvard University Press, 2003.

Sun Tzŭ. *Art of War.* Translated by Lionel Giles. 1910. https://www.gutenberg.org/files/132/132-h/132-h.htm.

Sweet, Louise E. "Camel Raiding of North Arabian Bedouin: A Mechanism of Ecological Adaptation." *American Anthropologist* 67, no. 5 (1965): 1132–50. http://www.jstor.org/stable/668360.

Swift, Jonathan. *Gulliver's Travels into Several Remote Nations of the World.* Vol. 1. Paris: A. and W. Gaglignani, 1826.

Tanczer, Leonie Maria. "50 Shades of Hacking: How IT and Cybersecurity Industry Actors Perceive Good, Bad, and Former Hackers." *Contemporary Security Policy* 41, no. 1 (January 2, 2020): 108–28. https://doi.org/10.1080/13523260.2019.1669336.

Taylor, Philip M. *Munitions of the Mind: A History of Propaganda.* Manchester: Manchester University Press, 1995.

Tellis, Ashley J. "China's Military Space Strategy." *Survival,* 49, no. 3 (2007): 41–72. https://doi.org/10.1080/00396330701564752.

Tenet, George. *At the Center of the Storm: My Years at the CIA.* New York: HarperCollins, 2007.

Thayer, Albert S. "Adverse Possession." *Journal of the Society of Comparative Legislation* 13, no. 3 (1913) 582–602. http://www.jstor.org/stable/752305.

Thompson, Paul. "Tempo and Mode in Evolution: Punctuated Equilibria and the Modern Synthetic Theory." *Philosophy of Science* 50, no. 3 (1983): 432–52. http://www.jstor.org/stable/187858.

Thomson, Jennifer. "Biocentrism and the Health of the Wild." In the *Wild and the Toxic: American Environmentalism and the Politics of Health,* 71–97. Durham, NC: University of North Carolina Press, 2019. http://www.jstor.org/stable/10.5149/9781469651668_thomson.7.

"To Stave Off Arab Spring Revolts, Saudi Arabia and Fellow Gulf Countries Spend $150 Billion." Knowledge at Wharton, September 21, 2011. https://knowledge.wharton.upenn.edu/article/to-stave-off-arab-spring-revolts-saudi-arabia-and-fellow-gulf-countries-spend-150-billion/.

Townsend, Mark. "Being Shot at While Trying to Disarm a Bomb—Just Another Day at Work," *UK Guardian,* November 7, 2009.

Treverton, Gregory F. "Intelligence Analysis: Between 'Politicization' and Irrelevance." In *Analyzing Intelligence: Origins, Obstacles, and Innovation,* edited by Roger Z. George and James B. Bruce, 93. Washington, DC: Georgetown University Press, 2008.

Trombetta, Maria Julia. "Environmental Security and Climate Change: Analysing the Discourse." *Cambridge Review of International Affairs* 21, no. 4 (2008): 585–602.

Tse, Wicky W. K. "Cutting the Enemy's Line of Supply: The Rise of the Tactic and Its Use in Early Chinese Warfare." *Journal of Chinese Military History* 6 (2017): 131–56.

Tsouras, Peter. *Warriors' Words: A Dictionary of Military Quotations*. London: Cassell Arms and Armour, 1994.

Tuchman, Barbara W. *March of Folly: From Troy to Vietnam*. New York: Knopf, 1984.

Tung, Liam. "Microsoft: This Is How the Sneaky Solarwinds Hackers Hid Their Onward Attacks for So Long—the Solarwinds Hackers Put in 'Painstaking Planning' to Avoid Being Detected on the Networks of Hand-Picked Targets." *ZDNet*, January 21 2021. https://www.zdnet.com/article/microsoft-this-is-how-the-sneaky-solarwinds-hackers-hid-their-onward-attacks-for-so-long/.

Tversky, Amos, and Daniel Kahneman. "Framing of Decisions and the Psychology of Choice." *Science* 211, no. 4481 (January 30, 1981): 453.

Tzabar, Shimon. *White Flag Principle: How to Lose a War and Why*. London: Penguin, 1972.

Ungar, Michael, ed. *Social Ecology of Resilience: A Handbook of Theory and Practice*. New York: Springer, 2012.

Union of Concerned Scientists (UCS). *UCS Satellite Database*. January 1, 2023. https://www.ucsusa.org/resources/satellite-database.

United Nations. "Collective Action Now, the Only Way to Meet Global Challenges, Guterres Reaffirms in Annual Report." *UN News*, September 23, 2019. https://news.un.org/en/story/2019/09/1047042.

———. "Our Aim." Together: Respect, Safety and Dignity for All campaign. Accessed September 4, 2023. https://together.un.org/our-aim.

———. "Tolerance Is a Commitment 'to Seek in Our Diversity the Bonds that Unite Humanity'—UN." *UN News*, November 16, 2016. https://news.un.org/en/story/2016/11/545462-tolerance-commitment-seek-our-diversity-bonds-unite-humanity-un.

United Nations Educational, Scientific and Cultural Organization. *Declaration of Principles on Tolerance*. November 16, 1995. https://www.refworld.org/docid/453395954.html.

United Nations General Assembly. *Charter of the United Nations and Statute of the International Court of Justice*. 1945. United Nations Treaty Series 993. https://treaties.un.org/doc/publication/ctc/uncharter.pdf.

———. *Multilateral Convention against Torture and Other Cruel, Inhuman or Degrading Treatment or Punishment*. December 10, 1984. United Nations Treaty Series 1465. https://treaties.un.org/doc/publication/unts/volume%201465/volume-1465-i-24841-english.pdf.

———. Resolution 41/128. *Declaration on the Right to Development*. December 4, 1986. https://www.ohchr.org/sites/default/files/rtd.pdf.

United Nations General Assembly. Resolution 217 A (III). *Universal Declaration of Human Rights*. December 10, 1948. United Nations Treaty Series. https://www.un.org/sites/un2.un.org/files/2021/03/udhr.pdf.

United Nations Human Rights, Office of the High Commissioner. "Human Rights Council Holds Panel Discussion on Promoting Tolerance, Inclusion, Unity and Respect for Diversity." March 19, 2018. https://www.ohchr.org/EN/HRBodies/HRC/Pages/NewsDetail.aspx?NewsID=22845&LangID=E.

———. "Status of Ratification Interactive Dashboard." Accessed September 4, 2023. https://indicators.ohchr.org/.US Air Force. *Air Force Doctrine: Annex 3-14 Coun-

terspace Operations. Maxwell AFB, AL: Curtis E. Lemay Center for Doctrine Development and Education, 2018. https://www.doctrine.af.mil/Doctrine-Annexes/Annex-3-14-Counterspace-Ops/.

US Army. *Army Leadership*. Field Manual 22-100. Washington, DC: US Government Printing Office, August 31, 1999.

———. *Battalion Commander's Handbook*. Carlisle, PA: US Army War College, 1996. https://apps.dtic.mil/sti/pdfs/ADA309539.pdf.

———. Field Manual 6-22. *Developing Leaders*. Washington, DC: US Government Printing Office, November 2022.

———. Field Manual 6-22. *Leader Development*. Washington, DC: US Government Printing Office, June 2015.

US Central Intelligence Agency. *Comprehensive Report of the Special Advisor to the DCI on Iraq's WMD, with Addendums* (Duelfer Report). Washington, DC: Central Intelligence Agency, April 25, 2005.

US Department of Defense. *Defense Space Strategy: Summary*. June 2020. https://media.defense.gov/2020/Jun/17/2002317391/-1/-1/1/2020_DEFENSE_SPACE_STRATEGY_SUMMARY.PDF?fbclid=IwAR2TYXPZVQm0o-ybSXM3UHOYK-sRsdlvaETKChf1raiGaFMYR64QkMxu_9o.

———. "DoD News Briefing—Secretary Rumsfeld and Gen. Myers." September 12, 2002. https://georgewbush-whitehouse.archives.gov/news/releases/2003/04/20030401-5.html.

US National Air and Space Agency. "'Oumuamua Overview." December 19, 2019. https://solarsystem.nasa.gov/asteroids-comets-and-meteors/comets/oumuamua/in-depth/.

US War Department. *War Dogs*. War Department Technical Manual 10-396. Washington, DC: US War Department, 1943.

Van Creveld, Martin. *Supplying War: Logistics from Wallenstein to Patton*. Cambridge, MA: Cambridge University Press, 1977.

van der Kolk, B. A., and M. S. Greenberg. "The Psychobiology of the Trauma Response: Hyperarousal, Constriction, and Addiction to Traumatic Reexposure." In *Psychological Trauma*, edited by B. A. van der Kolk, 63–87. Washington, DC: American Psychiatric Press, 1987.

Van Dyke, C., N. J. Zilberg, and J. A. McKinnon. "Posttraumatic Stress Disorder: A Thirty-Year Delay in a World War II Veteran." *American Journal of Psychiatry* 142, no. 9 (1985): 1070–73.

Vandepeer, Charles. "Intelligence and Knowledge Development: What Are the Questions Intelligence Analysts Ask?" *Intelligence and National Security* 33, no. 6 (2018): 785–86.

Vandewalle, Dirk. *A History of Modern Libya*. 2nd ed. New York: Cambridge University Press, 2012.

Varner, John. *Dogs of the Conquest*. Norman: University of Oklahoma Press, 1983.

Varshney, Ashutosh. "Nationalism, Ethnic Conflict, and Rationality." *Perspectives on Politics* 1, no. 1 (March 2003): 85–99.

Vatikiotis, P. J. *Islam and the State*. New York: Croom Helm, 1987.

Vaughan-Nichols, Steven J. "Solarwinds: The More We Learn, the Worse It Looks." *ZDNET Online*, January 4 2021. https://www.zdnet.com/article/solarwinds-the-more-we-learn-the-worse-it-looks/.

Vernon, Richard. "Unintended Consequences." *Political Theory* 7, no. 1 (1979): 57–73. http://www.jstor.org/stable/190824.

Violino, Bob. "Phishing Attacks Are Increasing and Getting More Sophisticated: Here's How to Avoid Them." *CNBC*, January 10, 2023. https://www.cnbc.com/2023/01/07/phishing-attacks-are-increasing-and-getting-more-sophisticated.html.

Vogt, George L. "When Posters Went to War: How America's Best Commercial Artists Helped Win World War I." *Wisconsin Magazine of History* 84, no. 2 (2000–2001): 38–47.

Walker, C., and J. Hall, eds. "Technical Document: Impact of Satellite Constellations on Optical Astronomy and Recommendations Toward Mitigations." NoirLab, National Science Foundation (NSF), August 25, 2020. https://noirlab.edu/public/products/techdocs/techdoc003/.

Walt, Stephen M. "Is the Cyber Threat Overblown?" *Foreign Policy*, March 30 2010.

Waltz, Kenneth. *Theory of International Politics*. Reading, MA: Addison-Wesley, 1979.

Warner, Michael. "A Brief History of Cyber Conflict." In *Ten Years In: Implementing Strategic Approaches to Cyberspace*, edited by Jacquelyn Schneider, Emily O Goldman, and Michael Warner, 13–30. Newport, RI: Naval War College Press, 2020.

Warnke, Paul C. "Apes on a Treadmill." *Foreign Policy*, no. 18 (Spring 1975): 12–29. https://doi.org/10.2307/1147960.

Warren, Bill. *Keep Watching the Skies: American Science Fiction Films of the Fifties*. 2 vols. Jefferson, NC: McFarland, 2009.

———. "Moral Identity and Protest Cascades in Syria." *British Journal of Political Science* 48, 4 (October 2018): 877–901.

Wedeen, Lisa. "Acting 'As If': Symbolic Politics and Social Control in Syria." *Comparative Studies in Society and History* 40, no. 3 (July 1998): 503–23.

Weinstock, Robert A. "The Lorax State: Parens Patriae and the Provision of Public Goods." *Columbia Law Review* 109, no. 4 (May 2009): 798–843.

Weinzierl, Matthew, and Mehak Sarang. "The Commercial Space Age Is Here." *Harvard Business Review*, February 28, 2021. https://hbr.org/2021/02/the-commercial-space-age-is-here.

Wells, Mark. "Aviators and Air Combat: A Study of the U.S. Eighth Air Force and R.A.F. Bomber Command." PhD dissertation, Kings College London, July 1992.

Werner, Wouter G. "The Curious Career of Lawfare." *Case Western Reserve Journal of International Law* 43 (2010): 61–72.

West, Mark. "Dr. Seuss's Responses to Nazism: Historical Allegories or Political Parables." *Jabberwocky* 19, no. 1 (July 2016): 1–4.

Westacott, Emrys. "Moral Relativism." *Internet Encyclopedia of Philosophy* (ISSN 2161-0002). Accessed February 2, 2021. https://iep.utm.edu/moral-re/.

Wezeman, Pieter D., Dr Aude Fleurant, Alexandra Kuimova, Dr Diego Lopes da Silva, Dr Nan Tian, and Siemon T. Wezeman. "Trends in International Arms

Transfers, 2019." Stockholm International Peace Research Institute, March 2020. https://www.sipri.org/publications/2020/sipri-fact-sheets/trends-international-arms-transfers-2019.

Whaley, Barton. *Stratagem: Deception and Surprise in War*. Cambridge: Massachusetts Institute of Technology Center for International Studies, 1969.

White House. Executive Order [14028] on Improving the Nation's Cybersecurity, May 12, 2021. https://www.federalregister.gov/documents/2021/05/17/2021-10460/improving-the-nations-cybersecurity.

———. *National Cybersecurity Strategy*. March 1, 2023. https://www.whitehouse.gov/wp-content/uploads/2023/03/National-Cybersecurity-Strategy-2023.pdf.

———. *National Security Strategy of the United States*. December 2017. https://trumpwhitehouse.archives.gov/wp-content/uploads/2017/12/NSS-Final-12-18-2017-0905-2.pdf.

———. "Statement by the Press Secretary on the 'NotPetya' Cyber-Attack." February 15, 2018. https://www.presidency.ucsb.edu/documents/statement-the-press-secretary-the-notpetya-cyber-attack.

Widom, C. S., and R. L. Shepard. "Accuracy of Adult Recollections of Childhood Victimization: Part 1, Childhood Physical Abuse." *Psychological Assessment 8*, no. 4 (1996): 412–21.

Wilkins, H. A. "The Soldier of Auvergne." In *Selections from Canadian Poets: With Occasional Critical and Biographical Notes and an Introductory Essay on Canadian Poetry*, edited by Edward Hartley Dewart, and Douglas Lochhead, 126–28. Toronto: University of Toronto Press, 1973.

Wilkinson, Charles F., and John M. Volkman. "Judicial Review of Indian Treaty Abrogation: 'As Long as Water Flows, or Grass Grows upon the Earth'; How Long a Time Is That?" *California Law Review* 63, no. 3 (1975): 601–61. https://doi.org/10.2307/3479850.

Williams, K. L., and R. V. Galliher. "Predicting Depression and Self-Esteem from Social Connectedness, Support and Competence." *Journal of Clinical Social Psychology* 25 (2006): 855–74.

Williams, Paul. "Protection, Resilience and Empowerment: United Nations Peacekeeping and Violence against Civilians in Contemporary War Zones." *Politics* 33 (2013): 287–98.

Williams, Robert. *The "China, Inc." Challenge to Cyberspace Norms*. Stanford, CA: Hoover Institution, 2018.

Willmott, H. P. *World War I*. New York: Dorling Kindersley, 2003.

Wilmes, John. "Dr. Seuss's Forgotten Anti-War Book Made Him an Enemy of the Right." *Outline*, July 30, 2018. https://theoutline.com/post/5601/dr-seuss-the-butter-battle-book-history?zd=1&zi=h5webi5x .

Wilson, Peter H. "The Causes of the Thirty Years War 1618–48." *English Historical Review* 123, no. 502 (2008): 554–86. http://www.jstor.org/stable/20108541.

Winship, Michael. "Yertle the Commander in Chief: Dr. Seuss Shows Us How Protest Can Sometimes Topple a Tyrant; A Children's Classic Rings True Today and Offers an Answer to Authoritarian Rule." *Salon*, January 16, 2017. https://www

.salon.com/2017/01/16/yertle-the-commander-in-chief-dr-seuss-shows-us-how-protest-can-sometimes-topple-a-tyrant_partner/.

Winters, Dick. Interview by Mark Cowen in *We Stand Alone Together*, 55:10. HBO, 2001.

Winters, Harold. *Battling the Elements: Weather and Terrain in the Conduct of War*. Baltimore, MD: Johns Hopkins University Press, 1998.

Witts, L. J. "Chlorosis in Males." *Guy's Hospital Report* 10, no. 4 (1930): 417–20.

Wohlstetter, Roberta. *Pearl Harbor: Warning and Decision*. Stanford, CA: Stanford University Press, 1962.

Wolosky, Shira. "Democracy in America: By Dr. Seuss." *Southwest Review* 85, no. 2 (Spring 2000): 167–83.

Womack, James K. "Soviet Correlation of Forces and Means: Quantifying Modem Operations." Master's thesis, US Army Command and General Staff College, 1990.

Wood, Hayley. "Interview with Filmmaker Ron Lamothe about *The Political Doctor Seuss*." *Massachusetts Foundation for the Humanities—News & Events*, September 2007. https://web.archive.org/web/2007091604421/http://www.mfh.org/lamotheinterview/.

Wood, Paul. "Aetiology of Da Costa's Syndrome." *British Medical Journal* 1, no. 4196 (1941): 845–51.

Yang, Li-chung. "Globetrotters and Exotic Creatures: The Imaginary Others in Dr. Seuss." *Wenshan Review of Literature and Culture* 12, no. 2 (June 2019): 165–86.

Young, Shalanda D. "Memorandum for the Heads of Executive Departments and Agencies." Office of Management and Budget, Executive Office of the President, January 26, 2022. https://www.whitehouse.gov/wp-content/uploads/2022/01/M-22-09.pdf.

Zetter, Kim. *Countdown to Zero Day: Stuxnet and the Launch of the World's First Digital Weapon*. New York: Broadway Books, 2014.

———. "The Untold Story of the Boldest Supply-Chain Hack Ever." *Wired*, May 2, 2023. https://www.wired.com/story/the-untold-story-of-solarwinds-the-boldest-supply-chain-hack-ever/.

Zetter, Kim, and Huib Modderkolk. "Revealed: How a Secret Dutch Mole Aided the U.S.-Israeli Stuxnet Cyberattack on Iran." *Yahoo News*, September 2, 2019. https://news.yahoo.com/revealed-how-a-secret-dutch-mole-aided-the-us-israeli-stuxnet-cyber-attack-on-iran-160026018.html.

Zicht, Jennifer. "In Pursuit of the Lorax," *EPA Journal* 17, no. 4 (September 1991): 27–30.

Zipes, Jack. *Irresistible Fairy Tale: The Cultural and Social History of a Genre*. Princeton, NJ: Princeton University Press, 2012. http://www.jstor.org/stable/j.ctt7sknm.

Zittrain, Jonathan. "'Netwar': The Unwelcome Militarization of the Internet Has Arrived." *Bulletin of the Atomic Scientists* 73, no. 5 (2017): 300–304.

Zunes, Stephen. "The Iranian Revolution: 1977–1979." International Center for Non-Violent Conflict, April 2009. https://www.nonviolent-conflict.org/wp-content/uploads/2016/02/The-Iranian-Revolution-1.pdf.

Zurbriggen, E. L., R. L. Gobin, and L. A. Kaehler. "Trauma, Attachment, and Intimate Relationships." *Journal of Trauma Dissociation* 13 (2012): 127–33.

Index

acceptance, 274–75
adaptation, 27
advanced persistent threat (APT), 114–16
Advanced Research Projects Agency Network (ARPANET), 102
Afghanistan, 35, 182, 187, 218
Air Force, US, 51–52, 139, 217
Alexander, Keith, 187
Allenby, Edmund, 219
alliance politics, 92
Allied Forces. *See* World War II
altruism, 160
Anderson, Lisa, 151
And to Think That I Saw It on Mulberry Street (Dr. Seuss), 12, 213
animal abuse, 279n1
Animal Farm (Orwell), 34–35
Anonymous, 115
the Anthropocene, 86, 97
anti-isolationism, 30–32
anti-Semitism, 63
Any Bonds Today? (film), 124
APT. *See* advanced persistent threat
Arab Spring, 147–50, 152–57, 161–62
Arendt, Hannah, 213–14
Armchair General! (Heatherly), 6
arms race, 238–40, 251, 254n13

Army, US, 4, 128–29; Geisel, T., in, 7–10, 63–64; leadership in, 214; recruits, 268; Signal Corps, 125; soldiers in, 137–40; US Department of Defense and, 132; in Vietnam War, 263–64; War College, 222
ARPANET. *See* Advanced Research Projects Agency Network
art, 10–12
al-Asad, Hafez, 152
attribution problem, 110–13
Australia, 48
Austria-Hungary, 244
authority: anti-authoritarianism, 132–35; authoritarianism, 118, 131–32, 146–47, 152–57, 161–62, 166n59; deviance and, 115–16; dissidents of, 157–61; to Geisel, T., 74–75, 152, 157, 159, 161; keeping, 148–52; military, 132–33; in national security, 185–86; political, 162n2, 174
Axis powers, 8–9. *See also* Germany; Japan

Baiocchi, David, 49
Bakshi, Ralph, 236
Baldwin, Alec, 116

banking, 49
Bank of America, 49
Bartholomew and the Oobleck (Dr. Seuss), 5, 10, 23, 152, 160
Bartrum, Giulia, 11–12
Bateson, Gregory, 173, 188n8
Battle of the Bulge, 9–10, 32, 63, 212–13, 215, 262, 263–64
Battle of Waterloo, 228
Bayat, Asef, 166n59
Bay of Pigs invasion, 31
al-Behairy, Galal, 157–58
Bellin, Eva, 155–56
Belt-and-Road Initiative, 48
Ben Ali, Zine al Abedine, 149, 153, 158
Berlin, Irving, 124
Berlin Wall, 236, 252
Bezos, Jeff, 50
Biden, Joseph, 206
Bill of Rights, 17
bin Laden, Osama, 178–79, 221–22
The Birds (Hitchcock), 34
Blanc, Mel, 125
Blood Rites (Eherenreich), 95
"Bluebeard" (Perrault), 7
Blue Origin, 50
Boko Haram, 226
Booby Traps (film), 128
Bosnia, 244
Bouazizi, Mohammad, 156, 158–59, 168n97, 169n120
Bozeman, Adda, 175
Bradley, Omar, 9
Brands, Hal, 55
Broderick, Matthew, 115
brotherhood, 67
Brothers Grimm, 7
Brown, L. Carl, 153, 225
Browning, Miles, 225
Budapest Agreement, 236
Bunyan, John, 24
Burma, 226
business, 43, 49–50
The Butter Battle Book (Dr. Seuss): arms race in, 238–40; Cold War and, 16; criticism of, 237–38; deterrence in, 6, 235–36, 252–53; hesitation in, 240–41; nonoccurrence in, 245–47; punishment in, 247–50; rationality in, 241–43; renegades in, 243–44; war termination and, 250–52

Caddick-Adams, Peter, 228
Cambodia, 75, 226
Candide (Voltaire), 31
capacity, 86–87
Capra, Frank, 124–25
carrying capacity, 86–87
Carter, Jimmy, 174, 238
cartoons, 7–8, 62–64, 124–25, 147, 160
Cartwright, James, 187
Castro, Fidel, 244
The Cat in the Hat (Dr. Seuss): attribution problem and, 110–13; authority in, 115–16; cyber operations and, 15–16; illustrations in, 121n46, 121n48; for Navy Seals, 5; offense-defense balance and, 113–14; racism in, 12–13; scholarship on, 3, 102, 116–18, 129, 132–33, 136; staged operations in, 109; success of, 28; symbolic meaning in, 11; unintended consequences in, 107–8; unwitting cooperation in, 103–7
"The Cat is Out of the Bag" (Ishizuka and Stephens), 12–13
Cat's Quizzer (Dr. Seuss), 12
Cedar Revolution, 152
Censored (film), 137
Central Intelligence Agency (CIA), 174, 178–80, 183
character agency, 38
children: assumptions about, 12; bigotry and, 13; Dr. Seuss to, 155, 161–62, 209; education of, 42, 129–30; to Geisel, T., 24, 128–29, 146, 149–50; to Houghton Mifflin, 28; humanity to, 17, 95; lessons for, 16, 212–15; parents and, 6–7; sexual abuse of, 271

Children's Illustrated Clausewitz (Fitz Gerald), 230
Chile, 74
China: Belt-and-Road Initiative in, 48; Boxer Rebellion, 225–26; in Cold War, 208; Guam and, 110; Japan and, 50; military of, 52; US and, 206
choices, 43–44
The Chow Hound (film), 140
Churchill, Winston, 80n77, 230
CIA. *See* Central Intelligence Agency
Cicero, 234n79
"Cinderella" (Perrault), 7
citizenship, 74, 81n86
civil society, 139–42
Civil War, US, 268
Clark, Roger S., 6
Clarke, Richard, 179
Clausewitz, Carl von: friction to, 24–25, 33; philosophy of, 44, 214–15, 229; reputation of, 3, 7, 230; Sun Tzu and, 27
claustrophobia, 34
climate studies, 98
Cold War: China in, 208; Cuba in, 244; deterrence in, 237–39; disarmament in, 254n13; to Geisel, T., 247–48; hesitation in, 240–41; INF, 236; MAD in, 243, 245, 247–48, 251, 256n36; nuclear weapons in, 115; politics, 47–48, 236–38; rationality in, 241–43; science in, 239–40; to Soviet Union, 187; spies in, 255n16; START, 236; strategy in, 248–50; termination of, 250–52; theories, 16; US in, 70. *See also The Butter Battle Book*
Cole, USS, 186
collapse, 86–87
collective action, for human rights, 62, 73–74
collective bargaining, 80n82
combat tactics, 18n8
Coming Home (film), 142n1
commercialization, 43, 49–50

common pool resource dilemmas, 90–92
community, 275–77
compassion, 130
compeerment, 53–55
competition, 25–26, 49
comradery, 226–27
concentration camps, 72
Constitution, US, 17, 89
constructivism, 92–93
Convention against Torture and Other Cruel, Inhuman or Degrading Treatment or Punishment, 71
Cook, Steven A., 154, 159
Cooper, Gary, 31
Côte d'Ivoire, 226
counterspace race, 53
courage, 130
Coutin, Susan Bibler, 74
Crimean War, 268
Cuba, 31, 37, 244
Culora, Thomas, 5
customary international law, 71–72
cyber warfare: attribution problem with, 110–13; cyber hygiene, 113; cybersecurity, 103–5, 121n48; democracy in, 208–9; deterrence in, 114; deviance in, 115–16; in Europe, 206; hacking and, 16, 102, 108, 110, 113–18, 121n46; to military, 15–16; National Cyber Security Centre, 110; *National Cybersecurity Strategy*, 106; offense-defense balance and, 113–14; Office of the National Cyber Director, 106; propaganda and, 15; by Russia, 207–8; staged operations in, 109; strategy for, 102, 116–18; unintended consequences in, 107–8; unwitting cooperation in, 103–7

Dartmouth College, 7
The Day After (film), 251
DDoS attack. *See* denial of service attack
deception: in democracy, 207, 211n39; failed, 205–6; false flags, 199–203;

in international relations, 16, 195–99, 203–5; military, 7, 271; in national security, 184–85
Declaration of Principles of Tolerance (UNESCO), 68
deep ecology, 97–98
deforestation, 87
De Matos, Jose, 218
demobilization, 80n80
democracy: American, 141–42; in cyber warfare, 208–9; deception in, 207, 211n39; democratization, 43, 47–49; fascism and, 124; Geisel, T., and, 93, 123; in Islam, 166n61; propaganda and, 127–32, 144n62; western, 89–90
denial, 247–50
denial of service (DDoS) attack, 107–8, 111
Department of Defense, US, 132, 239–40, 248
Design for Death (documentary), 63–64
determination, 36–37
deterrence, 114; in Cold War, 237–39; escalation and, 6, 235–36, 252–53; MAD, 256n36; nonoccurrence and, 245–47; with nuclear weapons, 6, 208, 236–45, 247–53; psychology of, 246–47; to UN, 245–46
Development as Freedom (Sen), 70
deviance, 115–16
Diagnostic and Statistical Manual (DSM), 262, 266, 269, 272
Did I Ever Tell You How Lucky You Are? (Dr. Seuss): Battle of the Bulge and, 10; comradery in, 226–27; military intelligence in, 220–22; soldiers and, 14, 218–19, 223–26; survival in, 229–30; warfare environment in, 227–29; war technology in, 215–17; World War II and, 212–15
dignity, 78n50
diplomacy, 254n12
discrimination, 65
Disney Walt, 124

dissimulation, 175–76
Dr. Seuss: career of, 213, 237–38; to children, 155, 161–62, 209; illustrations by, 220–21; in military art and science, 4–7; philosophy of, 16–17; reputation of, 93–94, 128, 138, 142, 188; scholarship on, 3, 13–16, 91, 279n13; symbolic meaning to, 10–13; themes of, 154; in World War II, 4. *See also specific topics*
Dow, Christopher, 129, 141
Drengson, Alan, 97–98
DSM. *See Diagnostic and Statistical Manual*
du Picq, Ardant, 227
al-Duwish, Mohammad, 153

Earle, Edward Mead, 44
Eastman, Philip D., 125
ecology, 86–90, 97–98
education: of children, 42, 129–30; to Culora, 5; to Geisel, T., 17, 115; military education curriculum, 4–5; NEA, 12; pedagogy, 141; propaganda and, 130–31, 141–42; for tolerance, 139; UNESCO, 68; in US, 6–7; US Army Information and Education division, 8
Egypt, 148–49, 154, 157–59
Ehrenreich, Barbara, 95
Eisenhower, Dwight D., 127, 224
Eldredge, Wentworth, 185
Elers, George, 216–17
Ellul, Jacques, 130–32, 141, 144n62
El Salvador, 74
end states, 33–36
enlistment, in World War II, 7–10
environment, luck and, 227–29
environmental laws, 97
environmental security: capacity and, 86–87; common pool resource dilemmas with, 90–92; deep ecology and, 97–98; long-term consequences of, 88–90; mass extinction and, 95–97; scholarship on, 15, 85–86, 98–

99; survival and, 93–95; vulnerability with, 92–93
escalation: in arms race, 238–40; deterrence and, 6, 235–36, 252–53; nationalism and, 240–41
escape, 7
ethical obligation, 77n37
Europe: collective bargaining in, 80n82; cyber warfare in, 206; decimation of, 10; fairy tales in, 7; forests in, 234n79; Nazis in, 7–8, 215; Soviet Union to, 248; US and, 48; in World War II, 8–9, 75
evasion, 7
Every War Must End (Ikle), 251–52
Executive Order on Improving the Nation's Cybersecurity, 106–7

failed deception, 205–6
fairy tales, 7
false flags, 199–203
fascism: cartoons about, 7–8; democracy and, 124; to Geisel, T., 138; imperialism and, 131; totalitarianism and, 30–31
Federal Bureau of Investigation (FBI), 221–22
Fighting Tools (film), 138
Fingar, Tom, 172
Finland, 50
Fitz Ferald, Caitlin, 230
The 500 Hats of Bartholomew Cubbins (Dr. Seuss), 151–52, 159
fog of war, 7
Fort Fox, 124, 128
Four Feathers (Mason), 278
Fox in Socks (Dr. Seuss), 4
Fox Studios, 124, 128
framing, 173–74, 188n8
France: Cambodia and, 226; Geisel, T., and, 227–28; Germany and, 9; Napoleon and, 228–29; Spain and, 11; United Kingdom and, 256n47; in World War I, 217
Franco, Francisco, 11

Franz Ferdinand, 244
freedom, 135–38, 140
Freleng, Friz, 125
friction, 24–25, 33
Friedrich, Carl Joachim, 129–30
Fritz, Stephan, 216
Froese, Katrin, 140–41

Gaddis, John Lewis, 46
Gaia hypothesis, 98
Gas (film), 138
Geisel, Helen, 28–29, 63–64
Geisel, Theodor: adaptation to, 27; authority to, 74–75, 152, 157, 159, 161; in Battle of the Bulge, 32; career of, 39n4, 140–41, 161–62; character agency to, 38; children to, 24, 128–29, 146, 149–50; Cold War to, 247–48; criticism of, 12–13; death of, 236; democracy and, 93, 123; education to, 17, 115; environment to, 99; fascism to, 138; forests to, 234n79; France and, 227–28; Great Balancing Act to, 44–45, 54–55; Grinch and, 262–63, 278; illustrations by, 167n74; isolationism to, 31; military deception and, 7, 271; nationalism to, 241; national security to, 171; philosophy of, 16–17, 37, 237, 242–43, 253; political authority to, 162n2, 174; political cartoons by, 62–64; preparedness to, 18n8; race and, 40n17, 76n25; resilience to, 274–75, 277; scholarship on, 3, 13–14; on snap judgments, 163n10; suicide to, 28–29; symbolic meaning to, 10–13; themes of, 154; trauma of, 261–62, 264, 266–67; in US Army, 7–10, 63–64; utopia to, 34–35; on warfare, 23; in World War II, 4, 14, 95, 122–25, 128, 131, 142, 212–15, 221, 239, 263; World War I to, 244; writing process of, 61–64, 91, 162n8; writing style of, 76n28. *See also* Dr. Seuss
genocide, 5

geology, 96
Germany: Axis powers and, 8–9; Berlin Wall, 236, 252; to Churchill, 230; concentration camps in, 72; culture of, 30–31; France and, 9; Hitler for, 228; Japan and, 125, 141, 254n12; Operation Fortitude and, 185; reconstruction in, 93; US and, 7–8, 217, 221, 227; in World War II, 11, 80n77, 131
Gewirtzman, Doni, 154
Gitlin, Todd, 189n15
Goffman, Erving, 173, 189n15
Going Home (film), 142n1
The Goldbrick (film), 122
Golding, William, 34–35
Goldman Sachs, 49
Good Samaritan laws, 77n31
Gorbachev, Mikhail, 236
grand strategy, 43–46, 53–55
Graveyard of Clerics (Menoret), 161
Great Balancing Act, 44–45, 54–55
Green Eggs and Ham (Dr. Seuss), 5, 12, 129
Gripes (film), 137
Guam, 110
Guccifer, 115
Guernica (Picasso), 10–11
Guillain-Barré syndrome, 28
Gulf of Tonkin incident, 30
Gulf War, 35
Gulliver's Travels (Swift), 253, 256n47
Gusentine, Robert, 5

hacking, 16, 102, 108, 110, 113–18, 121n46. *See also* cyber warfare
Halsey, William, 225
"Hansel and Gretel" (Brothers Grimm), 7
Happy Birthday to You! (Dr. Seuss), 94
Hardin, Garrett, 91
Harris, William, 185
Hart, B. H. Liddell, 44
al-Hathloul, Loujain, 157
Heart of Altruism (Monroe), 160

Heatherly, Chris, 6
Henry VIII (king), 256n47
Hesford, Wendy, 77n38
hesitation, 240–41
Heuer, Richard J., Jr., 183
Hicks, Louis, 216
High Noon (film), 31
Hill, A. P., 224
Hitchcock, Alfred, 34
Hitchens, Theresa, 52
Hitler, Adolf, 30, 80n77, 131, 147, 160, 228, 250–51
Hobbes, Thomas, 95
Hollywood, 123–25
Holmes, Richard, 225
The Homefront (film), 139–40
Horton and the Kwuggerbug and other Lost Stories (Dr. Seuss): deception in, 16, 199–203, 205–6; international relations in, 195–99, 203–5; writing process for, 62–64
Horton Hatches an Egg (Dr. Seuss), 124, 145n83
Horton Hears a Who (Dr. Seuss): collective action in, 73–74; criticism of, 12–13; development in, 69–71; ethical obligation in, 77n37; humanity in, 17; human rights laws and, 16, 65–67, 74–75; listening in, 64–65; mass extinction in, 277; peace in, 79n62; PTSD in, 275–76; race and, 163n10; scholarship on, 61–62, 129, 135–36, 145n83; tolerance in, 67–69; torture in, 71–73
Hot Spot (film), 135
Houghton Mifflin, 28
How the Grinch Stole Christmas (Dr. Seuss): as allegory, 262–63; PTSD in, 15, 261–62, 264–72, 277–78; resilience in, 272–77; wartime trauma in, 263–64
Hugo, Victor, 228
human intelligence, 172, 187
humanity, 17, 50–51, 54, 71, 95–96, 155, 245

Humanity at Sea (Mann), 66
human rights: citizenship and, 74,
    81n86; collective action for, 62,
    73–74; development of, 69–71;
    dignity as, 78n50; Good Samaritan
    laws as, 77n31; history of, 77n32;
    international law for, 61–62, 65–67,
    74–75; laws, 16; listening and,
    64–65, 77; moral duty and, 77n37;
    organizations, 73–74; tolerance
    of, 67–69; torture and, 16, 71–73;
    *Universal Declaration of Human
    Rights*, 64–65, 67–68, 77n38, 79n67;
    violations, 72–73
Hunches in Bunches (Dr. Seuss), 3;
    deception in, 184–85; decision-
    making in, 172–73, 177–78, 182–84;
    framing in, 173–74; intelligence
    analysis in, 174–77; intelligence
    technology in, 186–87; military
    intelligence in, 181–82; national
    security and, 16, 170–71, 187–88;
    political intelligence in, 179–80
Huntington, Samuel, 147
Hussein, Saddam, 149, 151, 184
Huston, John, 124
Huxley, Aldous, 41n23

I Can Read with My Eyes Shut (Dr.
    Seuss), 17
IEDs. *See* improvised explosive devices
If I Ran the Zoo (Dr. Seuss), 12, 147
I Had Trouble Getting to Solla Sollew
    (Dr. Seuss), 14; end states in, 33–36;
    isolationism/anti-isolationism in,
    30–32; military leadership in, 32;
    misadventure in, 29–30; plot of,
    23–24; preparedness in, 18n8;
    psychological resilience and, 27–29;
    scholarship on, 38–39; strategic
    pitfalls in, 24–27; utility of force
    and, 36–38
Ikle, Fred, 251–52
imagination, 3, 8, 128, 130, 138, 266–67
imperialism, 85, 131

improvised explosive devices (IEDs),
    218–19
India, 48
individual freedom, 135–38, 140
indoctrination strategies, 141
INF. *See* Intermediate-Range Nuclear
    Forces
The Infantry Blues (film), 139–40
information. *See* intelligence
Ingersoll, Ralph, 9, 76n6, 185
Inoue, Yuichi, 97–98
"Instructions for the Last Night,"
    175–77
intelligence: deception and, 184–85;
    human, 172, 187; in international
    relations, 205–6; military, 103,
    113–14, 117–18, 181–82, 220–22;
    national security and, 170–71, 174–
    77, 181–82, 187–88; ODNI, 185–86;
    political, 179–80; psychology of,
    180–81, 183; signals, 172, 178–79,
    187; technology, 186–87
intercontinental ballistic missiles, 249
Intermediate-Range Nuclear Forces
    (INF), 236
international law: customary, 71–72;
    Hesford on, 77n38; for human rights,
    61–62, 65–67, 74–75; modern, 67–
    68; politics of, 48–49; for weapons
    of mass destruction, 6
international relations: deception in,
    16, 195–99, 203–5; diplomacy in,
    199–203; intelligence in, 205–6
International Space Station, 50
international system: international
    political system, 95; military and,
    16–17; nation-states in, 44
International Union for the Conservation
    of Nature (IUCN), 87
intimidation, 29–30
Iran, 105, 155–56, 174, 182
Iraq, 35–36, 149, 183–84, 187, 219, 224
Irish Republican Army, 232n48
Irwin, Harvey, 224
Ishizuka, Katie, 12–13

Islam, 150–51, 152–53, 166n61, 175–77
Island (Huxley), 41n23
isolationism: anti-isolationism and, 30–32; totalitarianism and, 7–8
It Happened One Night (film), 125
IUCN. *See* International Union for the Conservation of Nature

Japan: China and, 50; Germany and, 125, 141, 254n12; India and, 48; Navy of, 134; Orientalism and, 147; Pearl Harbor Invasion, 124; propaganda and, 63–64; racism against, 128; technology in, 51–52; US and, 40n17, 76n25; women in, 224; World War II and, 163n10
Jasper, James, 158–60
Johnson-Freese, Joan, 52
Jones, Chuck, 125
Jurassic Park (film), 96–97

Kant, Immanuel, 66–67
Karmān, Tawakkol, 158
Kay, David, 183
Kenney, George, 224
Kent, Sherman, 175, 180
Kenya, 91
Khanh, Nguyen, 30
Khashoggi, Jamal, 118
Khomeini (ayatollah), 155
The King's Stilts (Dr. Seuss), 151–52, 162n2
Kippenberg, Hans G., 177
Kleinberg, Mindy, 222
Korean War, 75
Kowalski, Dean, 79n62
Kurzman, Charles, 155

Leaf, Munro, 125
Lebanon, 152
legal meaning, 66
legal principles, 71
Levi, Primo, 72
Leviathan (Hobbes), 95

Libya, 151
Life (magazine), 64
Lindbergh, Charles, 76n6
listening, 64–65, 77
"Literary Rape in America" (Strong), 13
Living Planet Report, UN, 96
lone dissidents, 157–61, 168n102
long-term consequences, of environmental security, 88–90
Looney Tunes, 125
The Lorax (Dr. Seuss), 85–86, 90–92, 97, 98
Lord of the Flies (Golding), 34–35
low earth orbit broadband companies, 51
luck: environment and, 227–29; to military, 218–19; military intelligence and, 220–22; to soldiers, 14, 212–15, 223–26; statistics and, 231n12; survival and, 229–30; war technology and, 215–17
Lynch, Marc, 158, 168n97

MacDonald, J. Fred, 130–31
MAD. *See* mutually assured destruction
Maersk, 108
Mann, Itamar, 66, 77n32
Mansoor, Peter, 55
Marine Corps Reserve, 14, 218
Mars, 48–49, 51, 54
Marshall, George C., 123–24
Marshall Plan, 93
Mason, A. E. W., 278
mass extinction, 95–97, 277
mass protest, 147, 152–57, 168n88
McCarthy, Kevin, 12
McClintock, Mike, 213
McElligot's Pool (Dr. Seuss), 5, 12
McManus, John C., 225
McMorris, Charles H., 225
McNamara, John, 227
McNamara, Robert S., 239, 248–49
Menoret, Pascal, 161
Merck, 108
Microsoft, 108, 110, 206

Middle East: Arab Spring in, 147–50, 152–57, 161–62; culture of, 147–52, 157–61; US in, 177–78
Milevski, Lukas, 53
military: art and science, 4–7; authority, 132–33; battles, 38–39; of China, 52; combat tactics, 18n8; cyber warfare to, 15–16; deception, 7, 271; decisions, 47; demobilization, 80n80; education curriculum, 4–5; intelligence, 103, 113–14, 117–18, 181–82, 220–22; international system and, 16–17; Joint Publication 3.0 for, 126–27; leadership, 32; lessons, 14; luck to, 218–19; means, 45; militarization, 43, 51–53; military information support operations, 127; NCOs in, 145n72; philosophy, 3; punishment in, 272; strategic pitfalls, 24–25; strategy, 14, 38–39, 228–29; in Syria, 155–56; theory, 3–4; training, 6, 133–35; veterans, 118, 198, 220, 229, 264, 269–75; warfare environment and, 234n83
Miller, William, 225
Milley, Mark, 216
misadventure, 29–30
Mr. Smith Goes to Washington (film), 125
modern international law, 67–68
Monroe, Kristen, 160
moral duty, 77n37
Morgan, Judith, 62
Morgan, Neil, 62
Morgan Stanley, 49
Morgenstern, Oskar, 241–42
Morgenthau Plan, 93
Mubarak, Hosni, 149, 158
Munch, Edward, 11–12
Munitions of the Mind (Taylor), 125
Murphy, Audie, 220
Murray, Williamson, 45, 55
Musk, Elon, 49–50
Muslims. *See* Islam

Mussolini, Benito, 30
mutually assured destruction (MAD), 243, 245, 247–48, 251, 256n36
Myanmar, 156
Myers, Mike, 116

Napoleon Bonaparte, 228–29
NASA, 50
National Cyber Security Centre, 110
National Cybersecurity Strategy, 106
National Education Association (NEA), 12
nationalism, 240–41
national security: affairs, 3, 6; authority in, 185–86; deception in, 184–85; decision-making, 16, 172–73, 177–78, 182–84; freedom and, 138; intelligence and, 170–71, 174–77, 181–82, 187–88; obstruction of, 180–81; policymakers and, 179–80; strategy, 30–32; to US, 131, 171, 175–81, 185–86, 191n64
nation-states, 44–46
Native Americans, 90
NATO. *See* North Atlantic Treaty Organization
Navy, US: reputation of, 139; Seals, 5, 182; submarines, 223; US Naval Academy, 231n29; US Naval War College, 4–5, 15–16
Nazis: concentration camps by, 72; defeat of, 8–9; diplomacy with, 254n12; in Europe, 7–8, 215; Hitler and, 131; psychology of, 228; retaliation by, 230; to United Kingdom, 80n77; to US, 30–31
NCOs. *See* Non-commissioned Officers
NEA. *See* National Education Association
Nel, Philip, 130, 145n83
Nevin, Patrick, 226
New Space, 46–47
New Zealand, 48
Ngalyike, Ibrahim Tada, 226

Night (Wiesel), 278
9/11 attacks, 175–76, 182, 186, 221–22
No Buddy Atoll (film), 134
Non-commissioned Officers (NCOs), 145n72
North Atlantic Treaty Organization (NATO), 92, 111, 126
North Korea, 110
NotPetya event, 108
Novy, Ron, 151
nuclear weapons: in Cold War, 115; deterrence with, 6, 208, 236–45, 247–53; to Iran, 105, 109, 172; Iraq and, 183–84; war and, 37
Nye, Gerald, 76n6

Ocean's Eleven (film), 117–18
ODNI. *See* Office of the Director of National Intelligence
offense-defense balance, 113–14
Office of the Director of National Intelligence (ODNI), 185–86
Office of the National Cyber Director, 106
Office of War Information (OWI), 123–24
Oh, the Places You'll Go! (Dr. Seuss), 3; Europe and, 10; grand strategy in, 43–46, 53–55; space race and, 14–15, 42–43; uncertainty in, 46–53
Olympic Summer Games, 8
On Beyond Zebra! (Dr. Seuss), 3, 6, 12, 147
One Fish, Two Fish, Red Fish, Blue Fish (Dr. Seuss), 5, 129
On War (Clausewitz), 214
open-source intelligence, 172
Operation Fortitude, 185
Operation Snafu (film), 128
Orientalism, 147
ornithophobia, 34
Orwell, George, 34–35
Outer Space Treaty, 49
Outpost (film), 134
overstretch, 45–56

OWI. *See* Office of War Information
Oxford University, 7

Pakistan, 182, 222
parents, 4, 6–7, 12, 115, 155, 252, 275
Patton, George S., 185, 225
Pay Day (film), 137–38
peace, 54–55, 79n62
Pearl Harbor (Wohlstetter), 178
Pearl Harbor invasion, 124
Pearlman, Wendy, 156–57, 161, 168n88
pedagogy, 141
Pegasus, 118
Permian extinction, 96
Perrault, Charles, 7
perseverance, 55
Philippines, 48
philosophy: of Clausewitz, 44, 214–15, 229; of Geisel, T., 16–17, 37, 237, 242–43, 253; of Kant, 66–67; military, 3; political, 8, 16–17; of power, 34–35; utilitarianism, 136; of Voltaire, 31
phishing, 104
Picasso, Pablo, 10–11
Pilgrim's Progress (Bunyan), 24
Pillar, Paul, 180, 222
Pinochet, Augusto, 74
policymakers, 26, 170–71, 179–80
politics: alliance, 92; Cold War, 47–48, 236–38; hacking in, 116–17; of international law, 48–49; international political system, 95; of nation-states, 44; political authority, 162n2, 174; political cartoons, 7–8, 62–64, 124–25, 147; political intelligence, 179–80; political philosophy, 8, 16–17; political propaganda, 122–23; political resistance, 168n97, 169n120; of writing, 37
Popular Front, US, 125
post-traumatic stress disorder (PTSD): diagnosis of, 264–72; resilience

and, 272–77; soldiers and, 261–62, 277–78; symptoms of, 15; wartime trauma and, 263–64
power, 34–35, 74–75, 117–18, 148–52
precautionary principle, 77n39
preparedness, 18n8
Princip, Gavrilo, 244
prisoner's dilemma, 242
Private Snafu (Dr. Seuss), 8, 63; anti-authoritarianism in, 132–35; civil society in, 139–42; democracy and, 127–32; Hollywood and, 123–25; individual freedom in, 135–38; as propaganda, 122–23, 125–27, 144n69; World War II and, 15
probability. *See* luck
problem-solving, 25–26, 38–39
propaganda: for anti-authoritarianism, 132–35; cyber warfare and, 15; democracy and, 127–32, 144n62; education and, 130–31, 141–42; Ellul on, 131–32, 141, 144n62; freedom and, 135–38; Japan and, 63–64; to soldiers, 136–37; in US, 125–27, 139–42, 144n69; in World War II, 15, 123–25, 142n1, 145n83
Propaganda (Ellul), 130–32
property law, 89
provocateurs, 243–44
Pruncken, Hank, 206
psychology: acceptance, 274–75; of animal abuse, 279n1; of choices, 43–44; claustrophobia, 34; of decision-making, 174; denial, 247–50; of deterrence, 246–47; of diplomacy, 181–82; DSM, 262, 266, 269, 272; framing, 173–74; of intelligence, 180–81, 183; intimidation, 29–30; of Nazis, 228; perseverance, 55; prisoner's dilemma, 242; of problem-solving, 38–39; psychological resilience, 27–29; psychological warfare, 127; rationality, 241–43; selfishness, 88–89; of soldiers, 263–72; of solidarity, 74; of utopias, 23–24, 29, 34–35, 39, 41n23; withdrawal, 270
Psychology of Intelligence Analysis (Heuer Jr.), 183
PTSD. *See* post-traumatic stress disorder
punishment, 247–50, 272

Qaddafi, Muammar, 151
al-Qaeda, 175–77, 182

Rabe, Tish, 116
race, 12–13, 31–32, 40n17, 76n25, 147
racism, 12–13, 63, 69, 128
RAND. *See* research and development center
rationality, 241–43
Read Across America program, 12
Reagan, Ronald, 236
realism, 93–95
Rebel Without a Cause (film), 37–38
religion, 68
renegades, 243–44
research and development center (RAND), 240
resilience, 207, 272–77
resistance: authoritarianism and, 146–47, 161–62; lone dissidents, 157–61, 168n102; mass protest, 152–57, 168n88; political, 168n97, 169n120; power and, 148–52
Rice, Condoleezza, 179
Roosevelt, Franklin D., 63, 70
Rosen, Lawrence, 153–54
Rousseau, Jean Jacques, 140
royalties, 37
Rumsfeld, Donald, 178
Russia: cyber warfare by, 207–8; DDos attack and, 111; Outer Space Treaty to, 49; Russian Federation, 236; Sandworm in, 108, 110; SolarWinds and, 112; spies from, 255n16; US and, 47–48, 104, 206. *See also* Soviet Union

Sadiki, Larbi, 153
Sandworm, 108, 110
Sassoon, Siegfried, 220
satellites, 50
Saudi Arabia, 150–51, 157–58, 161
Saving Private Ryan (film), 278
Scrambled Eggs Super! (Dr. Seuss), 12–13, 147
The Scream (Munch), 11–12
security studies, 4, 15, 98. *See also* environmental security
segregation, 63
selfishness, 88–89
Sen, Amartya, 70
Serbia, 244
sexism, 12–13, 69
sexual abuse, 271
Shelling, Thomas, 242
Shia Islam, 176
Short, Martin, 116
short-term interests, 88–90
Sierra Leone, 226
signals intelligence, 172, 178–79, 187
"Sir Gawain and the Green Knight" (poem), 267
Skocpol, Theda, 155
Slaughter, Joseph, 72
Sleep Book (Dr. Seuss), 18n8, 94
The Sneetches (Dr. Seuss), 5, 139, 252
Snow, David, 173–74
social connectedness, 273–74
social norms, 92–93, 140
social sciences, 276
society, 65, 139–42
sociology, 173–74, 189n15, 251–52
SolarWinds, 104, 106, 112, 117–18, 206, 207–8
soldiers, 95; comradery with, 226–27; luck to, 14, 212–15, 223–26; propaganda to, 136–37; psychology of, 263–72; PTSD and, 261–62, 277–78; in US Army, 137–40
solidarity, 74
South Africa, 48

Soviet Union: in Budapest Agreement, 236; Cold War to, 187; to Europe, 248; nuclear weapons to, 239; in space race, 102; US and, 47–48, 52, 247–50, 252–53. *See also* Russia
space race: commercialization of, 43, 49–50; democratization in, 43, 47–49; grand strategy in, 43–46, 53–55; militarization with, 43, 51–53; Soviet Union in, 102; symbolic meaning with, 42–43; US in, 14–15; US Space Force and Space Command, 52
SpaceX, 49–51
Spain, 10–11
Special Operations Command's Strategic Leaders International Course, US, 5
Spielberg, Steven, 278
spies, 255n16
Spies (film), 128, 137
staged operations, 109
Stalin, Joseph, 250–51
Stalling, Carl, 125
Standard Oil, 7
START. *See* Strategic Arms Reduction Treaty
Stephens, Ramón, 12–13
stereotypes, 64
Stillman, Richard, 229
The Story of Ferdinand (film), 140
Strategic Arms Reduction Treaty (START), 236
strategy: in Cold War, 248–50; for cyber warfare, 102, 116–18; for end states, 33–36; for friction, 33; grand, 43–46, 53–55; indoctrination strategies, 141; military, 14, 38–39, 228–29; misadventure and, 29–30; national security, 30–32; psychological resilience and, 27–29; strategic pitfalls, 24–27, 38–39; US Special Operations Command's Strategic Leaders International

Course, 5; utility of force, 36–38; in Vietnam, 34
Strong, E., 13
Stuxnet, 105, 109, 117–18
suicide, 28–29
Sunni Islam, 152–53, 176
Sun Tzu, 3–4, 7, 18n8, 27
survival, 93–95, 229–30
Swift, Jonathan, 253, 256n47, 263
symbolic meaning, 10–14, 29, 42–43
Syria, 149, 152, 155–56

Taliban, 182, 218
Tang, USS, 223
taqiyya (dissimulation), 175–77
Taylor, Philip M., 125–26
technology, 51–52, 186–87, 215–17, 231n29. *See also specific technology*
Tenet, George, 178–81
terrorism, 175–79, 182, 186–87, 221–22
Tet Offensive, 181–82
Thailand, 224
Thatcher, Margaret, 232n48
There's a Wocket in My Pocket (Dr. Seuss), 94
Thidwick the Big-Hearted Moose (Dr. Seuss): collapse in, 86–87; common pool resource dilemmas and, 90–92; deep ecology in, 97–98; environmental security and, 15, 85–86, 98–99; mass extinction in, 95–97; realism in, 93–95; security consequences in, 15; short-term interests in, 88–90; social norms in, 92–93, 140; tolerance in, 139
Threads (film), 251
Thucydides, 3, 7, 246
Together campaign, 78n48
Tojo, Hedeki, 30
tolerance, 67–69, 139–40
torture, 7, 16, 71–73
totalitarianism: fascism and, 30–31; isolationism and, 7–8; symbolic meaning and, 14

"The Tragedy of the Commons" (Hardin), 91
training, military, 6, 133–35
triggers, for PTSD, 269–70
Truman, Harry, 36
trust, 27
Tullibee, USS, 223
Tunisia, 149, 154
Turkey, 48
Twiggs, Charles, 229

Ukraine, 108, 236, 264
UN. *See* United Nations
uncertainty, 46–53
UNESCO. *See* United Nations Educational, Scientific, and Cultural Organization
unintended consequences, 107–8
United Arab Emirates, 48
United Kingdom, 50; in Battle of the Bulge, 63; in Crimean War, 268; France and, 256n47; Irish Republican Army to, 232n48; National Cyber Security Centre, 110; Nazis to, 80n77; US and, 251, 254n12; in World War I, 220
United Nations (UN): in Abidjan, 226; Charter, 68, 78n47; *Convention against Torture and Other Cruel, Inhuman or Degrading Treatment or Punishment*, 71; *Declaration on the Right to Development*, 70; deterrence to, 245–46; establishment of, 64–65; Living Planet Report, 96; Subcommittee on Prevention and National Preventive Mechanisms, 72; Together campaign, 78n48; *Universal Declaration of Human Rights*, 64–65, 67–68, 77n38, 79n67
United Nations Educational, Scientific, and Cultural Organization (UNESCO), 68
United States (US): Afghanistan and, 218; Air Force, 51–52; China

and, 206; Civil War, 268; in Cold War, 70; Constitution, 17, 89; in counterspace race, 53; Cyber Command, 115; democracy in, 141–42; Department of Defense, 132; education in, 6–7; El Salvador and, 74; environmental laws in, 97; Europe and, 48; Executive Order on Improving the Nation's Cybersecurity, 106–7; Germany and, 7–8, 217, 221, 227; government, 32, 38; in Gulf War, 35; intercontinental ballistic missiles in, 249; Iran and, 174; in Iraq, 183–84; isolationism in, 7–8; Japan and, 40n17, 76n25; Joint Publication 3.0 by, 126–27; MAD to, 247; Marine Corps Reserve, 14; in Middle East, 177–78; national security to, 131, 171, 175–81, 185–86, 191n64; Native Americans to, 90; Nazis to, 30–31; 9/11 attacks in, 175–76, 182, 186, 221–22; *Office of the National Cyber Director*, 106; OWI in, 123–24; Pacific Northwest of, 91; peace to, 54–55; Popular Front, 125; propaganda in, 125–27, 139–42, 144n69; race riots in, 31–32; Russia and, 47–48, 104, 206; segregation in, 63; Soviet Union and, 47–48, 52, 247–50, 252–53; Space Force and Space Command, 52; in space race, 14–15; Special Operations Command's Strategic Leaders International Course, 5; terrorism to, 175–79, 182, 186–87, 221–22; United Kingdom and, 251, 254n12; in Vietnam War, 30–31, 36–37, 75, 126, 181–82; war technology in, 231n29; in World War II, 17, 223. *See also specific topics*
unwitting cooperation, 103–7
US. *See* United States
utilitarianism, 136
utility of force, 36–38
utopias, 23–24, 29, 34–35, 39, 41n23

Venus, 49
veterans, military, 118, 198, 220, 229, 264, 269–75
Vietnam War: strategy in, 34; US Army in, 263–64; US in, 30–31, 36–37, 75, 126, 181–82; veterans of, 274
viruses, 117
Voltaire, 31
Volt Typhoon, 110
von Neumann, John, 241–42
vulnerability, 92–93

war. *See* specific topics
warfare environment, 227–29, 234n83
WarGames (film), 115
Warner Brothers, 124–25
Warnke, Paul, 238
war technology, 215–17, 231n29
war termination, 250–52
wartime trauma, 263–64
Washington, George, 16–17, 125–26
weapons of mass destruction, 6, 183–84
Wedeen, Lisa, 151
Wessler, William, 49
West, Mark, 161–62
western democracy, 89–90
Whitworth, Colin, 219
Why We Fight (film), 124
Wiesel, Elie, 278
The Wild One (film), 37–38
Winning Your Wings (film), 124
Winters, Dick, 220
Winters, Harold, 228
withdrawal, 270
Wohlstetter, Roberta, 178
women, 157–58, 224
Wood, Paul, 268
Wood, R. Mark, 220
World War I, 126, 216, 217, 220, 238–39, 244
World War II: aerial bombardment in, 10, 217; anti-Semitism in, 63; Battle of the Bulge, 9–10, 32, 63, 212–13, 215, 262, 263–64; concentration camps in, 72; culture after, 70,

139–40; enlistment in, 7–10; Europe in, 8–9, 75; Geisel, T., in, 4, 14, 95, 122–25, 128, 131, 142, 212–15, 221, 239, 263; Germany in, 11, 80n77, 131; history before, 126; IEDs in, 219; imperialism in, 85; infrastructure after, 93; interventionism in, 32; Japan and, 163n10; medicine in, 268; Middle East in, 147; Operation Fortitude in, 185; propaganda in, 15, 123–25, 142n1, 145n83; US in, 17, 223
writing, 37

writing process, 61–64, 91, 162n8
writing style, 76n28

Yertle the Turtle (Dr. Seuss): authoritarianism in, 146–47, 161–62; lone dissidents in, 157–61; mass protest in, 152–57; power in, 148–52; tolerance in, 139; totalitarianism in, 14
Your Job in Japan (film), 63–64

The Zax (Dr. Seuss), 94
Zulu War, 229

About the Contributors

Katherine Blue Carroll received her doctorate from the University of Virginia with a concentration in the politics of the Middle East. She is associate professor of political science at Vanderbilt University, where she teaches courses on the Middle East, comparative politics, and the US military. Since 2009 she has also served as director or associate director of Vanderbilt's undergraduate program in public policy studies. From 2008 to 2009 she worked as a social scientist in the Human Terrain System in Baghdad, Iraq, where she provided cultural and political information to brigade commanders on the ground. Most recently she is the coeditor (with William B. Hickman) of an undergraduate textbook, *Understanding the U.S. Military* (Routledge, 2023).

Chris C. Demchak is Grace Hopper Chair of Cyber Security and Senior Cyber Scholar at the US Naval War College. She holds a PhD, an MA, and other degrees in engineering, economics, comparative complex organizations, and political science. Her current research on cyberspace as a global, insecure, conflict-prone "substrate" applies a sociotechnical-economic systems approach to comparative institutional evolution with emerging technologies, cyber's offspring in adversaries' cyber/AI/ML campaigns, virtual wargaming for learning, and national/enterprise resilience against complex systems surprise. Her books include *Designing Resilience* (University of Pittsburgh Press, 2010), *Wars of Disruption and Resilience* (University of Georgia Press, 2011), and forthcoming *Cyber Warfare and Navies*, co-edited with Samual J. Tangredi, US Naval Institute Press, 2024).

Antulio J. Echevarria II holds the General MacArthur Chair of Research at the US Army War College and has authored six books on strategic thinking:

War's Logic (Cambridge University Press, 2021), *Military Strategy: A Very Short Introduction* (Oxford, 2017), *Imagining Future War* (2007), *Clausewitz and Contemporary War* (Oxford University Press, 2007), *Reconsidering the American Way of War* (Georgetown University Press, 2014), and *After Clausewitz* (University Press of Kansas, 2001). He completed a NATO Fulbright Fellowship, a Visiting Research Fellowship at Oxford University (2011–12 and 2022–23), a Senior Research Fellowship at the Foreign Policy Research Institute (2017–19), and an Adjunct Fellowship at the Modern War Institute (2018–19). He is editor in chief of the US Army War College Press and *Parameters*.

Kevin P. Eubanks received his PhD in comparative literature from the University of North Carolina at Chapel Hill in 2010 and has held faculty positions at Wake Forest University and the US Military Academy at West Point. He is currently associate director of the Writing Center and associate professor of writing and humanities at the US Naval War College in Newport, Rhode Island, where he teaches writing as well as courses on ideology, the war film, and war and conflict in art and literature. In addition to composition and rhetoric, his areas of expertise include film studies, critical theory, and comparative modernisms. His academic research has appeared in a variety of scholarly journals, including *German Studies Review*, *Seminar*, and *College Literature*.

John Hursh is an attorney-adviser specializing in international human rights, international humanitarian law, and the use of force. Hursh was program director for Democracy for the Arab World Now (DAWN), policy analyst (Sudan) at the Enough Project, and director of research, Stockton Center for International Law and editor in chief, *International Law Studies*, at the US Naval War College. He was a visiting scholar at the Center for Human Rights and Humanitarian Studies at Brown University in 2020 and 2021. In addition to his academic scholarship, he has written for *African Arguments*, *Foreign Policy*, *Just Security*, and *World Politics Review*.

Genevieve Lester is former De Serio Chair of Strategic Intelligence at the US Army War College. Her areas of interest are intelligence, accountability, leadership, and decision-making. She is also an associate fellow for strategic intelligence at the International Institute for Strategic Studies. She holds a PhD and MA in political science from the University of California, Berkeley; an MA in international economics/international law and organizations from the Johns Hopkins University, School of Advanced International Studies; and a BA in history from Carleton College. She was also a Fulbright Scholar at the Technical University in Berlin. She has held positions at Georgetown

University, National Defense University, Chatham House, and the International Institute for Strategic Studies. She began her career at the Survivors of the Shoah Visual History Foundation, Steven Spielberg's visual history project. She is the author of *When Should State Secrets Stay Secret? Accountability, Democratic Governance, and Intelligence* and numerous other publications on intelligence and related matters. She is currently coediting a volume on comparative covert action, as well as completing a monograph on the consumption of strategic intelligence.

Jon R. Lindsay is associate professor at the School of Cybersecurity and Privacy and the Sam Nunn School of International Affairs at the Georgia Institute of Technology. He is the author of *Information Technology and Military Power* (Cornell University Press, 2020) and coauthor of *Elements of Deterrence: Strategy, Technology, and Complexity in Global Politics* (Oxford University Press, 2024). He holds a PhD in political science from the Massachusetts Institute of Technology and an MS in computer science and BS in symbolic systems from Stanford University. He has also served in the US Navy.

Montgomery McFate is professor at the US Naval War College in Newport, Rhode Island. She holds a PhD in anthropology from Yale University and a JD from Harvard Law School. She is the author and/or editor of *Considering Anthropology and Small Wars* (Routledge, 2020), *Military Anthropology: Soldiers, Scholars and Subjects at the Margins of Empire* (Oxford University Press, 2018), and *Social Science Goes to War* (Oxford University Press, 2015), among others. Her articles have appeared in such journals as *Defense and Security Analysis*, *Journal of Information Warfare*, *Journal of Small Wars & Insurgencies*, and *Joint Forces Quarterly*. She was a key contributor to US Army Field Manual 3-24, *Counterinsurgency*, and is a lifetime member of the Council on Foreign Relations.

John Nagl is professor of warfighting studies at the US Army War College. A retired Army officer, he served in both Iraq wars. Nagl is the author of *Learning to Eat Soup with a Knife* (University of Chicago Press, 2004) and of *Knife Fights: A Memoir of Modern War* (Penguin, 2014). He helped write the 2006 *U.S. Army/Marine Corps Counterinsurgency Field Manual*. Nagl is a Senior Fellow at the Foreign Policy Research Institute and at the Center for a New American Security, which he previously served as its second president and a Professorial Lecturer at George Washington University. He is a fellow of the Irregular Warfare Institute and is a life member of the Council on Foreign Relations and the Veterans of Foreign Wars.

Saadia M. Pekkanen is Job and Gertrud Tamaki Endowed Professor of International Studies, adjunct professor of political science, adjunct professor of law, and founding director of the Space Law, Data, and Policy Program (SPACE LDP) at the University of Washington in Seattle. She earned master's degrees from Columbia University and Yale Law School and a doctorate from Harvard University in government. She works at the intersection of international relations and international law, engaging broader themes of states, strategy, alliances, and industrial policy in the world order. In addition to her half dozen books, she has published in journals such as *International Studies Quarterly, Japanese Journal of Political Science, International Relations of the Asia-Pacific, American Journal of International Law Unbound*, and *International Security*. Most recently, she is coeditor of the special issue "Space Diplomacy: The Final Frontier of Theory and Practice," *The Hague Journal of Diplomacy* (2023), and coeditor of *The Oxford Handbook of Space Security* (forthcoming, 2024).

Rebecca Pincus is director of the Polar Institute at the Wilson Center in Washington, DC. Previously she was on the faculty at the US Naval War College in the Center for Naval Warfare Studies, where her research focused on environmental security and geopolitics. She also worked in the Office of the Secretary of Defense for Policy as arctic and climate strategy adviser. In 2015 she was a Fulbright Fellow in Iceland, conducting research on Arctic small states and security. She also worked on polar strategy and policy issues for the US Coast Guard at its Center for Arctic Study and Policy and served on the faculty at the US Coast Guard Academy. She is currently a contributing author for the 5th National Climate Assessment. Her research on arctic security issues has been published in academic journals (*Polar Geography, War and Society*, and others) as well as in popular outlets like *War on the Rocks*, the BBC, and others. In 2015, Yale University Press released her coedited book *Diplomacy on Ice: Energy and the Environment in the Arctic and Antarctic*. Dr. Pincus earned her PhD in 2013 from the University of Vermont.

Michael Poznansky is associate professor in the Strategic and Operational Research Department and a faculty member of the Cyber & Innovation Policy Institute at the US Naval War College. He is the author of *In the Shadow of International Law: Secrecy and Regime Change in the Postwar World* (Oxford University Press, 2020). Dr. Poznansky has held fellowships with the Belfer Center, the Dickey Center, and the Modern War Institute. His work focuses mainly on the role that secrecy and deception play in world politics. He holds a PhD from the University of Virginia.

Sam J. Tangredi is Leidos Chair of Future Warfare Studies and professor of national, naval, and maritime strategy in the Center for Naval Warfare Studies of the US Naval War College. He has wide experience as a strategic planner and director of strategic planning teams. Sam has published seven books, more than two hundred journal articles, and numerous reports and presentations for a wide range of government and academic organizations. His latest book, *Algorithms of Armageddon: The Impact of Artificial Intelligence on Future Wars* (coauthored with George Galdorisi) was published in March 2024. His book *Anti-Access Warfare: Countering A2/AD Strategies*—widely considered the definitive work on that subject—was rereleased by Naval Institute Press in paperback in 2023. Dr. Tangredi is a retired US Navy captain who held command at sea.

Erich Henry Wagner is colonel in the US Marine Corps Reserve, currently assigned as the Director of the Marine Corps' Field History Branch, where he leads a detachment of Marines tasked to record, preserve, and disseminate the cumulative, operational, and institutional experience of the force. He earned his BS in history from the US Naval Academy, an MA in international relations from the Fletcher School at Tufts University, and an MA in strategic studies from the US Army War College. He has published articles in the *Marine Corps University Journal*, *The Journal for the Anglo-Zulu War Historical Society*, *Joint Center for Operational Analysis*, and *Combating Terrorism Exchange*. His most recent publications include "Turning Bullets into Water: Invulnerability Notions & Irregular Armed Groups," in *Extremisms in Africa* (Good Governance Africa, 2020) and "Ingenuity, Excess, Incompetence and Luck: Air-Resupply Anecdotes in Military History," in the *Journal of Military History*.

www.ingramcontent.com/pod-product-compliance
Lightning Source LLC
Chambersburg PA
CBHW070818250426
43672CB00031B/2770